VOLUME SIX HUNDRED AND THIRTY THREE

METHODS IN ENZYMOLOGY

Chemical and Synthetic Biology Approaches To Understand Cellular Functions - Part C

METHODS IN ENZYMOLOGY

Editors-in-Chief

ANNA MARIE PYLE
Departments of Molecular, Cellular and Developmental Biology and Department of Chemistry
Investigator, Howard Hughes Medical Institute
Yale University

DAVID W. CHRISTIANSON
Roy and Diana Vagelos Laboratories
Department of Chemistry
University of Pennsylvania
Philadelphia, PA

Founding Editors

SIDNEY P. COLOWICK and NATHAN O. KAPLAN

VOLUME SIX HUNDRED AND THIRTY THREE

METHODS IN ENZYMOLOGY

Chemical and Synthetic Biology Approaches To Understand Cellular Functions - Part C

Edited by

ARUN K. SHUKLA
Department of Biological Sciences and Bioengineering, Indian Institute of Technology, Kanpur, India

ELSEVIER

ACADEMIC PRESS
An imprint of Elsevier

Academic Press is an imprint of Elsevier
50 Hampshire Street, 5th Floor, Cambridge, MA 02139, United States
525 B Street, Suite 1650, San Diego, CA 92101, United States
The Boulevard, Langford Lane, Kidlington, Oxford OX5 1GB, United Kingdom
125 London Wall, London, EC2Y 5AS, United Kingdom

First edition 2020

Notices
Knowledge and best practice in this field are constantly changing. As new research and experience broaden our understanding, changes in research methods, professional practices, or medical treatment may become necessary.

Practitioners and researchers must always rely on their own experience and knowledge in evaluating and using any information, methods, compounds, or experiments described herein. In using such information or methods they should be mindful of their own safety and the safety of others, including parties for whom they have a professional responsibility.

To the fullest extent of the law, neither the Publisher nor the authors, contributors, or editors, assume any liability for any injury and/or damage to persons or property as a matter of products liability, negligence or otherwise, or from any use or operation of any methods, products, instructions, or ideas contained in the material herein.

ISBN: 978-0-12-819128-6
ISSN: 0076-6879

For information on all Academic Press publications
visit our website at https://www.elsevier.com/books-and-journals

Publisher: Zoe Kruze
Acquisitions Editor: Ashlie M. Jackman
Editorial Project Manager: Shellie Bryant
Production Project Manager: James Selvam
Cover Designer: Alan Studholme
Typeset by SPi Global, India

Contents

Contributors

Yimon Aye
Institute of Chemical Sciences and Engineering, École Polytechnique Fédérale de Lausanne, Lausanne, Switzerland

Mithu Baidya
Department of Biological Sciences and Bioengineering, Indian Institute of Technology, Kanpur, India

Jerome Bonnet
Centre de Biochimie Structurale (CBS), INSERM, CNRS, Univ. Montpellier, Montpellier, France

Thomas Böttcher
Department of Chemistry, Konstanz Research School Chemical Biology, Zukunftskolleg, University of Konstanz, Konstanz, Germany

Yana Cen
Department of Pharmaceutical Sciences, Albany College of Pharmacy and Health Sciences, Colchester, VT, United States

Hung-Ju Chang
Centre de Biochimie Structurale (CBS), INSERM, CNRS, Univ. Montpellier, Montpellier, France

Alexandra V. Chatzikonstantinou
Laboratory of Biotechnology, Department of Biological Applications and Technology; Department of Chemistry, Section of Organic Chemistry and Biochemistry, University of Ioannina, Ioannina, Greece

Ian Cohen
Department of Pharmaceutical Sciences, Albany College of Pharmacy and Health Sciences, Colchester, VT, United States

Emily M. Dieter
Department of Chemistry, University of Washington, Seattle, WA, United States

Hemlata Dwivedi-Agnihotri
Department of Biological Sciences and Bioengineering, Indian Institute of Technology, Kanpur, India

Felix Faschinger
Institute of Biophysics, Johannes Kepler University, Linz, Austria

Aleksandra Filipovska
Harry Perkins Institute of Medical Research, Nedlands; The University of Western Australia Centre for Medical Research; School of Molecular Sciences, The University of Western Australia, Crawley, WA, Australia

Douglas M. Fowler
Department of Genome Sciences; Department of Bioengineering, University of Washington, Seattle, WA, United States

Marco Fragai
Magnetic Resonance Center (CERM), Department of Chemistry "Ugo Schiff"; Consorzio Interuniversitario Risonanze Magnetiche di Metallo Proteine, University of Florence, Sesto Fiorentino, Italy

Ioannis P. Gerothanassis
Department of Chemistry, Section of Organic Chemistry and Biochemistry, University of Ioannina, Ioannina, Greece

Hermann J. Gruber
Institute of Biophysics, Johannes Kepler University, Linz, Austria

Vesa P. Hytönen
Faculty of Medicine and Health Technology and BioMediTech, Tampere University; Fimlab Laboratories, Tampere, Finland

Günther Knör
Institute of Inorganic Chemistry, Johannes Kepler University, Linz, Austria

Bingbing X. Li
Program in Chemical Biology, Department of Chemical Physiology and Biochemistry, Knight Cancer Institute, Oregon Health and Science University, Portland, OR, United States

Marcus J.C. Long
Beverley East Riding of Yorkshire, United Kingdom

Claudio Luchinat
Magnetic Resonance Center (CERM), Department of Chemistry "Ugo Schiff"; Consorzio Interuniversitario Risonanze Magnetiche di Metallo Proteine, University of Florence, Sesto Fiorentino, Italy

Dustin J. Maly
Department of Chemistry; Department of Biochemistry, University of Washington, Seattle, WA, United States

James C. Mathews
Harry Perkins Institute of Medical Research, Nedlands; The University of Western Australia Centre for Medical Research, Crawley, WA, Australia

Elke Oberbichler
Institute of Biophysics, Johannes Kepler University, Linz, Austria

Giacomo Parigi
Magnetic Resonance Center (CERM), Department of Chemistry "Ugo Schiff"; Consorzio Interuniversitario Risonanze Magnetiche di Metallo Proteine, University of Florence, Sesto Fiorentino, Italy

Michaela Prothiwa
Department of Chemistry, Konstanz Research School Chemical Biology, Zukunftskolleg, University of Konstanz, Konstanz, Germany

Oliver Rackham
Harry Perkins Institute of Medical Research, Nedlands; School of Pharmacy and Biomedical Sciences; Curtin Health Innovation Research Institute, Curtin University, Bentley, WA, Australia

Enrico Ravera
Magnetic Resonance Center (CERM), Department of Chemistry "Ugo Schiff"; Consorzio Interuniversitario Risonanze Magnetiche di Metallo Proteine, University of Florence, Sesto Fiorentino, Italy

Sina Rütschlin
Department of Chemistry, Konstanz Research School Chemical Biology, Zukunftskolleg, University of Konstanz, Konstanz, Germany

Eva Schlögl
Institute of Biophysics, Johannes Kepler University, Linz, Austria

Louis H. Scott
Harry Perkins Institute of Medical Research, Nedlands; The University of Western Australia Centre for Medical Research, Crawley, WA, Australia

Arun K. Shukla
Department of Biological Sciences and Bioengineering, Indian Institute of Technology, Kanpur, India

Ashish Srivastava
Department of Biological Sciences and Bioengineering, Indian Institute of Technology, Kanpur, India

Haralambos Stamatis
Laboratory of Biotechnology, Department of Biological Applications and Technology, University of Ioannina, Ioannina, Greece

Jessica Stangl
Institute of Biophysics, Johannes Kepler University, Linz, Austria

Antonis D. Tsiailanis
Department of Chemistry, Section of Organic Chemistry and Biochemistry, University of Ioannina, Ioannina, Greece

Andreas G. Tzakos
Department of Chemistry, Section of Organic Chemistry and Biochemistry, University of Ioannina, Ioannina, Greece

Daniel A. Urul
Institute of Chemical Sciences and Engineering, École Polytechnique Fédérale de Lausanne, Lausanne, Switzerland; Department of Chemistry and Chemical Biology, Cornell University, Ithaca, NY, United States

Cindy T. Wei
Molecular and Cellular Biology Program; Department of Chemistry; Department of Genome Sciences, University of Washington, Seattle, WA, United States

Maria Wiesauer
Institute of Inorganic Chemistry, Johannes Kepler University, Linz, Austria

Jessica Wohlfahrt
Department of Pharmaceutical Sciences, Albany College of Pharmacy and Health Sciences, Colchester, VT, United States

Xiangshu Xiao
Program in Chemical Biology, Department of Chemical Physiology and Biochemistry, Knight Cancer Institute, Oregon Health and Science University, Portland, OR, United States

Song Zheng
Department of Pharmaceutical Sciences, Albany College of Pharmacy and Health Sciences, Colchester, VT, United States

Preface

Investigating the basic structure, function, and dynamics of cellular proteins is crucial to develop a complete understanding of cellular and physiological processes. In this context, development of novel methods and tools, and improvement of existing ones are of immense importance and broad interest. Over the last few years, a series of developments in the area of chemical and synthetic biology have surfaced that have allowed us to generate novel insights into a wide range of cellular processes. Many of these recent advances are now pushing the boundaries of our knowledge in various domains of cellular biology not only at the molecular and systems level.

In these three volumes of *Methods in Enzymology*, we cover a broad spectrum of topics including protein design and engineering, novel labeling approaches, emerging biochemical and biophysical methods, design of novel genetic circuits, genome engineering, and computational techniques. Some of the chapters in these volumes provide step-by-step experimental protocols that can be directly utilized in different laboratories, while others provide an overview of a specific method or a particular topic for a comprehensive understanding.

I take this opportunity to express my sincere gratitude to all the authors who readily agreed to contribute to this volume and delivered their work in stipulated time frame. These volumes would not be possible without, and I also want to thank, the members of the editorial and production team for doing a wonderful job on getting these chapters ready for publication and efficiently coordinating the entire publication process.

On behalf of the entire *Methods in Enzymology* team, I present to you these three volumes of *Methods in Enzymology* on Chemical and Synthetic Biology and sincerely hope that you find the topics covered in these volumes very useful for your research. As always, we look forward to your comments and feedback.

Arun K. Shukla, PhD
Associate Professor, Joy Gill Chair Professor and
EMBO Young Investigator
Intermediate Fellow, Wellcome Trust DBT India Alliance
Department of Biological Sciences and Bioengineering
Indian Institute of Technology, Kanpur, India

CHAPTER ONE

Competitive binding assay for biotin and biotin derivatives, based on avidin and biotin-4-fluorescein

Elke Oberbichler[a,†], Maria Wiesauer[b,†], Eva Schlögl[a], Jessica Stangl[a], Felix Faschinger[a], Günther Knör[b], Hermann J. Gruber[a], Vesa P. Hytönen[c,d,*]

[a]Institute of Biophysics, Johannes Kepler University, Linz, Austria
[b]Institute of Inorganic Chemistry, Johannes Kepler University, Linz, Austria
[c]Faculty of Medicine and Health Technology and BioMediTech, Tampere University, Tampere, Finland
[d]Fimlab Laboratories, Tampere, Finland
*Corresponding author: e-mail address: vesa.hytonen@tuni.fi

Contents

† These authors contributed equally.

Methods in Enzymology, Volume 633
ISSN 0076-6879
https://doi.org/10.1016/bs.mie.2019.10.028

Abstract

Biotinylated molecules are extensively employed in bioanalytics and biotechnology. The currently available assays for quantification of biotin groups suffer from low sensitivity, low accuracy, or provide highly variable responses for different biotin derivatives. We developed a competitive binding assay in which avidin was pre-blocked to different extents by the biotinylated analyte and a constant amount of biotin-4-fluorescein (B4F) was added, resulting in strong quenching of the B4F. The assay was robust and the shape of the titration curve immediately revealed whether the data were reliable or perturbed by steric hindrance in case of large biotin derivatives. These advantages justified well the $10\times$ higher sample consumption ($\sim$0.6 nmol) compared to single point assays. The assay was applied to a representative set of small biotin derivatives and validated by cross-control with the well-established 2-anilinonaphthalene-6-sulfonic acid (2,6-ANS) binding assay. In comparison to the 2,6-ANS binding assay, the lower precision ($\pm$10%) was compensated by the 100-fold higher sensitivity and the deviations from the ANS assay were $\leq$5%. In comparison to the more sensitive biotin group assays, the new assay has the advantage of minimal bias for different biotin derivatives. In case of biotinylated DNA with 30 nucleotides, steric hindrance was found to reduce the accuracy of biotin group determination; this problem was overcome by partial digestion to $n \leq 5$ nucleotide residues with a 3′-exonuclease. The newly proposed biotin group assay offers a useful compromise in terms of sensitivity, precision, trueness, and robustness.

1. Introduction

Biotin is the prosthetic group of many carboxyl transferases in prokaryotes and eukaryotes. At the same time, biotinylated molecules are popular tools in bioanalytics and medical diagnostics (Dundas, Demonte, & Park, 2013), due to their high affinity and almost irreversible binding toward (strept)avidin (Green, 1990). For these reasons a number of assays have been developed by which biotin, its metabolites, and biotinylated molecules can be quantified over a wide range of concentrations, as reviewed in Livaniou et al. (2000) and further exemplified by recent studies (Ikeda, Miyao, & Sueda, 2014; Indyk, Gill, & Woollard, 2014; Rispens & Ooijevaar-de Heer, 2016).

Bioassays (Baur, Suormala, Bernoulli, & Baumgartner, 1998) and solid-phase assays with radiolabels (Baur et al., 1998; Groman, Rothenberg, Bayer, & Wilchek, 1990; Mock, 1990) or marker enzymes (Bayer & Wilchek, 1990; Mock, 1997; Staggs, Sealey, McCabe, Teague, & Mock, 2004) provide for very high sensitivity but they are time-consuming and their precision is moderate. In comparison, homogenous biotin assays are less sensitive but more convenient and usually more precise. In the classical example (Green, 1965), avidin is mixed with an excess of 4′-hydroxyazobenzene-2-carboxylic acid (HABA, $K_d = 5.8\,\mu M$) which occupies most biotin-binding sites whereupon it shows high absorbance at 500 nm. Step-wise titration of this mixture with biotin results in a linear decrease of A_{500}, up to the point where all binding sites are occupied with biotin and A_{500} remains constant upon further addition of biotin (Green, 1965). Due to its linear dose-response, high accuracy, and simplicity, the HABA assay is still used in most commercial kits for biotin determination, whereby the biotin consumption has been reduced from ~40 to ~4 nmol biotin groups by switching from cuvettes to microtiter plates (MTP) and by measuring only the initial linear slope (e.g., in the biotin quantitation kits from Abcam or from Pierce (Thermo Fischer Scientific)). Alternatively, HABA has been replaced by the fluorescent probe 2-anilinonaphthalene-6-sulfonic acid (ANS), providing for a fluorescence assay which closely resembles the HABA assay with respect to its linear dose-response, its high accuracy, and its moderate sensitivity (Mock & Horowitz, 1990).

A linear signal decrease is also observed if the endogenous tryptophan fluorescence of (strept)avidin is used to monitor binding of biotin (Der-Balian, Gomez, Masino, & Parce, 1990; Lin & Kirsch, 1977, 1979). The sensitivity is ~100 × higher than in the HABA and ANS assays but (because of fluorescence excitation at 290 nm) this assay cannot be used when significant amounts of other proteins are present. Moreover, it is not known whether it works equally well with various derivatives of biotin.

The most sensitive fluorescence assay for biotin is based on the discovery that avidin is preferentially labeled by fluorescein-isothiocyanate (FITC) next to the four biotin-binding sites and that the bound FITC residues are significantly quenched in this location (Al-Hakiem, Landon, Smith, & Nargessi, 1981). Binding of biotin abolishes the quenching and results in >100% fluorescence increase (Al-Hakiem et al., 1981; Nargessi & Smith, 1986). FITC-labeled streptavidin is even more responsive, showing >160% fluorescence enhancement upon binding of biotin (Barbarakis, Smith-Palmer, Bachas, Chen, & Van Der Meer, 1993). The most dramatic effect is seen with Alexa488-labeled avidin if the Alexa488 labels are additionally quenched by bound HABA; thereby the addition of biotin causes displacement of HABA and results in an almost 20-fold fluorescence increase

(Batchelor, Sarkez, Cox, & Johnson, 2007). These assays are simple, precise and can determine few picomoles of biotin. Unfortunately, though, even closely similar biotin derivatives (such as biocytin) give very different responses from biotin. Therefore, the assay must always be calibrated with the biotin derivative of interest (Barbarakis et al., 1993; Batchelor et al., 2007) and mixtures of different biotin derivatives can only be analyzed by a combination of reversed-phase chromatography and post-column detection with FITC-streptavidin (Przyjazny, Hentz, & Bachas, 1993; Rao, Anderson, & Bachas, 1997). Another drawback of FITC-avidin-based assays is their non-linear dose-response and lack of a defined breakpoint associated to the state where exactly four biotin groups are added per (strept)avidin tetramer. This behavior is obviously due to the imperfect site-specificity of (strept) avidin labeling with FITC. No such problems were seen in case of site-specific incorporation of non-natural amino acids like methoxycoumarine-alanine or –glutamic acid in position 120 of streptavidin (Murakami, Hohsaka, Ashizuka, Hashimoto, & Sisido, 2000) but unfortunately the engineered streptavidins with the methoxycoumarine labels are not commercially available.

The goal of the present study was to develop a homogeneous fluorescence assay for biotin derivatives which meets the following criteria: (i) The reagents for the assay should be commercially available. (ii) The sample consumption should not be higher than in the tryptophan quenching assay (~0.5 nmol). (iii) In contrast to the latter, the wavelengths of excitation and emission should be in the visible region to minimize background fluorescence. (iv) The dose-response should be linear, with a clear-cut breakpoint when all biotin-binding sites in avidin are saturated with biotinylated analyte molecules.

Being in need of such an assay, we remembered a published feasibility test which had met criteria (i)–(iv) (Ebner et al., 2008). The assay principle takes advantage of the irreversible binding of biotin (or a biotin derivative) to avidin, and of the fact that subsequently added B4F occupies the remaining sites and gets quenched, while the excess of unbound B4F molecules shows undiminished fluorescence (see Fig. 1).

In the present study we screened for the optimal assay parameters and checked whether the same linear dose-response is also found with various kinds of biotin derivatives. We also addressed steric problems of larger biotinylated molecules that might prevent saturation of all four binding sites per avidin tetramer under the given assay conditions. The new assay was validated by comparison with the ANS binding assay, taking advantage of its high accuracy at higher analyte concentrations.

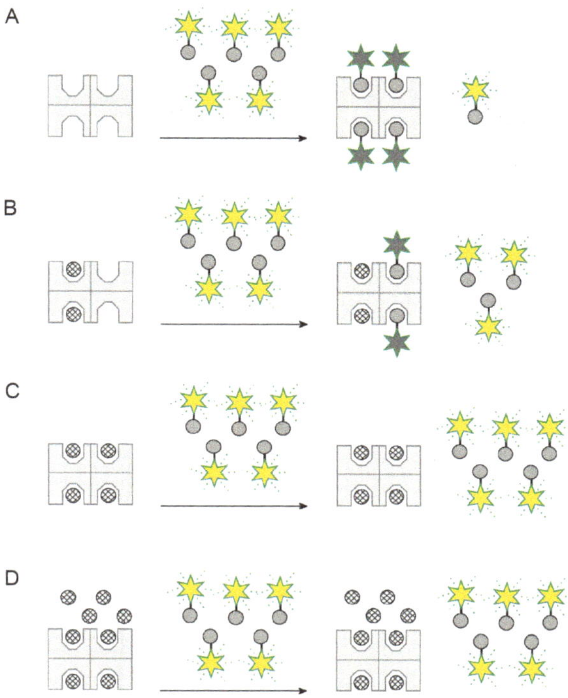

Fig. 1 Principle of the competitive B4F binding assay. Identical amounts of avidin are irreversibly blocked with increasing amounts of a biotin derivative and then always mixed with the same slight excess of B4F (~10%). (A) In absence of a biotin derivative, ~90% of the B4F is bound and quenched; only ~10% is free and fluoresces strongly. (B) Partial pre-blocking of avidin with a biotin derivative reduces both binding and quenching of B4F. (C) Saturation of avidin with equimolar biotin derivative prevents binding and quenching of all B4F molecules. (D) Due to irreversible and tight binding, an excess of biotin derivative has the same effect as the equimolar amount in (C).

2. Materials and methods

UV/vis measurements were conducted on a Hitachi U-3010 spectrophotometer, all fluorescence measurements were performed on a Hitachi F-2700 fluorimeter. Details on used materials and procedures can be found in the Supplementary Material in the online version at https://doi.org/10.1016/bs.mie.2019.10.028.

2.1 Buffers

Buffer A (100 mM NaCl, 50 mM Na_2HPO_4, 1 mM EDTA, pH adjusted to 7.5 with NaOH) and buffer B (1 mg mL^{-1} BSA in buffer A) were used in all titration experiments, as specified.

2.2 Stock solutions of fluorescent probes

ANS was dissolved in DMSO to give a 5 mM stock solution. HABA was dissolved in alkaline aqueous solution to give a 20 mM stock solution. B4F was dissolved in DMSO to obtain a concentration of approximately 500 μM ("nominal concentration by weight," NCBW). Further dilution with buffer A yielded an approximately 20 μM stock solution (NCBW). The effective concentration (EC) of the B4F stock solution was determined to be 16.0 μM (Fig. S2 in Supplementary Material in the online version at https://doi.org/10.1016/bs.mie.2019.10.028) (Kada, Falk, & Gruber, 1999). The nominal concentration of B4F deviates from the effective concentration because of imperfect purity of commercial B4F (Kada, Falk, & Gruber, 1999). B4F was diluted to 0.160 μM (EC) with buffer B immediately before its use in a B4F binding assay.

2.3 Stock solutions of biotin and biotinylated compounds

Except for biotinylated DNA, all concentration values stated in this section are "nominal concentrations by weight" (NCBW), with the provisional assumption that the compound had a purity of 100%. All stock solutions were diluted with buffer A to 400 μM for the ANS assay and with buffer B to 0.2 μM for the competitive B4F assay, respectively.

D-biotin was dissolved in an aqueous solution of 100 mM Tris base to yield a 4 mM stock.

Biocytin was dissolved in an acetic acid-water mixture and, if necessary, diluted with buffer A to a final stock concentration of 1 or 4 mM.

Biotin-PEG$_8$-amine was converted into the trifluoroacetic acid (TFA) salt and taken to dryness, while biotin-C$_3$-amine was already available as TFA salt. Both compounds were dissolved in DMSO at a concentration of 4 mM.

The TFA salts of biotin-PEG$_1$-amine and biotin-PEG$_2$-amine were both purchased as DMSO stock solutions with a declared concentration of 25 mg mL^{-1}. For the competitive B4F binding assay, aliquots of the DMSO stock solution were first diluted with 100 mM NaH_2PO_4 to a concentration of 50 or 100 μM, respectively, ensuring protonation by this acidic buffer.

5′-Biotin-DNA was dissolved in water to give a 100 μM solution. Further dilution with buffer A yielded a 4 μM stock solution. Determining the absorption A_{260} of a 10-fold diluted sample gave a concentration value of 0.413 μM in the diluted sample ($\varepsilon_{260} = 306 \times 10^3\,M^{-1}\,cm^{-1}$, OligoCalc (Kibbe, 2007)), thus a concentration of 4.13 μM in the stock solution.

A sample of 5′-biotin-DNA was subjected to treatment with exonuclease I in order to receive a digested sample (see Section S4 in Supplementary Material in the online version at https://doi.org/10.1016/bs.mie.2019.10.028).

2.4 Stock solutions of avidin

For the HABA and ANS assays, avidin concentration was 3 μM tetramers (NCBW, batch 1). For determination of the optimal assay conditions (Section 2.5), avidin "working solution No. 1" (0.105 μM (EC)) was used. For the competitive binding assays, avidin "working solutions No. 2" (final concentration: 34.8 nM (EC) in batch 2 and 33.4 nM (EC) in batch 3) were used.

2.5 Titration of B4F with avidin to determine the optimal competitive B4F assay conditions

Triplicates of 12 polystyrene tubes (5 mL) were prepared containing increasing amounts of avidin (working solution No. 1) and a constant amount of B4F (0.160 μM) in a constant final volume of 1 mL. For this purpose, $1000 - x - 100$ μL buffer B was mixed with x μL of avidin working solution No. 1 (x = 0, 10, 20, 30, 40, 50, 60, 80, 100, 120, 140, 160 μL). Then, 100 μL of 0.160 μM B4F was added to each tube and each tube was vortexed immediately after the addition. After incubation for 10 min in the dark, the fluorescence at 525 nm (5 nm slit) was measured while exciting at 490 nm (2.5 or 5 nm slit). The fluorescence values were plotted against the added volume of avidin working solution No. 1. A linear fit was applied to the segment which showed a steep initial drop, as well as to the moderate subsequent increase of the titration curve (see Fig. 4 for batch 2 and Fig. S7 in Supplementary Material in the online version at https://doi.org/10.1016/bs.mie.2019.10.028 for batch 3). The intersection point of the two lines gave the volume of the avidin working solution (in μL) at which all B4F was bound to avidin.

The purpose of the above described B4F titration with avidin was to find out how much avidin was necessary to bind 90% of the B4F molecules in each assay tube; this was the desired working condition for the competitive B4F binding assay.

In batch 2 (Fig. 4) the breakpoint occurred at 37.01 μL of 0.105 μM avidin tetramer (3.87 pmol). From this result, the desired concentration of the avidin "working solution No. 2" was calculated as 0.9×3.87 pmol/100 μL = 34.8 nM. In batch 3 (Fig. S7 in Supplementary Material in the online version at https://doi.org/10.1016/bs.mie.2019.10.028), the break-point was at 35.6 μL of 0.105 μM avidin tetramer (3.72 pmol), thus concentration of "working solution No. 2" was calculated as $0.9 \times 3.72\,\text{pmol}/100\,\mu\text{L} = 33.4\,\text{nM}$ avidin tetramer.

2.6 Competitive binding assay with biotin-4-fluorescein

Typically, triplicates of 17 polystyrene tubes (5 mL) were prepared containing increasing amounts of D-biotin (or a biotin derivative) and constant amounts of avidin and B4F. First, $1000 - x - 200\,\mu\text{L}$ buffer B was mixed with x μL of 0.2 μM biotin-reagent ($x = 0$, 5, 10, 15, 20, 25, 30, 40, 50, 60, 70, 80, 90, 100, 110, 120, 130 μL). Then, 100 μL of "avidin working solution No. 2" was added and each tube was vortexed immediately after the addition of avidin. After incubation for 10 min, 100 μL of 0.160 μM B4F was added to each tube. After incubation for 10 min in the dark, the fluorescence of the samples was measured at 525 nm (5 nm slit) while irradiating at 490 nm (5 nm slit).

The fluorescence values were plotted against the volume (in μL) of the added biotin-reagent solution (Figs. 5 and 6 and Fig. S8 in Supplementary Material in the online version at https://doi.org/10.1016/bs.mie.2019.10.028 with avidin batch 2, Fig. S9 in Supplementary Material with avidin batch 3). A linear fit was applied to the two linear segments, the steep initial rise and the plateau at high biotin concentration using the piecewise fitting function "PWL2" of the software Origin® (OriginLab, Northhampton). The obtained intersection point (x_i) indicated the volume at which all binding sites were saturated by biotin residues. When multiple titration curves were measured, the mean value of the intersection point was calculated and the error was computed as the standard deviation of the obtained x_i values.

3. Results and discussion

3.1 Use of the ANS assay for accurate characterization of the biotinylated test compounds

The goal of the present study was to establish the B4F-based biotin assay illustrated in Fig. 1 and to test it with a variety of biotin derivatives. Besides several small molecules, a longer polyethylene glycol (PEG) derivative and a 30-nucleotide biotinylated single-stranded DNA molecule were chosen. Due to the imperfect purity of most test samples, it was first necessary to determine their true biotin content.

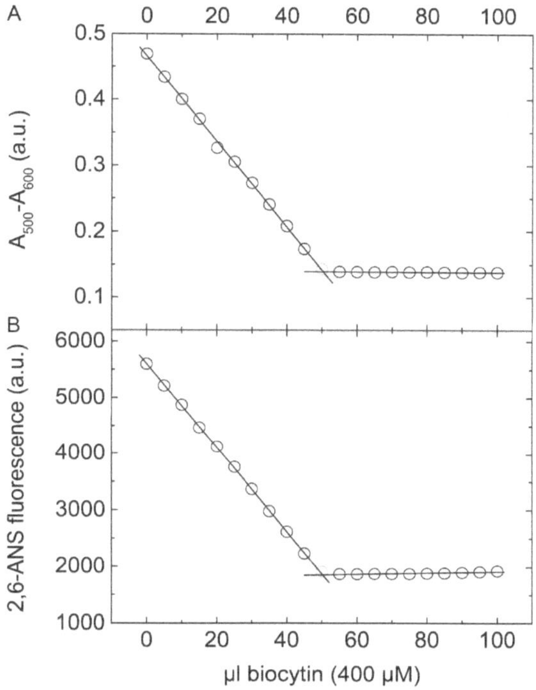

Fig. 2 Comparison of HABA (A) and ANS (B) as reporters for the titration of avidin with biocytin. 2 mL 3 μM avidin tetramer (NCBW) was mixed with HABA (20 μL, 20 mM) or ANS (25 μL, 5 mM) in a stirred cuvette and titrated with 400 μM biocytin (NCBW). The absorbance values (A_{500}–A_{600}) or the fluorescence signals (at 340 nm) were corrected for the dilution upon biocytin addition and plotted vs the consumption of biocytin. Data shown in gray was excluded from linear fitting. The intersection points (50.4 μL biocytin in (A) and 50.2 μL biocytin in (B)) reflected the amount of biocytin necessary for saturation of all binding sites in avidin sample.

The most accurate methods for biotin group determination are the HABA assay (Green, 1965) and the ANS assay (Mock & Horowitz, 1990; Mock, Lankford, & Horowitz, 1988). Testing their performance (Fig. 2), breakpoints occurred at the same biocytin consumption in the HABA and ANS assay (50.4 and 50.2 μL, respectively), proving full equivalence of the two assays.

Except for biotin-5′-DNA due to lack of material, the biotin content of all biotin derivatives was then tested by the ANS assay (Fig. 3). As expected for samples with different purity, the breakpoint did not occur at the same consumption of biotin derivative solution (see Table 1 and Table S1 in the online version at https://doi.org/10.1016/bs.mie.2019.10.028 in

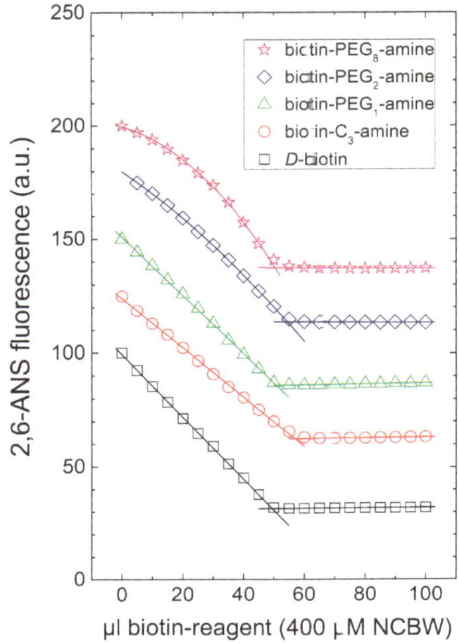

Fig. 3 ANS assay for D-biotin and biotin derivatives. As described in Fig. 2B, a 2 mL aliquot from the identical avidin solution was mixed with ANS and titrated with 400 µM D-biotin or biotin derivative (NCBW). The data sets were processed as in Fig. 2B and the initial fluorescence of each curve was renormalized to 100 units. Except for D-biotin (□), all titration profiles were vertically displaced by multiples of 25 fluorescence units to avoid overlap of the data. The steep segments for biotin-PEG$_8$-amine and biotin-PEG$_2$-amine were analyzed by a polynomial fit ($y = A + B\,x + C\,x^2$), all other fits were strictly linear.

Supplementary Material in the online version at https://doi.org/10.1016/bs.mie.2019.10.028). The lowest consumption was observed with nominally 400 µM D-biotin (49.6 ± 0.6 µL, second column in Table 1 and Table S1 in Supplementary Material in the online version at https://doi.org/10.1016/bs.mie.2019.10.028), indicating that D-biotin was of highest purity among the tested samples. For this reason, D-biotin was selected as standard in the ANS assay.

The biotin content of the different compounds was calculated by dividing the breakpoint for the biotin standard (46.9 µL) by the breakpoint of the particular biotin derivative (third column in Table S1 Supplementary Material in the online version at https://doi.org/10.1016/bs.mie.2019.10.028). Biocytin is a preferred standard when determining the biotin group content of biotinylated proteins (Oberoi et al., 2014; Przyjazny et al., 1993; Rao et al., 1997). D-biotin is less convenient to use as standard because it can sometimes degrade on a time scale of hours at room temperature

Table 1 Determination of the biotin group contents of biotin derivatives by the ANS and B4F assay.

Biotin derivative	Breakpoint[a] (μL)	Biotin content ANS assay[b] (%)	Biotin content B4F assay[c] (%)	Recovery in B4F assay [d] (%)
D-biotin	49.6 ± 0.6	$\equiv 100.0 \pm 1.2$[e]	n.d.	n.d.
Biocytin	50.7 ± 0.9	97.7 ± 1.8	$\equiv 98 \pm 7$[f]	$\equiv 100 \pm 7$[f]
Biotin-C_3-amine	56.9	87.1	86 ± 3	99 ± 3
Biotin-PEG_1-amine	50.8	97.6	93 ± 3	95 ± 3
Biotin-PEG_2-amine	55.6[g]	89.1[g]	94 ± 4	105 ± 5
Biotin-PEG_8-amine	51.2[g]	96.9[g]	97 ± 3	100 ± 3

[a]Calculated from the data in Table S1 in Supplementary Material.
[b]Calculated from the data in Table S1 in Supplementary Material, referring to D-biotin as standard.
[c]Calculated from the B4F assay data (Table 2) as $100\% \times$ CBB/NCBW.
[d]The recovery in the B4F assay was calculated from the ratio CBB/CBA (see last two columns of Table 2).
[e]D-Biotin was used as standard in the ANS assay. Therefore, its "biotin content" was set to 100% by definition, as indicated by the triple bar (≡).
[f]Biocytin was used as standard for the B4F assay. Therefore, its biotin content was set to 97.7% (as determined in the ANS assay) and its recovery was 100% by definition, as indicated by the triple bar (≡).
[g]Average value from two titrations.

(Ebner et al., 2008) and it is completely destroyed during the HCl digestion of proteins (unpublished results). For these reasons we decided to henceforth use biocytin as standard, taking into account its biotin content of 97.7%.

3.2 Concept of the new competitive B4F binding assay

The main drawback of quantifying the biotin content by ANS displacement is the high consumption of sample. One measurement requires ~100 μL of approximately 400 μM biotinylated compound (i.e., 40 nmol). We therefore set out to establish a more sensitive homogeneous assay.

The principles of the new biotin assay are illustrated in Fig. 1. In a series of tubes (or wells), avidin is always mixed with exactly the same amount of biotin-4-fluorescein (B4F). The amount of B4F exceeds the amount of binding sites by ~10%. This means that in absence of a biotinylated analyte (panel A) ~90% of B4F molecules are bound and strongly quenched (Kada, Falk, & Gruber, 1999). In presence of a small amount of biotinylated analyte (panel B), only the remaining binding sites in avidin can be occupied by B4F, leading to a higher fraction of free and fluorescent B4F. Panel C illustrates the situation when the sample contains just the right amount of biotinylated analyte to saturate all binding sites in avidin. Here, the addition of B4F leads to maximal fluorescence because all B4F molecules are unbound and unquenched. No further fluorescence increase is seen when mixing avidin with higher amounts of biotinylated analyte before adding B4F (panel D).

A feasibility test of this competitive B4F binding assay had been reported with D-biotin as the analyte (Ebner et al., 2008). In the present study the assay conditions and the data evaluation method were standardized and the general applicability to different biotin derivatives was tested as described below.

3.3 Adjustment of the working conditions for the competitive B4F binding assay

The reasons for choosing a 10% excess of free B4F over the number of avidin's ligand-binding sites in the assay mixture are the following: (i) It is essential to avoid any excess of binding sites over B4F because otherwise small amounts of a biotinylated analyte would not cause a response of the assay. (ii) It would be difficult to reproducibly adjust an exact match between binding sites and B4F in each assay tube. (iii) The 10% fraction of unbound B4F in the blank sample is important to ensure fast association of B4F with avidin within the standard incubation time (10 min).

The method for finding the proper assay condition is shown in Fig. 4. In parallel tubes, a constant amount of B4F (16 pmol) was mixed with

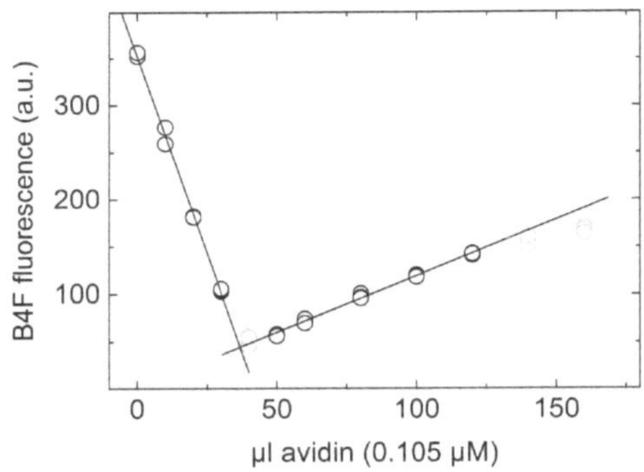

Fig. 4 Determination of optimal assay conditions by non-cumulative titration of a constant amount of B4F with increasing amounts of avidin. In a series of tubes, the specified volume (µL) of the 0.105 µM avidin working solution No. 1 (batch 2) was mixed with buffer B at a final volume of 900 and 100 µL 16.0 µM B4F was added. The resulting B4F fluorescence was plotted vs avidin consumption per tube. Data shown in gray was excluded from linear fitting. The intersection point (37.01 µL × 0.105 µM = 3.87 pmol avidin tetramer) indicated saturation of all binding sites by B4F. From this result, the desired concentration of the avidin "working solution No. 2" (batch 2) was calculated as 0.9 × 3.87 pmol/100 µL = 34.8 nM.

increasing amounts of avidin in a final volume of 1 mL. As expected (Kada, Kaiser, Falk, & Gruber, 1999), this titration resulted in a linear decrease of B4F fluorescence, with a sharp minimum at 37.0 μL when the number of binding sites matched the number of B4F molecules in the sample.

It is of utmost importance to use the same amount of B4F (16 pmol per tube) in the pre-test (Fig. 4) as in the final assay (Figs. 5 and 6), as well as a constant amount of avidin which binds 90% of B4F. According to Fig. 4, the proper volume of the "avidin working solution No. 1" was $0.9 \times 37.0 = 33.3$ μL. Using 100 μL volume for convenience and accuracy, the "working solution No. 1" was diluted 100/33.3, resulting in "working solution No. 2."

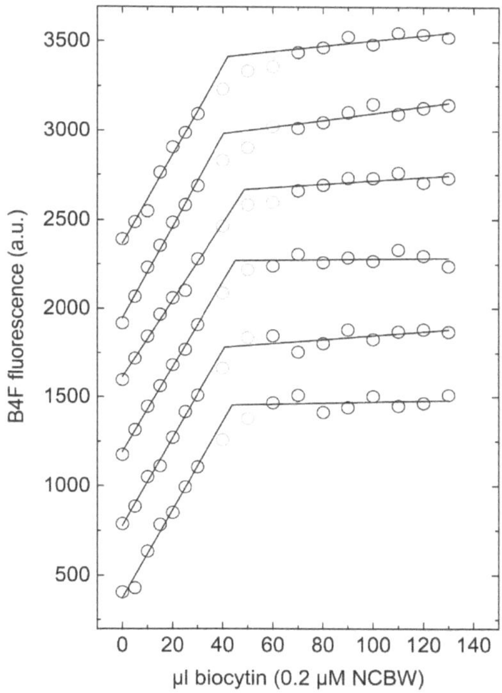

Fig. 5 Competitive B4F binding assay for determination of the biotin group content in the biocytin standard. In separate tubes, the specified volume of a nominally 0.2 μM biocytin solution was mixed with buffer B at a final volume of 800 μL. 100 μL of avidin working solution No. 2 (batch 2, 34.8 nM according to Fig. 4) was added. After 10 min, 100 μL 160 nM B4F was added and after another 10 min the fluorescence of B4F was measured. Data shown in gray was excluded from the linear fits. The intersection point (mean value 43.44 ± 3.10 μL) indicated saturation of all binding sites by avidin. In order to ensure better visibility, the individual titration curves were vertically stacked with an increment of 400 units.

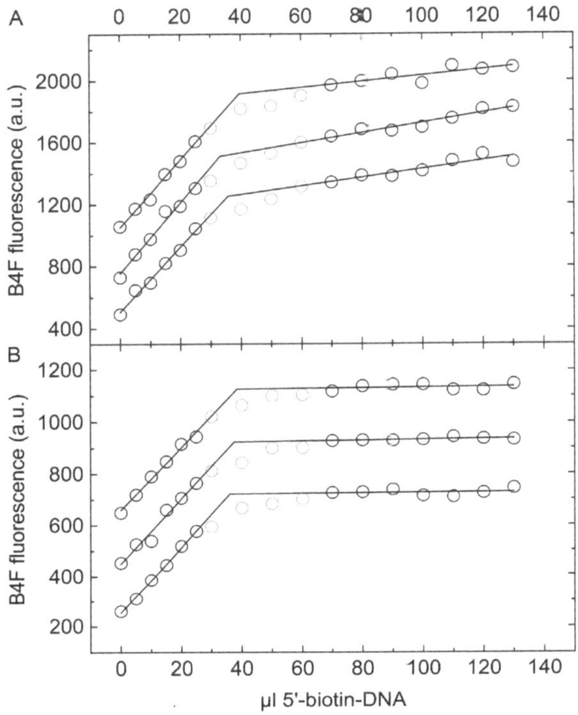

Fig. 6 Application of the competitive B4F binding assay to 5′-biotin-DNA (A) before digestion and (B) after digestion with exonuclease I. Undigested or digested 5′-biotin-DNA was mixed with buffer B at a final concentration of 0.200 or 0.207 µM, respectively (according to A_{260}). These working solutions were used for the analogous titration experiment as shown for biocytin in Fig. 5. All other details are the same as in Fig. 5. In order to ensure better visibility, the individual titration curves were vertically stacked with an increment of 300 units in A and 200 units in B, respectively.

Of note, the "effective concentration" values of the avidin (34.8 nM) and B4F (160 nM) solutions are only used for adjustment of the proper working conditions. These numbers have no influence on the evaluation of the final B4F binding assay because the latter is calibrated with a certified standard.

3.4 Determination of biotin groups by the competitive B4F binding assay

The calibration of the B4F binding assay is presented in Fig. 5. In each titration experiment, a constant amount of avidin was mixed with increasing amounts of biocytin, followed by the addition of a constant amount of B4F which represented a 10% excess over the number of binding sites.

In absence of biocytin the fluorescence was minimal, as expected for maximal binding of B4F. Increasing amounts of biocytin caused a linear rise in B4F fluorescence up to complete saturation of avidin with biocytin, followed by a constant high fluorescence level when the amount of biocytin exceeded the number of binding sites in the sample. The linear rise and the horizontal plateau were fitted by straight lines and the intersection point was determined using the piecewise fitting procedure of Origin®.

The biocytin solution used in Fig. 5 had a nominal concentration of 0.200 μM, corresponding to an effective concentration of 0.195 μM according to the ANS assay data (Table 2). Consequently, the average breakpoint for six titration experiments at 43.4 ± 3.1 μL of 0.195 μM biocytin standard indicated that with this batch of avidin working solution No. 2 the breakpoint always occurred when 8.49 pmol of a biotin derivative are present in the sample (Table S2 in Supplementary Material in the online version at https://doi.org/10.1016/bs.mie.2019.10.028).

Subsequently, the same titration as in Fig. 5 was performed with nominally 0.200 μM solutions of biotin-PEG_1-amine and biotin-PEG_8-amine (Fig. S8 in Supplementary Material in the online version at https://doi.org/10.1016/bs.mie.2019.10.028). Division of 8.49 pmol by the volume

Table 2 Measurement of the biotin content of various biotin derivatives by the competitive B4F binding assay.

Biotin derivative	NCBW[a] (μM)	Conc. from A_{260} (μM)	CBA[b] (μM)	CBB[c] (μM)
Biocytin	0.200	–	0.195 ± 0.004	$\equiv 0.195 \pm 0.014$[d]
Biotin-C_3-amine	0.200	–	0.174	0.173 ± 0.006
Biotin-PEG_1-amine	0.200	–	0.195	0.186 ± 0.006
Biotin-PEG_2-amine	0.200	–	0.178	0.187 ± 0.009
Biotin-PEG_8-amine	0.200	–	0.194	0.195 ± 0.006
5′-biotin-DNA	n.d.[e]	0.200	n.d.[e]	0.235 ± 0.021
5′-biotin-DNA, digested	n.d.[e]	0.207	n.d.[e]	0.227 ± 0.007

[a]"Nominal concentration by weight."
[b]"Concentration by ANS assay," calculated by multiplication of NCBW with the biotin content according to the ANS assay (Table 1).
[c]"Concentration by the B4F assay," determined as described in Table S2 in Supplementary Material.
[d]Biocytin with the known biotin content of 97.7% (according to the ANS assay) was used as standard in the B4F assay. Therefore, the concentration of nominally 0.200 μM biocytin in the B4F assay was 0.195 μM by definition, as indicated by the triple bar (≡).
[e]The amount of 5′-biotin-DNA was too little for preparing a known concentration by weight and for the ANS assay.

of biotin derivative consumed at the intersection point gave the effective biotin concentration in the nominally 0.200 μM biotin derivative solution (last column in Table 2, batch 2).

The analogous procedure was performed with a separate batch of the "avidin working solution No. 2." Here, the calibration gave a breakpoint at 40.8 ± 2.3 μL 0.195 μM biocytin (7.97 pmol, Fig. S9A and Table S2 in Supplementary Material in the online version at https://doi.org/10.1016/bs.mie.2019.10.028, batch 3) and this batch was used to determine the effective biotin content in nominally 0.200 μM solutions of biotin-C_3-amine and biotin-PEG_2-amine (Fig. S9 and Table S2 in Supplementary Material in the online version at https://doi.org/10.1016/bs.mie.2019.10.028).

As illustrated in Fig. 1, the B4F binding assay relies on blocking of all four binding sites in avidin by the biotinylated analyte. In this respect, problems were expected with large analytes, such as DNA. We thus tested single-stranded biotin-5′-DNA composed of 30 nucleotides (9199 g/mol, Fig. 6A). When avidin was titrated with this biotin-5′-DNA, a quasi-linear rise and a breakpoint were observed but the usual plateau after the breakpoint was no longer horizontal. We took this as an indication of mutual steric hindrance when four large biotin-5′-DNA molecules were to be bound per avidin molecule.

Digestion of an aliquot of the same biotin-5′-DNA with a 3′-exonuclease resulted in single strands with ≤5 nucleotides. The proper digestion conditions were found by digesting the analogous TMR-5′-DNA and analyzing the resulting fragment size by gel filtration (Figs. S3 and S4 in Supplementary Material in the online version at https://doi.org/10.1016/bs.mie.2019.10.028). After the digestion of biotin-5′-DNA, the B4F binding assay showed a much more horizontal plateau (Fig. 6B) but the consumption of biotin-5′-DNA was still very similar as before the digestion (Fig. 6A), as also reflected in the derived "concentration by B4F assay" (CBB) for undigested and digested biotin-5′-DNA (last column in Table 2). As expected, these numbers show that digestion of biotin-DNA gives a narrower random error in the B4F binding assay, although the extent of deviation in the breakpoint with undigested biotin-DNA was not as large as expected.

The locations of the breakpoints in Fig. 6A and B indicated binding of four, rather than only two, biotin-5′-DNA molecules per avidin tetramer. This can be concluded from the comparison between the nominal concentration of the biotin-5′-DNA solution (0.200 μM according to A_{260}, third column in Table 2) with the concentration measured by the B4F binding assay (0.235 and 0.227 μM, last column). The deviation from the concentration calculated from A_{260} is attributed to the fact that the molar absorptivity ($\varepsilon_{260} = 306 \times 10^3\,M^{-1}\,cm^{-1}$) obtained by OligoCalc is an estimate (Kibbe, 2007).

A comparison between the well-established ANS assay and the more sensitive B4F assay is presented in Table 1. Biocytin with a biotin content of 97.7% was used as the common standard, therefore its recovery in the B4F assay (i.e., the agreement with the ANS assay) was defined as 100% (last column in Table 1). For the other biotinylated analytes, the relative deviation of the B4F assay from the ANS assay was within the range of ±5%.

The major advantage of the new B4F-based titration assay was much lower sample consumption. One HABA or ANS titration requires 100 μL of 400 μM biotinylated analyte (40 nmol), whereas one complete B4F titration required <0.6 nmol of sample (17 × 3 tubes with an average dosage of <80 μL 0.2 μM biotin per tube, see Fig. 5). A further 10-fold reduction in sample consumption in the new B4F assay could be achieved by switching from cuvettes to the MTP format as realized in commercially available HABA assay kits. Alternatively, the sample consumption could further be reduced by working at lower avidin concentrations in all three above mentioned assays. This strategy, however, not only implies a lower signal-to-noise ratio but also a loss of the clear-cut breakpoint in case of larger biotinylated analytes (Kaiser, Marek, Haselgrübler, Schindler, & Gruber, 1997).

The tradeoff for higher sensitivity of the new B4F assay, as compared to the ANS assay, is the lower precision of the B4F assay (see Table 1). This is in part attributed to the multiple pipetting steps in the latter method.

Most importantly, the B4F assay data were in good agreement with the reliable results from the ANS assay, expressed as "recovery" in the last column of Table 1. The "trueness" of the B4F assay originates from the assay principle: As in the HABA and ANS titration assays, the amount of biotinylated analyte is not directly calculated from a measured signal but from the consumption required for saturation of all binding sites in avidin, which is the breakpoint in the fluorescence profile. This makes the B4F assay robust to very high background color and background fluorescence of the biotin sample, as already shown in the first feasibility test of this assay (Ebner et al., 2008).

4. Conclusions

The goal of this study was to establish a biotin group assay which requires much less sample than the HABA and ANS assay, while retaining the high robustness and reliability of these conventional biotin group assays. This goal has been successfully reached with the new titration assay where a standardized avidin solution is progressively saturated with the biotinylated analyte, and a constant amount of B4F is added to determine the fraction of unoccupied binding sites. The sample consumption is two orders of magnitude smaller than in the HABA or ANS titration assays but one order of magnitude higher than when using FITC-labeled (strept)avidin as the probe molecule. The new B4F binding assay shows a linear dose-response and a clear-cut breakpoint when all binding sites are saturated with analyte. Consequently, the new assay can be calibrated with biotin (or biocytin) and does not require calibration with the biotin derivative of interest. In conclusion, the new B4F assay provides for a balanced compromise in terms of sensitivity, precision, trueness, and easy calibration. Furthermore, the new assay is insensitive for the background fluorescence signal, making it attractive choice for colorful samples.

Acknowledgment

This work was supported by the Austrian Science Funds (FWF, project FWFW1250 "DK NanoCell").

References

Al-Hakiem, M. H., Landon, J., Smith, D. S., & Nargessi, R. D. (1981). Fluorimetric assays for avidin and biotin based on biotin-induced fluorescence enhancement of fluorescein-labeled avidin. *Analytical Biochemistry*, *116*(2), 264–267.

Barbarakis, M. S., Smith-Palmer, T., Bachas, L. G., Chen, S. Y., & Van Der Meer, B. W. (1993). Enhancement of the emission intensity of fluorophore-labeled avidin by biotin and biotin derivatives. Evaluation of different fluorophores for improved sensitivity. *Talanta*, *40*(7), 1139–1145.

Batchelor, R. H., Sarkez, A., Cox, W. G., & Johnson, I. (2007). Fluorometric assay for quantitation of biotin covalently attached to proteins and nucleic acids. *BioTechniques*, *43*(4), 503–507.

Baur, B., Suormala, T., Bernoulli, C., & Baumgartner, E. (1998). Biotin determination by three different methods: Specificity and application to urine and plasma ultrafiltrates of patients with and without disorders in biotin metabolism. *International Journal for Vitamin and Nutrition Research*, *68*(5), 300–308.

Bayer, E. A., & Wilchek, M. (1990). Protein biotinylation. *Methods in Enzymology*, *184*, 138–160.

Der-Balian, G. P., Gomez, B., Masino, R. S., & Parce, J. W. (1990). A fluorometric method for determining the degree of biotinylation of proteins. *Journal of Immunological Methods*, *126*(2), 281–285.

Dundas, C. M., Demonte, D., & Park, S. (2013). Streptavidin-biotin technology: Improvements and innovations in chemical and biological applications. *Applied Microbiology and Biotechnology*, *97*, 9343–9353.

Ebner, A., Marek, M., Kaiser, K., Kada, G., Hahn, C. D., Lackner, B., et al. (2008). Application of biotin-4-fluorescein in homogeneous fluorescence assays for avidin, streptavidin, and biotin or biotin derivatives. *Methods in Molecular Biology*, *418*, 73–88.

Green, N. M. (1965). A spectrophotometric assay for avidin and biotin based on binding of dyes by avidin. *The Biochemical Journal*, *94*, 23C–24C.

Green, N. M. (1990). Avidin and streptavidin. *Methods in Enzymology*, *184*, 51–67.

Groman, E. V., Rothenberg, J. M., Bayer, E. A., & Wilchek, M. (1990). Enzymatic and radioactive assays for biotin, avidin, and streptavidin. *Methods Enzymol*, *184*, 208–217.

Ikeda, T., Miyao, H., & Sueda, S. (2014). Ultrasensitive biotin assay of a noncompetitive format in a homogeneous solution based on resonance energy transfer induced by a protein–protein interaction. *Analytical Chemistry*, *86*(12), 5673–5677. https://doi.org/10.1021/ac501180n.

Indyk, H. E., Gill, B. D., & Woollard, D. C. (2014). Biotin content of paediatric formulae, early lactation milk and seasonal bovine milk powders by biosensor immunoassay. *International Dairy Journal*, *35*, 25–31. https://doi.org/10.1016/j.idairyj.2013.10.002.

Kada, G., Falk, H., & Gruber, H. J. (1999). Accurate measurement of avidin and streptavidin in crude biofluids with a new, optimized biotin–fluorescein conjugate. *Biochimica et Biophysica Acta (BBA)—General Subjects*, *1427*(1), 33–43. https://doi.org/10.1016/S0304-4165(98)00178-0.

Kada, G., Kaiser, K., Falk, H., & Gruber, H. J. (1999). Rapid estimation of avidin and streptavidin by fluorescence quenching or fluorescence polarization. *Biochimica et Biophysica Acta (BBA)—General Subjects*, *1427*(1), 44–48. https://doi.org/10.1016/S0304-4165(98)00177-9.

Kaiser, K., Marek, M., Haselgrübler, T., Schindler, H., & Gruber, H. J. (1997). Basic studies on heterobifunctional biotin – PEG conjugates with a 3-(4-pyridyldithio)propionyl marker on the second terminus. *Bioconjugate Chemistry*, *8*(4), 545–551. https://doi.org/10.1021/bc970086u.

Kibbe, W. A. (2007). OligoCalc: An online oligonucleotide properties calculator. *Nucleic Acids Research*, *35*(Web Server issue), W43–W46. https://doi.org/10.1093/nar/gkm234.

Lin, H. J., & Kirsch, J. F. (1977). A sensitive fluorometric assay for avidin and biotin. *Analytical Biochemistry*, *81*(2), 442–446.

Lin, H. J., & Kirsch, J. F. (1979). A rapid, sensitive fluorometric assay for avidin and biotin. *Methods in Enzymology*, *62*, 287–289.

Livaniou, E., Costopoulou, D., Vassiliadou, I., Leondiadis, L., Nyalala, J. O., Ithakissios, D. S., et al. (2000). Analytical techniques for determining biotin. *Journal of Chromatography. A*, *881*(1–2), 331–343.

Mock, D. M. (1990). Sequential solid-phase assay for biotin based on 125I-labeled avidin. *Methods in Enzymology*, *184*, 224–233.

Mock, D. M. (1997). Determinations of biotin in biological fluids. *Methods in Enzymology*, *279*, 265–275.

Mock, D. M., & Horowitz, P. (1990). Fluorometric assay for avidin-biotin interaction. *Methods in Enzymology*, *184*, 234–240.

Mock, D. M., Lankford, G., & Horowitz, P. (1988). A study of the interaction of avidin with 2-anilinonaphthalene-6-sulfonic acid as a probe of the biotin binding site. *Biochimica et Biophysica Acta*, *956*(1), 23–29.

Murakami, H., Hohsaka, T., Ashizuka, Y., Hashimoto, K., & Sisido, M. (2000). Site-directed incorporation of fluorescent nonnatural amino acids into streptavidin for highly sensitive detection of biotin. *Biomacromolecules*, *1*(1), 118–125.

Nargessi, R. D., & Smith, D. S. (1986). Fluorometric assays for avidin and biotin. *Methods in Enzymology*, *122*, 67–72.

Oberoi, P., Xiaoyi, S., Zhang, C.-J., Barrow, S.-w., Dzantiev, L., Davydov, I. V., et al. (2014). *Characterization of biotinylated antibodies and other proteins*. AAPS Abstracts.

Przyjazny, A., Hentz, N., & Bachas, L. (1993). Sensitive and selective liquid chromatographic postcolumn reaction detection system for biotin and biocytin using a homogeneous fluorophore-linked assay. *Journal of Chromatography. A*, *654*(1), 79–86.

Rao, S., Anderson, K., & Bachas, L. (1997). Determination of the extent of protein biotinylation by fluorescence binding assay. *Bioconjugate Chemistry*, *8*(1), 94–98.

Rispens, T., & Ooijevaar-de Heer, P. (2016). Quantification of the degree of biotinylation of proteins using proteinase K digestion and competition ELISA. *Journal of Immunological Methods*, 430, 61–63. https://doi.org/10.1016/j.jim.2016.01.009.

Staggs, C. G., Sealey, W. M., McCabe, B. J., Teague, A. M., & Mock, D. M. (2004). Determination of the biotin content of select foods using accurate and sensitive HPLC/avidin binding. *Journal of Food Composition and Analysis*, *17*(6), 767–776.

CHAPTER TWO

(Strept)avidin as a template for ligands other than biotin: An overview

Vesa P. Hytönen*
Faculty of Medicine and Health Technology and BioMediTech, Tampere University, Tampere, Finland
Fimlab Laboratories, Tampere, Finland
*Corresponding author: e-mail address: vesa.hytonen@tuni.fi

Contents

Abstract

Chicken avidin and bacterial streptavidin are workhorses in biotechnology. We have used avidin as a scaffold protein to develop avidin variants with novel ligand-binding affinity, so-called antidins. This article covers the strategy applied in the development of antidins. Using a phage display developed for avidin, immobilized ligands were used to select binders from a phage pool displaying avidin variants with randomized sequence in the protein loops. Antidins binding various ligands with nanomolar affinity were obtained. Antidins have already been demonstrated to be suitable for a diagnostic assay measuring serum progesterone levels and they offer a promising alternative to antibodies for the recognition of small molecules.

Chicken avidin purified from egg white (Green, 1975) and bacterial streptavidin originally isolated from *Streptomyces avidinii* (Chaiet & Wolf, 1964) are the best-known members of the avidin protein family, collectively called (strept)avidin. These biotin-binding proteins have been widely utilized in biotechnology (Laitinen, Nordlund, Hytönen, & Kulomaa, 2007). Later, numerous avidins were characterized structurally and functionally both from prokaryotes and eukaryotes. Rhizavidin, produced using the genetic information obtained from *Rhizobium etli*, is such an example, having a dimeric structure (Helppolainen et al., 2007) and therefore representing an atypical ligand-binding architecture compared to tetrameric (strept)avidin, where the ligand-binding affinity results from the structural contribution of two subunits (Hendrickson et al., 1989; Livnah, Bayer,

Methods in Enzymology, Volume 633
ISSN 0076-6879
https://doi.org/10.1016/bs.mie.2019.10.029

Wilchek, & Sussman, 1993). Later, efforts have been made to convert rhizavidin into a monomeric protein with high affinity toward biotin (Lee et al., 2016).

The tight interaction between (strept)avidin and biotin has motivated scientists to reveal the cornerstones of this extremely high affinity ($K_d \sim 10^{-15}$ M). We have learned with the help of structural biology, mutagenesis work and simulations, that the excellent structural complementarity between ligand and binding pocket (Weber, Ohlendorf, Wendoloski, & Salemme, 1989), a hydrogen bond network (Hyre et al., 2006; Klumb, Chu, & Stayton, 1998) and interactions mediated by hydrophobic residues (Freitag et al., 1998) all are required for the tight biotin binding. Detailed studies also revealed the contribution of particular amino acid residues to the binding affinity (Freitag et al., 1998; Klumb et al., 1998).

The development of novel receptors for small molecules is not straightforward. However, there is a need for selective, high-affinity capture of small molecules for instance in chemical synthesis and as diagnostic tools. Therefore, we speculated that avidin, possessing a ligand-binding site optimized for a small ligand, could be potentially utilized as scaffold protein in such a work. When equipped with a bacterial signal peptide, functional avidin can be produced in *E. coli* (Hytönen, Laitinen, et al., 2004).

To develop a method for the screening of novel binding activities, we created a M13 bacteriophage displaying functional avidin. Two different strategies were applied in the first study (Riihimäki, Hiltunen, et al., 2011; Riihimäki, Kukkurainen, et al., 2011) to display avidin on the surface of the M13 phage in fusion with the C-terminal region of the minor coat protein pIII (3 + 3 display mode used). In the first approach, avidin was produced solely fused with pIII, whereas in the avidin/avidin-pIII display strategy, free avidin subunits were produced in addition to the avidin-pIII fusion. These experiments revealed proteolytic cleavage of a significant portion of the avidin-pIII fusion proteins. Therefore production of the free avidin subunits was not absolutely necessary to achieve functional tetrameric avidins, allowing the use of DNA libraries to produce an array of avidin variants (Riihimäki, Hiltunen, et al., 2011; Riihimäki, Kukkurainen, et al., 2011). However, later we introduced a version of the expression cassette, where the amber stop codon was used between pIII and avidin to ensure the availability of free avidin subunits for the assembly of the tetramer on the phage surface (Niederhauser et al., 2012).

To achieve novel binding specificities, the previously determined structures of the avidin-biotin complex were analyzed, and the randomization of the avidin sequence was targeted to the loop regions between β-strands. The first study, where only the loop between β-strands 1 and 2 was manipulated, yielded highly promising results and the avidin variant sbAvd-1 (steroid-binding avidin) having binding affinity toward testosterone was obtained (Riihimäki, Hiltunen, et al., 2011; Riihimäki, Kukkurainen, et al., 2011). This binding, however, was rather weak ($K_d = 9 \times 10^{-6}$ M; determined using surface plasmon resonance), and the binding toward testosterone was inhibited by dehydroepiandrosterone, androstenedione and biotin, indicating poor specificity. Further randomized mutagenesis of the loop between β-strands 3 and 4 yielded the avidin variant sbAvd-2, possessing a substantially decreased affinity to biotin ($K_d = 6.6 \times 10^{-7}$ M), while the binding affinity toward testosterone was not affected significantly ($K_d = 1.1 \times 10^{-5}$ M; Riihimäki, Hiltunen, et al., 2011; Riihimäki, Kukkurainen, et al., 2011).

The developed avidin variants, also called antidins, showed high thermal stability ($T_m = 80.6$ °C for sbAvd-1; $T_m = 82.5$ °C for sbAvd-2), comparable to that of wild-type avidin ($T_m = 85.5$ °C; Riihimäki, Hiltunen, et al., 2011; Riihimäki, Kukkurainen, et al., 2011). This indicated that the avidin fold is robust and the loops of avidin can be manipulated without compromising the thermal stability. Although not systematically investigated in all previous studies, the vast amount of data obtained from (strept)avidin mutagenesis studies supports this conclusion (Laitinen, Hytönen, Nordlund, & Kulomaa, 2006).

The second generation of antidins were developed using three different DNA libraries (Lehtonen et al., 2016). In the first library, four residues within the loop between β-strands 5 and 6 were randomized. The second library randomized five residues in the loop between β-strands 1 and 2 as well as four residues within the loop between β-strands 5 and 6. The third library randomized four residues within the loop between β-strands 3 and 4. These libraries were then displayed on M13 bacteriophage and subjected to selection using magnetic nanoparticles functionalized with target ligands. Out of the seven ligands used (progesterone, testosterone, hydrocortisone, cholic acid, folic acid, ketoprofen, ibuprofen), all but ibuprofen yielded an increase in the specific response toward a ligand-HSA-coated surface compared to the HSA reference, indicating success in the selection of antidins with novel binding functionality (Lehtonen et al., 2016).

The antidin variants enriched by the phage selection protocol were then expressed with a C-terminal His_6 tag and characterized for ligand-binding affinity and thermal stability. The measured ligand-binding affinities were in the nano-micromolar range, for example, sbAvd-5 bound progesterone with $K_d = 108$ nM, fabAvd-a bound folic acid with $K_d = 188$ nM and hbAvd-1 bound hydrocortisone with $K_d = 1513$ nM (Lehtonen et al., 2016). These ligands were also found to stabilize the corresponding proteins as evidenced by differential scanning calorimetry. We also determined their oligomeric state using analytical gel filtration (Lehtonen et al., 2016).

Previous studies have suggested tetramer stability to be linked with the recombinant production yield of avidin (Hytönen, Määttä, et al., 2005). We therefore speculated that including the mutation I117Y, motivated by avidin-related protein 4 (Hytönen, Nyholm, et al., 2004), to engineered antidins might lead to higher yields and improved thermal and chemical stability (Määttä et al., 2011), making the novel avidin variants more attractive for various applications. As expected, a higher thermal stability was reached ($\Delta T_m = 15$–$35\,°C$) and we obtained two- to sixfold improved protein yields, but we also observed slight changes in the ligand-binding affinity (Lehtonen et al., 2016). This is not surprising, as the residue Asn118 is known to be in direct contact with biotin, and the mutation I117Y affecting the neighboring residue might cause slight changes in the surrounding protein structure, enhancing the interaction between avidin subunits (Hytönen, Määttä, et al., 2005). This increased stability could be one of the reasons why we were successful in determining the 3D structure of sbAvd-2(117Y) by X-ray crystallography (Agrawal et al., 2019).

The third generation of antidins were based on chemically synthesized DNA libraries generated through trinucleotide mutagenesis technology to obtain maximal diversity and good sequence coverage with a reasonable library size. The libraries were based on the monovalent phage display mode (3 + 3) and utilized Gateway recombination cloning (Lehtonen et al., 2015). Since we included an amber stop codon between the avidin variants and the C-terminal pIII, efficient surface display was achieved, as described in Niederhauser et al. (2012). Three selection rounds with an automated magnetic particle processor resulted in three novel antidin forms (sbAvd7-9 having dissociation constants of 5–13 nM for progesterone, Agrawal et al., 2019). We performed a more thorough biophysical characterization for one of the avidin forms, sbAvd-7, which revealed its tetrameric state and high thermal stability ($T_m = 85.84\,°C$; $T_m = 96.07\,°C$ with progesterone). Importantly, the biotin binding affinity of sbAvd-7 was clearly diminished and we observed negligible thermal stabilization upon addition of biotin ($\Delta T_m = 0.27\,°C$). The use of molecular docking revealed several potential reasons for the increased affinity of sbAvd-7 compared to sbAvd-2

(Agrawal et al., 2019), but the exact molecular origins of the enhanced affinity remain to be discovered.

These studies enable us to draw the following conclusions:

(1) Avidin is a suitable protein to be used as a molecular scaffold

The engineered avidin forms have high thermal stability and remain soluble, indicating proper protein folding. The tetrameric structure of avidin is a potential limitation for phage display, but the use of the amber stop codon appears to be a suitable solution to overcome this challenge (Lehtonen et al., 2015). Various antidin forms have been purified in reasonable yield.

(2) The avidin ligand-binding site has advantages over antibodies when developing binders for small molecules

The ligand-binding site of avidin is a deep cavity, which might be optimal for relatively small molecules (~200–300 Da). Larger molecules (such as folic acid, MW = 441 Da) might not be ideal targets for the development of high-affinity binders. An elongated shape of the ligand might be an optimal starting point.

(3) Predicting the ligand-binding affinity for novel variants is difficult

We have noticed that changes applied to the avidin sequence may cause "side effects," as exemplified by the mutation I117Y. Therefore, efficient screening methods are absolutely necessary.

Would antidins be the solution for addressing the needs for novel ligand binders? Most likely they cannot replace antibodies, but our results are promising. The progesterone-binding avidin variant was successfully used to assay serum progesterone from dog samples (Agrawal et al., 2019). Recent studies by Skerra and coauthors have shown that close relatives to avidin, lipocalins, can be converted to bind for instance vascular endothelial growth factor receptor 3 with picomolar affinity (Richter & Skerra, 2017), β-amyloid peptide with picomolar affinity while being capable of inhibiting amyloid formation (Rauth et al., 2016) and prostate-specific membrane antigen, detecting the target protein on live cells (Barinka et al., 2016). The first anticalins are entering clinical-stage research (Rothe & Skerra, 2018).

Another interesting approach to utilize (strept)avidin has been shown by Ward and colleagues, who built artificial metalloenzymes based on genetically engineered (strept)avidins (reviewed in Davis & Ward, 2019). In their recent example, streptavidin variants were used to obtain good activity and selectivity for the reduction of prochiral imines (Wu et al., 2019). While their approach was based on tandem dimers of streptavidin subunits, our experience with chicken avidin suggests that circular permutation could be a powerful tool in situations, where two or more binding functions need to be precisely positioned within the protein (Hytönen, Nordlund, et al., 2005; Hytönen et al., 2006; Leppiniemi et al., 2011; Määttä et al., 2008; Nordlund et al., 2004, 2005; Riihimäki, Hiltunen, et al., 2011; Riihimäki, Kukkurainen, et al., 2011).

(Strept)avidin-biotin technology is one of the cornerstones of biotechnology. We have demonstrated that avidin has the potential to become exploited for the creation of novel binding functions. The current molecular tool pack contains various ligand-binding specificities as well as solutions to control both the oligomeric assembly and the oligomeric state of the protein. The future of the avidin family appears therefore bright.

Acknowledgments

I would like to acknowledge all the colleagues and collaborators who contributed to these studies and enabled the progress made. My specific thanks go to Dr. Soili Lehtonen, who had a central role in the antidin research and also provided comments for the manuscript. I also thank Dr. Steffi Goffart for her comments for the manuscript draft.

References

Agrawal, N., Lehtonen, S. I., Uusi-Mäkelä, M., Jain, P., Viitala, S., Määttä, J. A. E., et al. (2019). Molecular features of steroid-binding antidins and their use for assaying serum progesterone. *PLoS One, 14* e0212339.

Barinka, C., Ptacek, J., Richter, A., Novakova, Z., Morath, V., & Skerra, A. (2016). Selection and characterization of anticalins targeting human prostate-specific membrane antigen (PSMA). *Protein Engineering, Design & Selection, 29*, 105–115.

Chaiet, L., & Wolf, F. J. (1964). The properties of streptavidin, a biotin-binding protein produced by streptomycetes. *Archives of Biochemistry and Biophysics, 106*, 1–5.

Davis, H. J., & Ward, T. R. (2019). Artificial metalloenzymes: Challenges and opportunities. *ACS Central Science, 5*, 1120–1136.

Freitag, S., Le Trong, I., Chilkoti, A., Klumb, L. A., Stayton, P. S., & Stenkamp, R. E. (1998). Structural studies of binding site tryptophan mutants in the high-affinity streptavidin-biotin complex. *Journal of Molecular Biology, 279*, 211–221.

Green, N. M. (1975). Avidin. *Advances in Protein Chemistry, 29*, 85–133.

Helppolainen, S. H., Nurminen, K. P., Määttä, J. A., Halling, K. K., Slotte, J. P., Huhtala, T., et al. (2007). Rhizavidin from rhizobium etli: The first natural dimer in the avidin protein family. *The Biochemical Journal, 405*, 397–405.

Hendrickson, W. A., Pähler, A., Smith, J. L., Satow, Y., Merritt, E. A., & Phizackerley, R. P. (1989). Crystal structure of core streptavidin determined from multiwavelength anomalous diffraction of synchrotron radiation. *Proceedings of the National Academy of Sciences of the United States of America, 86*, 2190–2194.

Hyre, D. E., Le Trong, I., Merritt, E. A., Eccleston, J. F., Green, N. M., Stenkamp, R. E., et al. (2006). Cooperative hydrogen bond interactions in the streptavidin-biotin system. *Protein Science, 15*, 459–467.

Hytönen, V. P., Hörhä, J., Airenne, T. T., Niskanen, E. A., Helttunen, K. J., Johnson, M. S., et al. (2006). Controlling quaternary structure assembly: Subunit interface engineering and crystal structure of dual chain avidin. *Journal of Molecular Biology, 359*, 1352–1363.
Hytönen, V. P., Laitinen, O. H., Airenne, T. T., Kidron, H., Meltola, N. J., Porkka, E. J., et al. (2004). Efficient production of active chicken avidin using a bacterial signal peptide in *Escherichia coli*. *The Biochemical Journal, 384*, 385–390.
Hytönen, V. P., Määttä, J. A., Nyholm, T. K., Livnah, O., Eisenberg-Domovich, Y., Hyre, D., et al. (2005). Design and construction of highly stable, protease-resistant chimeric avidins. *The Journal of Biological Chemistry, 280*, 10228–10233.
Hytönen, V. P., Nordlund, H. R., Hörhä, J., Nyholm, T. K., Hyre, D. E., Kulomaa, T., et al. (2005). Dual-affinity avidin molecules. *Proteins, 61*, 597–607.
Hytönen, V. P., Nyholm, T. K., Pentikäinen, O. T., Vaarno, J., Porkka, E. J., Nordlund, H. R., et al. (2004). Chicken avidin-related protein 4/5 shows superior thermal stability when compared with avidin while retaining high affinity to biotin. *The Journal of Biological Chemistry, 279*, 9337–9343.
Klumb, L. A., Chu, V., & Stayton, P. S. (1998). Energetic roles of hydrogen bonds at the ureido oxygen binding pocket in the streptavidin-biotin complex. *Biochemistry, 37*, 7657–7663.
Laitinen, O. H., Hytönen, V. P., Nordlund, H. R., & Kulomaa, M. S. (2006). Genetically engineered avidins and streptavidins. *Cellular and Molecular Life Sciences, 63*, 2992–3017.
Laitinen, O. H., Nordlund, H. R., Hytönen, V. P., & Kulomaa, M. S. (2007). Brave new (strept)avidins in biotechnology. *Trends in Biotechnology, 25*, 269–277.
Lee, J. M., Kim, J. A., Yen, T. C., Lee, I. H., Ahn, B., Lee, Y., et al. (2016). A rhizavidin monomer with nearly multimeric avidin-like binding stability against biotin conjugates. *Angewandte Chemie (International Ed. in English), 55*, 3393–3397.
Lehtonen, S. I., Taskinen, B., Ojala, E., Kukkurainen, S., Rahikainen, R., Riihimäki, T. A., et al. (2015). Efficient preparation of shuffled DNA libraries through recombination (Gateway) cloning. *Protein Engineering, Design & Selection, 28*, 23–28.
Lehtonen, S. I., Tullila, A., Agrawal, N., Kukkurainen, S., Kähkönen, N., Koskinen, M., et al. (2016). Artificial avidin-based receptors for a panel of small molecules. *ACS Chemical Biology, 11*, 211–221.
Leppiniemi, J., Määttä, J. A., Hammaren, H., Soikkeli, M., Laitaoja, M., Jänis, J., et al. (2011). Bifunctional avidin with covalently modifiable ligand binding site. *PLoS One, 6* e16576.
Livnah, O., Bayer, E. A., Wilchek, M., & Sussman, J. L. (1993). Three-dimensional structures of avidin and the avidin-biotin complex. *Proceedings of the National Academy of Sciences of the United States of America, 90*, 5076–5080.
Määttä, J. A., Airenne, T. T., Nordlund, H. R., Jänis, J., Paldanius, T. A., Vainiotalo, P., et al. (2008). Rational modification of ligand-binding preference of avidin by circular permutation and mutagenesis. *Chembiochem, 9*, 1124–1135.
Määttä, J. A., Eisenberg-Domovich, Y., Nordlund, H. R., Hayouka, R., Kulomaa, M. S., Livnah, O., et al. (2011). Chimeric avidin shows stability against harsh chemical conditions—Biochemical analysis and 3D structure. *Biotechnology and Bioengineering, 108*, 481–490.
Niederhauser, B., Siivonen, J., Määttä, J. A., Jänis, J., Kulomaa, M. S., & Hytönen, V. P. (2012). DNA family shuffling within the chicken avidin protein family—A shortcut to more powerful protein tools. *Journal of Biotechnology, 157*, 38–49.
Nordlund, H. R., Hytönen, V. P., Hörhä, J., Määttä, J. A., White, D. J., Halling, K., et al. (2005). Tetravalent single-chain avidin: From subunits to protein domains via circularly permuted avidins. *The Biochemical Journal, 392*, 485–491.
Nordlund, H. R., Laitinen, O. H., Hytönen, V. P., Uotila, S. T., Porkka, E., & Kulomaa, M. S. (2004). Construction of a dual chain pseudotetrameric chicken avidin

by combining two circularly permuted avidins. *The Journal of Biological Chemistry*, *279*, 36715–36719.

Rauth, S., Hinz, D., Börger, M., Uhrig, M., Mayhaus, M., Riemenschneider, M., et al. (2016). High-affinity anticalins with aggregation-blocking activity directed against the Alzheimer β-amyloid peptide. *The Biochemical Journal*, *473*, 1563–1578.

Richter, A., & Skerra, A. (2017). Anticalins directed against vascular endothelial growth factor receptor 3 (VEGFR-3) with picomolar affinities show potential for medical therapy and in vivo imaging. *Biological Chemistry*, *398*, 39–55.

Riihimäki, T. A., Hiltunen, S., Rangl, M., Nordlund, H. R., Määttä, J. A., Ebner, A., et al. (2011). Modification of the loops in the ligand-binding site turns avidin into a steroid-binding protein. *BMC Biotechnology*, *11*, 64.

Riihimäki, T. A., Kukkurainen, S., Varjonen, S., Hörhä, J., Nyholm, T. K., Kulomaa, M. S., et al. (2011). Construction of chimeric dual-chain avidin by tandem fusion of the related avidins. *PLoS One*, *6* e20535.

Rothe, C., & Skerra, A. (2018). Anticalin® proteins as therapeutic agents in human diseases. *BioDrugs*, *32*, 233–243.

Weber, P. C., Ohlendorf, D. H., Wendoloski, J. J., & Salemme, F. R. (1989). Structural origins of high-affinity biotin binding to streptavidin. *Science*, *243*, 85–88.

Wu, S., Zhou, Y., Rebelein, J. G., Kuhn, M., Mallin, H., Zhao, J., et al. (2019). Breaking symmetry: Engineering single-chain dimeric streptavidin as host for artificial metalloenzymes. *Journal of the American Chemical Society*, *141*, 15869–15878. https://doi.org/10.1021/jacs.9b06923.

CHAPTER THREE

Engineering siderophores

Sina Rütschlin, Thomas Böttcher*
Department of Chemistry, Konstanz Research School Chemical Biology, Zukunftskolleg, University of Konstanz, Konstanz, Germany
*Corresponding author: e-mail address: thomas.boettcher@uni-konstanz.de

Contents

Abstract

Siderophores have important functions for bacteria in iron acquisition and as virulence factors. In this chapter we will discuss the engineering of cyclic hydroxamate siderophores by various biochemical approaches based on the example of *Shewanella algae*. The marine gamma-proteobacterium *S. algae* produces three different cyclic hydroxamate siderophores as metabolites via a single biosynthetic gene cluster and one of them is an important key player in interspecies competition blocking swarming of *Vibrio alginolyticus*. AvbD is the key metabolic enzyme assembling the precursors into three different core structures and hence an interesting target for metabolic and biochemical engineering. Synthetic natural and unnatural precursors can be converted in vitro with purified AvbD to generate siderophores with various ring sizes ranging from analytical to milligram scale. These engineered siderophores can be applied, for example, as swarming inhibitors against *V. alginolyticus*. Here, we describe the synthesis of the natural and unnatural siderophore precursors HS[*X*]A and provide our detailed

Methods in Enzymology, Volume 633
ISSN 0076-6879
https://doi.org/10.1016/bs.mie.2019.10.030

protocols for protein expression of AvbD, conversion of HS[*X*]A with the enzyme to produce ring-size engineered siderophores and secondly for a biosynthetic feeding strategy that allows to extract engineered siderophores in the milligram scale.

1. Introduction

Iron is poorly soluble in its ferric form but essential for microbial growth. Consequently, most microbes produce and secrete siderophores as high-affinity chelators that make iron available to their cells. The competition for iron has spurred the evolution of a great diversity of different siderophore structures along with specialized receptors for uptake of their corresponding iron complexes. Siderophores also are important virulence factors in pathogenic bacteria (Fischbach, Lin, Liu, & Walsh, 2006) and have been target of intensive studies dissecting their biosynthesis and regulation (Bosello, Robbel, Linne, Xie, & Marahiel, 2011; Dale, Doherty-Kirby, Lajoie, & Heinrichs, 2003; Seyedsayamdost et al., 2012).

The hydroxamate siderophore desferrioxamine B even is used as drug for the treatment of iron overload in patients with beta-thalassemia and has been included on the World Health Organization's (WHO) model list of essential medicines (Aydinok et al., 2015). Furthermore, customized siderophores tuned to different metal ions are of major interest for biotechnological applications.

Numerous approaches have been reported to engineer siderophores biochemically or biosynthetically (Ankenbauer, Staley, Rinehart, & Cox, 1990; Franke, Ishida, & Hertweck, 2014). Commonly employed strategies are, for example, the heterologous expression of the corresponding biosynthetic gene clusters, the manipulation of regulatory genes, mutasynthesis and feeding of a bacterial strain with non-natural precursors (Bilyk & Luzhetskyy, 2016) (Fig. 1A). These strategies give access to complex natural product scaffolds without the need for laborious total synthesis. While great progress has been made with engineering non-ribosomal peptide-synthetases (NRPS) for the production of new natural products (Bozhüyük et al., 2018; Mootz, Schwarzer, & Marahiel, 2002), the siderophores described in this chapter are produced by the class of NRPS-independent siderophore (NIS) synthetases (Kadi, Oves-Costales, Barona-Gomez, & Challis, 2007; Kadi, Song, & Challis, 2008; Oves-Costales, Kadi, & Challis, 2009) (Fig. 1B). Established strategies to engineer cyclic siderophore structures have been mainly based on metabolic engineering with feeding of inhibitors

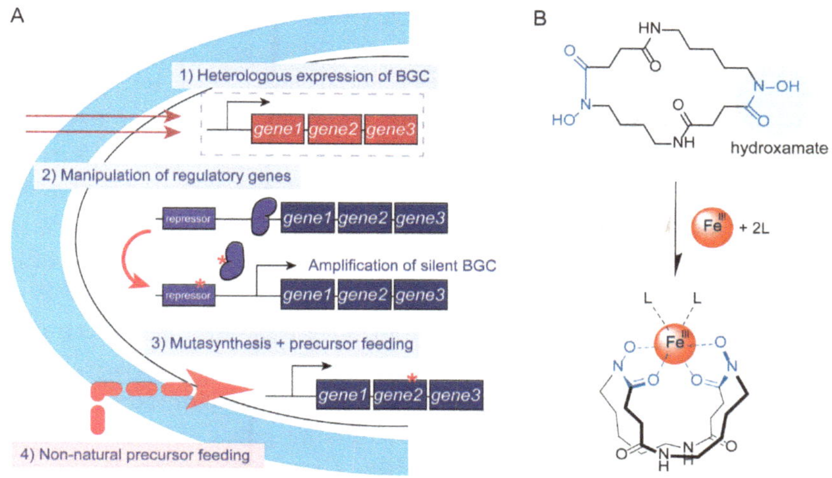

Fig. 1 Metabolic engineering of siderophores. (A) Biochemical and metabolic approaches described in literature to manipulate the biosynthesis of natural products. (B) Ferric iron chelation by hydroxamate siderophores discussed in this chapter.

and non-natural precursors (Richardson-Sanchez & Codd, 2018; Richardson-Sanchez, Tieu, Gotsbacher, Telfer, & Codd, 2017; Soe, Pakchung, & Codd, 2012; Soe, Telfer, Levina, Lay, & Codd, 2016). An alternative strategy is enzyme-free metal-templated synthesis. Hereby synthetic hydroxamate building blocks are preorganized as ligands of a metal ion and subsequently assembled to siderophores by peptide coupling (Lifa, Tieu, Hocking, & Codd, 2015; Sresutharsan, Tieu, Richardson-Sanchez, Soe, & Codd, 2017). In this chapter, we will showcase the manipulation of macrocyclic hydroxamate siderophore biosynthesis based on promiscuous enzymatic conversion of synthetic precursors by the NIS synthetase AvbD and feeding experiments in live cells of *Shewanella algae* B516. We will further describe detailed methods for the isolation of the ring-size engineered siderophores and highlight a screening platform to evaluate their potential as swarming inhibitors.

2. Production of three siderophores by one BGC in *S. algae* and its applications

2.1 Biosynthesis of siderophores by *S. algae*

The marine bacterium *S. algae* produces three structurally related cyclic hydroxamate siderophores by only one biosynthetic gene cluster

(Böttcher & Clardy, 2014; Rütschlin, Gunesch, & Böttcher, 2017). Two of them, bisucaberin (Kadi, Song, & Challis, 2008) and putrebactin (Kadi, Arbache, Song, Oves-Costales, & Challis, 2008), are macrocyclic homodimers of two *N*-hydroxy-*N*-succinyl-diamine precursors with either cadaverine (five methylene groups) or putrescine (four methylene groups) motifs in their structure, respectively. The third is the chimeric siderophore avaroferrin, with a heterodimeric structure encompassing one putrescine and one cadaverine motif (Fig. 2). Avaroferrin has been isolated as potent swarming inhibitor from the interspecies competition of *S. algae* and *Vibrio alginolyticus*. While avaroferrin effectively stalled motility of *V. alginolyticus*, putrebactin and bisucaberin were considerably less active. (Böttcher & Clardy, 2014). *V. alginolyticus* has specialized on the uptake of xenosiderophores that they do not produce, which is known as siderophore piracy (Pletzer, Braun, & Weingart, 2016; Traxler, Seyedsayamdost, Clardy, & Kolter, 2012). This behavior allows to compete for the limited resource of ferric iron while saving the metabolic cost of siderophore production. For *V. alginolyticus*, bioavailability of ferric iron was found to be a prerequisite of swarming motility and experimental evidence suggested that the asymmetric structure of avaroferrin thwarts siderophore piracy and thereby inhibits swarming (Böttcher & Clardy, 2014).

This provides an important advantage for *S. algae* in the competition with *V. alginolyticus* and may have favored the evolution of the chimeric siderophore. The avaroferrin genecluster is a mashup of known putrebactin and bisucaberin clusters and consists of four genes *avbABCD* encoding the biosynthesis enzymes and additional trailing genes *avtA-H* with transport

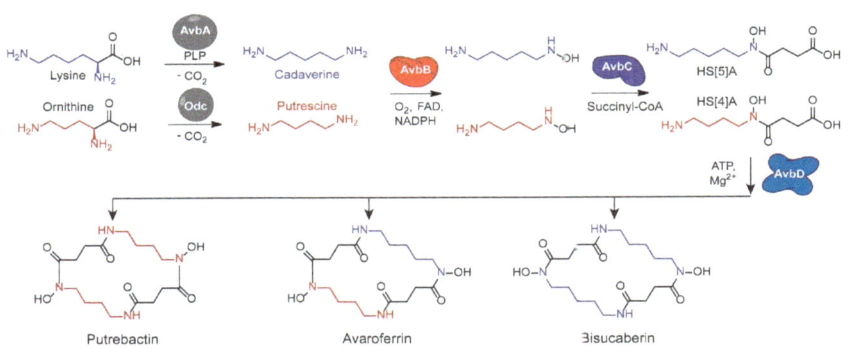

Fig. 2 Scheme of the siderophore biosynthesis pathway of *S. algae* leading to bisucaberin, putrebactin and avaroferrin production.

related functions. In a first step, the polyamine cadaverine is generated by decarboxylation of lysine via AvbA, while putrescine is produced from decarboxylation of ornithine (Ornithine decarboxylase, Odc) in the primary metabolism (Fig. 2). Putrescine can further be generated by metabolization of arginine (Shah & Swiatlo, 2008). The two diamines are then oxidized via the monooxygenase AvbB and subsequently coupled to succinyl-CoA by AvbC. The resulting *N*-hydroxy-*N*-succinyl-diamine precursors HS[4]A and HS[5]A are substrates of the NIS synthetase AvbD. This enzyme catalyzes the two-step dimerization and macrocyclization of HS[4]A and HS[5]A forming avaroferrin, as well as the homodimers putrebactin and bisucaberin (Rütschlin et al., 2017). Hereby, AvbD has a greater affinity for HS[5]A than for HS[4]A, favoring production of bisucaberin the siderophore with a larger ring-size. Yet, the cellular substrate pool of diamines is fine-tuned to maximized production of the chimeric avaroferrin which suggests a facile way for rapid adaptation of natural product ratios by evolution (Rütschlin et al., 2017). Furthermore, the biosynthetic promiscuity of AvbD in generating natural product diversity may be exploited for generating novel ring-size engineered macrocyclic hydroxamates (Fig. 2).

2.2 Manipulation of siderophore biosynthesis in *S. algae*

We thus synthesized a small library of unnatural *N*-hydroxy-*N*-succinyl-diamine HS[*X*]A precursors with different aliphatic chain lengths (X=3, 4, 5, 6, 8) ranging from three to eight methylene carbons (Rütschlin, Gunesch, & Böttcher, 2018). Indeed, all of the precursors were accepted by purified AvbD and converted to macrocyclic siderophores with homodimeric or heterodimeric architecture, depending on the combinations supplied. Hereby, macrocycles with 18- to 28-membered rings could be achieved. In the example presented here, the *avbD* gene of *S. algae* B516 with a sequence encoding a *N*-terminal Strep-tag was cloned into an expression vector and recombinant AvbD was produced in *Escherichia coli*. Overexpressed AvbD was then purified using an affinity column and used in a proper reaction buffer without further purification to convert combinations of synthetic HS[*X*]A precursors to dimeric macrocyclic siderophores (Kadi, Song, & Challis, 2008; Rütschlin et al., 2018). The produced siderophores were analyzed by LC-MS based methods. We describe below the synthesis of the synthetic precursors HS[*X*]A and a protocol for the enzymatic reactions catalyzed by AvbD.

2.3 In situ siderophore production in *S. algae*

Since the reactions with purified AvbD only resulted in siderophore yields in analytical scale, we sought to find a method to generate amounts suitable for biological testing. Upscaled conversion of AvbD with HS[*X*]A precursors resulted in a siderophore yield in milligram scale as previously published for homologous enzymes (Kadi, Arbache, et al., 2008; Kadi, Song, & Challis, 2008). For this experiment, the amount of purified AvbD and HS[*X*]A was multiplied by several orders and HPLC purification was necessary (Rütschlin et al., 2017). While this method generated suitable amounts of siderophores for biological evaluation, it has the drawback of high synthetic and protein purification effort. Therefore, we established a more comfortable and faster method based on biosynthetic manipulation and precursor feeding (Fig. 3). Hereby, the biosynthetic pathway of putrescine resulting from primary metabolism was impaired by addition of an ornithine decarboxylase inhibitor to the growth culture (Soe & Codd, 2014). Additionally, the culture was forced to produce more siderophores by addition of 2-2′-bipyridyl as iron chelator. The substrate pool was manipulated by feeding of 1,6-diaminohexane which was incorporated into the biosynthetic pathway replacing putrescine and thus results in the production of novel cyclic hydroxamate siderophores which could be isolated in preparative scale by natural product extraction (Rütschlin et al., 2018).

Bacteria are grown on NBE carrageenan plates mimicking the natural habitat of *S. algae*. Siderophores are extracted with isopropanol followed by several pre-purification and purification steps. Finally, the siderophores can be obtained in pure form by precipitation. After identification and confirmation by NMR and LC-HRMS, the siderophores can be stored as DMSO stocks for further applications.

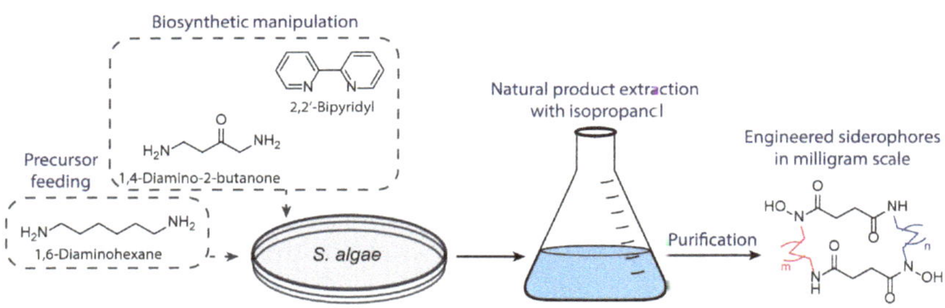

Fig. 3 Manipulation of *Shewanella algae* siderophore biosynthesis by feeding of an artificial precursor.

2.4 Swarming inhibition of *Vibrio alginolyticus*

Avaroferrin was discovered as potent inhibitor of the swarming behavior of pathogenic *V. alginolyticus* by escaping siderophore piracy and thereby blocking *Vibrio*'s access to ferric iron (Böttcher & Clardy, 2014). As the feeding experiments with 1,6-diaminohexane yielded novel homo- and heterodimeric siderophores, they were evaluated on their potential effects on swarming behavior of *V. alginolyticus*. They were tested with a standard swarming assay (Böttcher & Clardy, 2014) and especially the heterodimeric structures turned out to be inhibitors with a potency comparable to avaroferrin (Rütschlin et al., 2018).

3. Protocol

3.1 Materials

3.1.1 Equipment and supplies

- Liquid Chromatography system for preparative reversed-phase high-performance liquid chromatography (RP-HPLC)
- Liquid Chromatography system for semi-preparative reversed-phase high-performance liquid chromatograph (RP-HPLC)
- HPLC column Reprosil 100 C18, 10 μM, 250 × 20 mm (Dr. Maisch GmbH)
- HPLC column Reprospher 100 C18 AQ, 10 μm, 250 × 10 mm (Dr. Maisch GmbH)
- 600 MHz NMR (nuclear magnetic resonance) instrument
- LC-MS and LC-HRMS instrument
- Rotary evaporator
- Lyophilizer
- Magnetic stirrer
- Thermocycler instrument for PCR (e.g., peqSTAR thermal cycler (PEQLAB))
- Cell disruptor (e.g., Branson 450 Digital Sonifier with microtip)
- Microcentrifuge (cooled to 4 °C)
- Disposable 1.5 and 2 mL microcentrifuge tubes
- 1 L Erlenmeyer flasks for protein expression
- Shaking incubator (temperature set to 18, 30 or 37 °C)
- Plate incubator (temperature set to 30 or 37 °C)
- Puregene Yeast/Bacteria Kit B (Qiagen)
- Appropriate polymerase, primer and buffers for PCR (e.g., Phusion High-Fidelity PCR kit (ThermoFisher Scientific))

- ÄKTA™ start protein purification system (GE Healthcare Life Sciences)
- StrepTrap™ HP affinity column binding to strap tag II (GE Healthcare Life Sciences)
- Amicon® Ultra filtration column with a 50,000 MWCO membrane (Millipore)
- Large petri dishes (15 mm) for bacterial culturing
- Petri dishes (9 mm) for swarming assays
- Sep-Pak Vac 35 cc (10 g) C18 cartridge (Waters)
- Ice bath
- Hydrogen gas
- 5 L glass beaker for extraction

3.1.2 Reagents and solvents

- Solvents: dimethylformamide (DMF, dry), chloroform ($CHCl_3$), ethyl acetate (EE), hexane, diethyl ether (Et_2O), ethanol (EtOH), 1,4-dioxane, acetonitrile (HPLC-grade), isopropanol (HPLC-grade), methanol (MeOH), dichloromethane (CH_2Cl_2), pyridine (dry), *tert*-butanol (*tert*-BuOH), dimethyl sulfoxide (DMSO)
- Formic acid (CAS No.: 64-18-6)
- Potassium phthalimide (CAS No.: 1074-82-4)
- Sodium iodide (CAS No.: 7681-82-5)
- Sodium hydride 60% dispersion in mineral oil (CAS No.: 7646-69-7)
- *tert*-Butyl *N*-(benzyloxy)carbamate (CAS No.: 79722-21-7)
- Hydrazine hydrate (CAS No.: 10217-52-4)
- Benzyl chloroformate (CAS No.: 501-53-1)
- Trifluoroacetic acid (CAS No.: 76-05-1)
- Succinic anhydride (CAS No.: 108-30-5)
- Palladium on carbon, 10 wt% loading, matrix activated carbon support (Pd/C, Sigma)
- Sodium carbonate (Na_2CO_3)
- Magnesium sulfate (anhydrous)
- Pierce™ Coomassie Plus (Bradford) Assay Reagent (ThermoFisher Scientific)
- Carrageenan suitable for gel preparation (Sigma, CAS No.: 9000-07-1)
- 2,2′-Bipyridyl (CAS No.: 366-18-7)
- 1,4-Diamino-2-butanone dihydrochloride (DBO·2HCl, CAS No.: 3660-09-1)
- 1,6-Diaminohexane (CAS No.: 124-09-4)

- Adenosine 5′-triphosphate disodium salt hydrate (CAS No.: 34369-07-8)
- Silica gel for column chromatography (0.035–0.070 mm, mesh 60 Å, Roth, Germany)

3.1.3 Buffers and solutions

- Concentrated HCl solution
- 1% $NaHCO_3$ and sat. $NaHCO_3$ solution
- Brine
- 10% Trichloroacetic acid (TCA) solution
- Ampicillin 1 mg/mL (1000 ×) stock solution in sterile H_2O, stored at −20 °C
- Kanamycin 25 mg/mL (1000 ×) stock solution in sterile H_2O, stored at −20 °C
- Lysogenic Broth-Lennox (LB) medium and LB agar
- Nutrient Broth E (NBE) medium and NBE carrageenan agar
- Phosphate-buffered saline (PBS)
- StrepTrap binding buffer (100 mM Tris-HCl, 150 mM NaCl and 1 mM EDTA (pH = 8) in H_2O_{deion})
- StrepTrap elution running buffer (100 mM Tris-HCl, 150 mM NaCl and 1 mM EDTA (pH = 8), 2.5 mM desthiobiotin in H_2O_{deion})
- AvbD storage buffer (20 mM Tris-HCl (pH = 8.0), 100 mM NaCl and 10% glycerol)
- AvbD reaction buffer (4 mM ATP, 15 mM $MgCl_2$, 25 mM Tris-HCl (pH = 8.0))
- Anhydrotetracycline solution in DMF (2 mg/mL, CAS No.: 13803-65-1)

3.1.4 Bacterial cells, expression vectors and enzymes

- *S. algae* B516
- *V. alginolyticus* B522
- Chemically competent One Shot® TOP10 *E. coli* cells (Invitrogen)
- Chemically competent *E. coli* BL21 (DE3) cells (ThermoFisher Scientific)
- Gateway pDONR 201 donor vector (ThermoFisher Scientific)
- pDEST destination vector encoding streptavidin tag II and ampicillin resistance; or similar Gateway pDEST destination vector
- Gateway LR Clonase II Enzyme Mix (ThermoFisher Scientific)
- Gateway BP Clonase II Enyzme Mix (ThermoFisher Scientific)

3.2 Synthesis of AvbD substrates

For an overview of the synthesis (see Fig. 4).

3.2.1 Gabriel synthesis (step 1)

1. Add 40 mL dimethylformamide (DMF) and a halogenated alkane of desired chain length (50.3 mmol, 8.63 g, 1.0 eq) into a dried round-bottom flask equipped with a stirring bar under argon or nitrogen atmosphere
2. Add potassium phthalimide (50.0 mmol, 9.26 g, 1.0 eq) and stir for 20 h at room temperature
3. Add 100 mL of saturated brine and extract it with $CHCl_3$
4. Dry the organic layer over anhydrous magnesium sulfate
5. Remove the solvent under reduced pressure using a rotary evaporator
6. Perform preparative column chromatography on silica gel for purification (gradient: isocratic 30% EE, 70% hexane) to yield the product as white solid
7. Identify and characterize product purity by NMR and LC-HRMS, see Kadi, Song and Challis (2008), Lifa et al. (2015), Rütschlin et al. (2018)

3.2.2 Installing a benzyloxyamine (step 2)

1. Add 25 mL dimethylformamide (DMF) and *tert*-butyl *N*-(benzyloxy) carbamate (12.9 mmol, 2.90 g, 1.0 eq) into a dried round-bottom flask equipped with a stirring bar under argon or nitrogen atmosphere in an oil bath

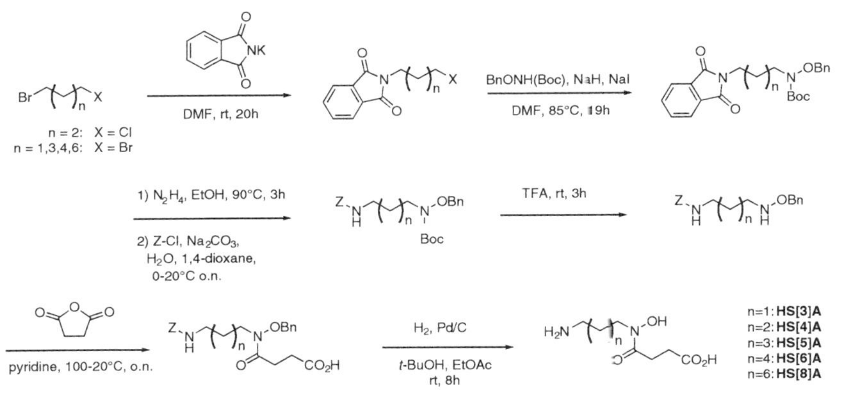

Fig. 4 Synthetic scheme for the preparation of *N*-hydroxy-*N*-succinyl-diamine precursors HS[*X*]A precursors according to Rütschlin et al. (2018).

2. Add sodium iodide (0.8 mmol, 123 mg, 0.1 eq) and 60% sodium hydrate dispersed in mineral oil (28.5 mmol, 685 mg, 2.2 eq) and stir for 15 min at 85 °C
3. Add the reaction product of step 1 (13.0 mmol, 3.50 g, 1.0 eq) and stir for 19 h at 85 °C
4. Let cool to room temperature and quench carefully with distilled water
5. Add 100 mL of Et_2O to extract the aqueous layer and wash the pooled Et_2O fractions with 1% $NaHCO_3$ solution, distilled water and brine
6. Dry the organic layer over anhydrous magnesium sulfate
7. Remove the solvent under reduced pressure using a rotary evaporator
8. Perform preparative column chromatography on silica gel for purification (gradient: isocratic 30% EE, 70% hexane) to yield the product as yellow oil
9. Identify and characterize product purity by NMR and LC-HRMS, see Kadi, Song and Challis (2008), Lifa et al. (2015), Rütschlin et al. (2018)

3.2.3 Hydrazinolysis and reprotection (step 3)

1. Add 20 mL ethanol (EtOH) and the reaction product of step 2 (7.5 mmol, 3.08 g, 1 eq) into a round-bottom flask equipped with a stirring bar
2. Add hydrazine hydrate (39.2 mmol, 19 mL, 5.2 eq) and heat to reflux for 3 h at 90 °C
3. Let cool to room temperature and filter the white precipitate
4. Wash the white precipitate with Et_2O and concentrate it in vacuo
5. Dissolve the residue in 48 mL 2:1 water/1,4-dioxane in a round-bottom flask under stirring
6. Add sodium carbonate (7.1 mmol, 755 mg, 0.9 eq) and cool to 0 °C in an ice bath
7. Add benzyl chloroformate (6.5 mmol, 930 μL, 0.9 eq) drop-wise and let warm to room temperature
8. Stir overnight at room temperature
8. Add saturated brine and extract it with CH_2Cl_2
9. Dry the organic layer over anhydrous magnesium sulfate
10. Remove the solvent under reduced pressure using a rotary evaporator
11. Perform preparative column chromatography on silica gel for purification (gradient: isocratic 30% EE, 70% hexane) to yield the product as yellow oil
12. Identify and characterize product purity by NMR and LC-HRMS, see Kadi, Song and Challis (2008), Lifa et al. (2015), Rütschlin et al. (2018)

3.2.4 Boc deprotection (step 4)

1. Add 30 mL of 1:4 TFA:CH_2Cl_2 and reaction product of step 3 (3.8 mmol, 1.58 g, 1.0 eq) into a round-bottom flask and stir for 3 h at room temperature
2. Remove the solvent under reduced pressure using a rotary evaporator
3. Add saturated sodium hydrogen carbonate solution and extract it with Et_2O (2 × 75 mL)
4. Dry the organic layer over anhydrous magnesium sulfate
5. Remove the solvent under reduced pressure using a rotary evaporator to yield the product as yellow oil
6. Identify and characterize product purity by NMR and LC-HRMS, see Kadi, Song and Challis (2008), Lifa et al. (2015), Rütschlin et al. (2018)

3.2.5 Succinate coupling (step 5)

1. Add 5 mL of pyridine and reaction product of step 4 (0.6 mmol, 0.19 g, 1.0 eq) into a dried round-bottom flask equipped with a stirring bar under argon or nitrogen atmosphere
2. Add succinic anhydride (0.6 mmol, 63 mg, 1.0 eq) and stir for 2 h at 100 °C
3. Let cool to room temperature and stir overnight
4. Remove the solvent under reduced pressure using a rotary evaporator
5. Add Et_2O and extract it with saturated sodium hydrogen carbonate solution
6. Acidify the combined aqueous layers with conc. HCl to pH 5
7. Add Et_2O and to extract the aqueous layer
8. Dry the organic layer over anhydrous magnesium sulfate
9. Remove the solvent under reduced pressure using a rotary evaporator to yield the product as pale pink oil
10. Identify and characterize product purity by NMR and LC-HRMS, see Kadi, Song and Challis (2008), Lifa et al. (2015), Rütschlin et al. (2018)

3.2.6 Reductive deprotection (step 6)

1. Add 1 mL of 1:1 *tert*-BuOH:EE (dry) and reaction product of step 5 (0.02 mmol, 7.00 mg, 1.0 eq) into a round-bottom flask equipped with a stirring bar
2. Add 10% Pd/C (1.5 mg) and stir under hydrogen atmosphere for 8 h at room temperature
3. Remove the catalyst by filtration and wash it with 1:1 MeOH/H_2O

4. Remove the solvent of the filtrate under reduced pressure using a rotary evaporator
5. Dissolve the residue in H_2O
6. Perform preparative HPLC for purification (gradient: T_{0min}: B=5%; T_{15min}: B=5%; T_{20min}: B=95%; T_{30min}: B=95%. A=H_2O (+0.1% FA), B=acetonitrile (+0.1% FA), Flow-rate: 15 mL/min) to yield the product as white solid
7. Identify and characterize product purity by NMR and LC-HRMS, see Kadi, Song and Challis (2008), Lifa et al. (2015), Rütschlin et al. (2018)

3.3 Overproduction of AvbD in *E. coli* and enzyme purification

1. Grow *S. algae* B516 in NBE medium overnight at 30 °C (shaking at 180 rpm) and extract the genome
2. Use appropriate primers and any cloning workflow (e.g., Invitrogen™ Gateway® Technology (ThermoFisher Scientific)) to generate the following expression vector with Amp selection marker: construct an expression vector coding for the enzyme AvbD of *S. algae* B516 (for Gateway technology generate pDEST/*avbD*, which contains additionally an *N*-terminal Strep-tag II for affinity purification)
3. Use any standard transformation protocol for transforming the constructed plasmid in competent *E. coli* cells
4. Prepare cryo-stock (8.5 parts glycerol/1.5 parts cells; frozen at −80 °C) of *E. coli* cells containing the expression vector for AvbD
5. Inoculate cells with a pipette tip from the cryo-stock into 5 mL LB medium supplemented with 100 μg/mL carbenicillin and incubate overnight at 37 °C and shaking at 180 rpm
6. On the next day, inoculate culture 1:50 in 1 L LB media containing 100 μg/mL carbenicillin
7. Incubate at 37 °C and shaking at 180 rpm until OD_{600} = 0.5
8. Induce expression using 0.2 μg/mL anhydrotetracycline stock solution
9. Shake overnight at 180 rpm and 18 °C
10. Harvest cells by centrifugation and resuspend the cell pellet in 30 mL PBS
11. Lyse cells with any cell disruptor with external cooling on ice
12. Centrifuge lysed cells (7000 rpm for 30 min at 4 °C) to remove cellular debris and purify the protein from the supernatant using a 5 mL StrepTrap™ HP affinity column binding to strap tag II on an ÄKTA™ protein purification system

13. Equilibrate the column with 5 CV (column volumes) StrepTrap binding buffer and add the lysed cell's supernatant. Remove unbound protein by washing with 5 CV of StrepTrap binding buffer and elute the protein with 6 CV of StrepTrap elution buffer
14. Concentrate the fractions in AvbD storage buffer by using an Amicon® Ultra filtration column with a 50,000 MWCO
15. Check protein concentration (Bradford assay) to obtain the protein content
16. Aliquote and freeze AvbD at −80 °C

3.4 Analytical scale AvbD reaction

1. Incubate purified AvbD (4 μM, preparation see Section 3.3) with one or combinations of two (1:1) of the synthesized HS[*X*]A (X = 3–8) (preparation see Section 3.2) (total 2 mM) in AvbD reaction buffer in a final volume of 140 μL for 4 h at room temperature

   Tip: Vary incubation time and temperature to find optimal conditions for the siderophore yield matching your application
2. Stop the reaction by quenching 40 μL of sample with 3 μL 10% TCA solution
3. Identify and characterize product purity by LC-HRMS or LC-MS analysis of the reaction mixture

   Tip: As the reaction buffer contains various salts which should not enter the mass spectrometer, make sure you desalt you sample online for good maintenance of the spectrometer

3.5 Upscaled siderophore production

3.5.1 Preparative scale AvbD reaction

1. Wash a reaction flask with concentrated HNO_3 and rinse it with H_2O_{deion}
2. Incubate purified AvbD (1.75 mM, preparation see Section 3.3) with one or combinations of two (1:1) of the synthesized HS[*X*]A (X = 3–8) (preparation see Section 3.2) (total 2 mM) in AvbD reaction buffer in a final volume of 7 mL for 15 h at 30 °C

   Tip: Vary incubation time and temperature to find optimal conditions for the siderophore yield matching your application
3. Stop the reaction by quenching with 200 μL of 7% TCA solution
4. Centrifuge the reaction mixture (10 min, 4000 rpm) and recover the supernatant

5. Remove the H_2O_{deion} under reduced pressure using a rotary evaporator or a lyophilizer
6. Perform preparative and semi-preparative HPLC chromatography to purify the siderophores (gradient: T_{0min}: B = 5%; T_{5min}: B = 5%; T_{25min}: B = 95%; T_{33min}: B = 95%, A = H_2O (+0.1% FA), B = acetonitrile (+0.1% FA)) yielding the products as white solids

   Tip: From up to a certain purity, the siderophores tend to precipitate readily from 1:1 acetonitrile:H_2O as white solids. Work up can be simplified by this method, as the precipitated siderophores can be centrifuged and washed with 1:1 acetonitrile:H_2O until homogeneity is reached
7. Remove the solvent under reduced pressure using a lyophilizer
8. Identify and characterize product purity by NMR and LC-HRMS
9. For application as swarming inhibitor follow instructions in Section 3.7

3.5.2 Metabolic feeding with 1,6-diaminohexane

1. Grow *S. algae* B516 in NBE medium overnight at 30 °C (shaking at 180 rpm)
2. Prepare about 50 large petri dishes (15 mm) filled each with 60 mL of NBE carrageenan supplemented with 100 μM 2,2′-bipyridyl, 10 mM DBO·2HCl and 10 mM 1,6-diaminohexane

   Tip: Add 2,2′-bipyridyl, DBO·2HCl and 1,6-diaminohexane after autoclaving into the warm media and shake vigorously to dissolve

   Tip: To test for feeding with other diamines, extract as initial experiment a small amount of plates to check, whether the diamine is incorporated into siderophore biosynthesis by *S. algae* at all. After soaking in isopropanol (see instructions in Section 3.6), dissolve the extract in H_2O_{deion} and subject to LC-MS analysis to check for the siderophores produced
3. Add 200 μL of an overnight culture of *S. algae* strain B516 in NBE medium onto each plate and distribute as fine film with a Drigalski spatula
4. Incubate for 4–5 days at 30 °C until nicely overgrown
5. For purification of the siderophores follow instructions in Section 3.6

3.6 Siderophore isolation from *S. algae* extracts

1. Chop plates into pieces and collect them in a 5 L glass container for extraction
2. Add 2.5 L EE and let the plate pieces soak overnight

3. Discard the EE by decanting and add 2.5 L isopropanol
4. Soak for 2 days
5. Decant isopropanol and remove the solvent under reduced pressure using a rotary evaporator

   Tip: For extraction of natural products, make sure you always use a water bath temperature of only 30 °C to ensure that substances are not degrading
6. Dissolve the residue in 8 mL H_2O_{deion} and a few drops of methanol. Pre-fractionate the solution with a Sep-Pak Vac 35 cc (10 g) C18 cartridge (Waters) and elute with water, 30% MeOH and 60% MeOH in water and MeOH
7. Remove the solvent under reduced pressure using a lyophilizer
8. Perform preparative and semi-preparative HPLC chromatography to purify the siderophores (gradient: T_{0min}: B = 5%; T_{5min}: B = 5%; T_{25min}: B = 95%; T_{33min}: B = 95%, A = H_2O (+0.1% FA), B = acetonitrile (+0.1% FA)) yielding the products as white solids

   Tip: From up to a certain purity, the siderophores tend to precipitate readily from 1:1 acetonitrile:H_2O as white solids. Work up can be simplified by this method, as the precipitated siderophores can be centrifuged and washed with 1:1 acetonitrile:H_2O until homogeneity is reached
9. For application as swarming inhibitor prepare 10 mM siderophore stock solutions in DMSO and store up at −20 °C. For screening as swarming inhibitor follow instructions in Section 3.7

3.7 Applications

3.7.1 Swarming inhibition assay with V. alginolyticus

1. Grow *V. alginolyticus* B522 in NBE medium overnight at 30 °C (shaking at 180 rpm)
2. Prepare petri dishes (9 mm) filled with 30 mL NBE carrageenan

   Tip: Pour the plates with hand-warm media to avoid excess condensing water, which could change the water content of the solid medium's surface. As the surface wetness is critical for swarming assays this step should be a critical one for the success of your swarming inhibition experiment
3. Add paper discs onto the plates with a sterile tweezer as outlined in Fig. 5 (distance 2 cm)
4. Inoculate 5 μL of *V. alginolyticus* B522 (V) overnight culture in NBE medium onto the central paper disc

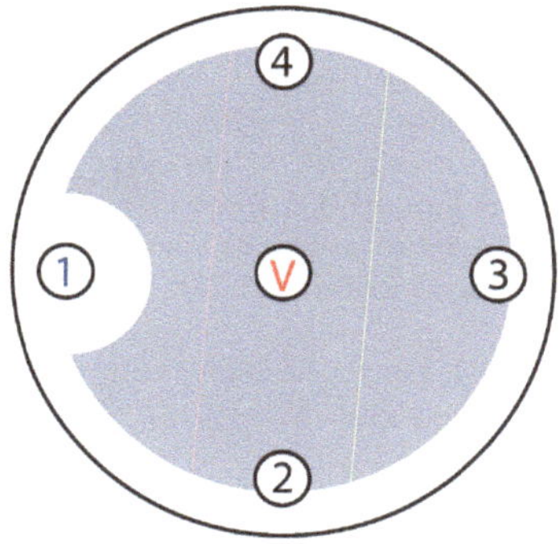

Fig. 5 Swarming inhibition test with siderophores. V: point of inoculation of swarming colony of *V. alginolyticus*. 1–4: test compounds and controls.

5. Incubate the plate for 12 h at 30 °C (lag phase before the colony initiates swarming)
6. Add 5 μL of 10 mM siderophore stock onto the surrounding paper discs (1–4). For negative control, add 5 μL of DMSO to test onto a paper disc
7. Incubate the plate for 6 h at 30 °C until visual inspection results in a clear inhibition zone

   Tip: Screen several readout times after inoculation of the siderophores to find the best condition for the visual readout

4. Summary

Biosynthetic and biochemical approaches are useful tools to generate ring-size engineered macrocyclic hydroxamate siderophores which otherwise require high synthetic effort in total syntheses. Recent work has demonstrated the utility of these strategies for the production of siderophores in suitable scale for analytical and biological investigations. This chapter describes the methods for the synthesis of the natural and artificial siderophore precursors HS[*X*]A (X = 3–8) as well as methods to produce engineered siderophores in analytical and preparative scale. We also provide a protocol for the application of the engineered siderophores as swarming inhibitors against *V. alginolyticus*.

Acknowledgments

This work was funded by the Emmy Noether program of the Deutsche Forschungsgemeinschaft (DFG), EU FP7 Marie Curie Zukunftskolleg Incoming Fellowship Program—University of Konstanz grant no. 291784, Fonds der Chemischen Industrie (FCI), and Konstanz Research School Chemical Biology (KoRS-CB). S.R. acknowledges support by a KoRS-CB PhD fellowship.

References

Ankenbauer, R. G., Staley, A. L., Rinehart, K. L., & Cox, C. D. (1990). Mutasynthesis of siderophore analogues by *Pseudomonas aeruginosa*. *Proceedings of the National Academy of Sciences of the United States of America*, *88*, 1878–1882.

Aydinok, Y., Kattamis, A., Cappellini, M. D., El-Beshlawy, A., Origa, R., Elalfy, M., et al. (2015). Effects of deferasirox-deferoxamine on myocardial and liver iron in patients with severe transfusional iron overload. *Blood*, *125*(25), 3868–3877. https://doi.org/10.1182/blood-2014-07-586677.

Bilyk, O., & Luzhetskyy, A. (2016). Metabolic engineering of natural product biosynthesis in actinobacteria. *Current Opinion in Biotechnology*, *42*, 98–107. https://doi.org/10.1016/j.copbio.2016.03.008.

Bosello, M., Robbel, L., Linne, U., Xie, X., & Marahiel, M. A. (2011). Biosynthesis of the siderophore rhodochelin requires the coordinated expression of three independent gene clusters in Rhodococcus jostii RHA1. *Journal of the American Chemical Society*, *133*(12), 4587–4595. https://doi.org/10.1021/ja1109453.

Böttcher, T., & Clardy, J. (2014). A chimeric siderophore halts swarming vibrio. *Angewandte Chemie International Edition*, *53*(13), 3510–3513. https://doi.org/10.1002/anie.201310729.

Bozhüyük, K. A. J., Fleischhacker, F., Linck, A., Wesche, F., Tietze, A., Niesert, C. P., et al. (2018). De novo design and engineering of non-ribosomal peptide synthetases. *Nature Chemistry*, *10*(3), 275–281. https://doi.org/10.1038/nchem.2890.

Dale, S. E., Doherty-Kirby, A., Lajoie, G., & Heinrichs, D. E. (2003). Role of siderophore biosynthesis in virulence of Staphylococcus aureus: Identification and characterization of genes involved in production of a siderophore. *Infection and Immunity*, *72*(1), 29–37. https://doi.org/10.1128/iai.72.1.29-37.2004.

Fischbach, M. A., Lin, H., Liu, D. R., & Walsh, C. T. (2006). How pathogenic bacteria evade mammalian sabotage in the battle for iron. *Nature Chemical Biology*, *2*(3), 132–138. https://doi.org/10.1038/nchembio771.

Franke, J., Ishida, K., & Hertweck, C. (2014). Evolution of siderophore pathways in human pathogenic bacteria. *Journal of the American Chemical Society*, *136*(15), 5599–5602. https://doi.org/10.1021/ja501597w.

Kadi, N., Arbache, S., Song, L., Oves-Costales, D., & Challis, G. L. (2008). Identification of a gene cluster that directs putrebactin biosynthesis in Shewanella species: PubC catalyzes cyclodimerization of N-hydroxy-N-succinylputrescine. *Journal of the American Chemical Society*, *130*(32), 10458–10459. https://doi.org/10.1021/ja8027263.

Kadi, N., Oves-Costales, D., Barona-Gomez, F., & Challis, G. L. (2007). A new family of ATP-dependent oligomerization-macrocyclization biocatalysts. *Nature Chemical Biology*, *3*(10), 652–656. https://doi.org/10.1038/nchembio.2007.23.

Kadi, N., Song, L., & Challis, G. L. (2008). Bisucaberin biosynthesis: An adenylating domain of the BibC multi-enzyme catalyzes cyclodimerization of N-hydroxy-N-succinylcadaverine. *Chemical Communications*, (41), 5119–5121. https://doi.org/10.1039/b813029a.

Lifa, T., Tieu, W., Hocking, R. K., & Codd, R. (2015). Forward and reverse (retro) iron(III) or gallium(III) desferrioxamine E and ring-expanded analogues prepared using metal-templated synthesis from endo-hydroxamic acid monomers *Inorganic Chemistry*, *54*(7), 3573–3583. https://doi.org/10.1021/acs.inorgchem.5b00141.

Mootz, H. D., Schwarzer, D., & Marahiel, M. A. (2002). Ways of assembling complex natural products on modular nonribosomal peptide synthetases. *Chembiochem*, *3*(6), 490–504. https://doi.org/10.1002/1439-7633(20020603)3:6<490::Aid-Cbic490>3.0.Co;2-N.

Oves-Costales, D., Kadi, N., & Challis, G. L. (2009). The long-overlooked enzymology of a nonribosomal peptide synthetase-independent pathway for virulence-conferring siderophore biosynthesis. *Chemical Communications*, (43), 6530–6541. https://doi.org/10.1039/b913092f.

Pletzer, D., Braun, Y., & Weingart, H. (2016). Swarming motility is modulated by expression of the putative xenosiderophore transporter SppR-SppABCD in *Pseudomonas aeruginosa* PA14. *Antonie van Leeuwenhoek International Journal of General and Molecular Microbiology, 109*(6), 737–753. https://doi.org/10.1007/s10482-016-0675-8.

Richardson-Sanchez, T., & Codd, R. (2018). Engineering a cleavable disulfide bond into a natural product siderophore using precursor-directed biosynthesis. *Chemical Communications, 54*(70), 9813–9816. https://doi.org/10.1039/c8cc04981e.

Richardson-Sanchez, T., Tieu, W., Gotsbacher, M. P., Telfer, T. J., & Codd, R. (2017). Exploiting the biosynthetic machinery of *Streptomyces pilosus* to engineer a water-soluble zirconium(iv) chelator. *Organic and Biomolecular Chemistry, 15*(27), 5719–5730. https://doi.org/10.1039/c7ob01079f.

Rütschlin, S., Gunesch, S., & Böttcher, T. (2017). One enzyme, three metabolites: Shewanella algae controls siderophore production via the cellular substrate pool. *Cell Chemical Biology, 24*(5), 598–604.e510. https://doi.org/10.1016/j.chembiol.2017.03.017.

Rütschlin, S., Gunesch, S., & Böttcher, T. (2018). One enzyme to build them all: Ring—Size engineered siderophores inhibit the swarming motility of Vibrio. *ACS Chemical Biology, 13*(5), 1153–1158. https://doi.org/10.1021/acschembio.8b00084.

Seyedsayamdost, M. R., Cleto, S., Carr, G., Vlamakis, H., Joao Vieira, M., Kolter, R., et al. (2012). Mixing and matching siderophore clusters: Structure and biosynthesis of serratiochelins from *Serratia* sp. V4. *Journal of the American Chemical Society, 134*(33), 13550–13553. https://doi.org/10.1021/ja304941d.

Shah, P., & Swiatlo, E. (2008). A multifaceted role for polyamines in bacterial pathogens. *Molecular Microbiology, 68*(1), 4–16. https://doi.org/10.1111/j.1365-2958.2008.06126.x.

Soe, C. Z., & Codd, R. (2014). Unsaturated macrocyclic dihydroxamic acid siderophores produced by Shewanella putrefaciens using precursor-directed biosynthesis. *ACS Chemical Biology, 9*(4), 945–956. https://doi.org/10.1021/cb400901j.

Soe, C. Z., Pakchung, A. A., & Codd, R. (2012). Directing the biosynthesis of putrebactin or desferrioxamine B in *Shewanella putrefaciens* through the upstream inhibition of ornithine decarboxylase. *Chemistry and Biodiversity, 9*(9), 1880–1890. https://doi.org/10.1002/cbdv.201200014.

Soe, C. Z., Telfer, T. J., Levina, A., Lay, P. A., & Codd, R. (2016). Simultaneous biosynthesis of putrebactin, avaroferrin and bisucaberin by Shewanella putrefaciens and characterisation of complexes with iron(III), molybdenum(VI) or chromium(V). *Journal of Inorganic Biochemistry, 162*, 207–215. https://doi.org/10.1016/j.jinorgbio.2015.12.008.

Sresutharsan, A., Tieu, W., Richardson-Sanchez, T., Soe, C. Z., & Codd, R. (2017). Dimeric and trimeric homo- and heteroleptic hydroxamic acid macrocycles formed using mixed-ligand Fe(III)-based metal-templated synthesis. *Journal of Inorganic Biochemistry, 177*, 344–351. https://doi.org/10.1016/j.jinorgbio.2017.07.033.

Traxler, M. F., Seyedsayamdost, M. R., Clardy, J., & Kolter, R. (2012). Interspecies modulation of bacterial development through iron competition and siderophore piracy. *Molecular Microbiology, 86*(3), 628–644. https://doi.org/10.1111/mmi.12008.

CHAPTER FOUR

Competitive profiling for enzyme inhibitors using chemical probes

Michaela Prothiwa, Thomas Böttcher*
Department of Chemistry, Konstanz Research School Chemical Biology, Zukunftskolleg, University of Konstanz, Konstanz, Germany
*Corresponding author: e-mail address: thomas.boettcher@uni-konstanz.de

Contents

Abstract

Enzyme inhibitors are central tools for chemical biology. In this chapter we will discuss the application of chemical probes for competitive profiling of inhibitors of the quinolone biosynthesis enzyme PqsD of *Pseudomonas aeruginosa*. The human pathogen *P. aeruginosa* produces a large diversity of 2-alkyl-4(*1H*)-quinolones and their derivatives as metabolites with major roles in quorum sensing, virulence, and interspecies competition. PqsD is a central enzyme in the biosynthesis of all of these quinolones and hence an interesting target for inhibitor discovery. Activity-based probes with an electrophilic warhead bind covalently to active site nucleophiles like cysteine or serine. An α-chloroacetamide probe with terminal alkyne tag allowed to selectively label the active site cysteine of PqsD and was demonstrated to be a useful tool for inhibitor discovery using competition experiments. Potent inhibitors bind to the active site and thereby prevent labeling of the enzyme by the probe. Labeling intensity is quantified

Methods in Enzymology, Volume 633
ISSN 0076-6879
https://doi.org/10.1016/bs.mie.2019.10.031

on polyacrylamide gels by the fluorescence of a reporter tag appended by bioorthogonal click chemistry. The competitive inhibitor profiling strategy has many advantages over traditional screening approaches and is applicable in vitro as well as in live cells. Here we describe the synthesis of an activity-based probe and provide our detailed protocols for target enzyme labeling as well as its application for the screening for potent enzyme inhibitors of PqsD by a competitive profiling strategy.

1. Introduction

Enzyme inhibitors are central tools in chemical biology allowing to dissect and modulate cellular functions (Wijdeven, Neefjes, & Ovaa, 2014). Discovering enzyme inhibitors usually involves in vitro screening approaches with purified enzymes, structure-guided rational design or customized cell-based assays (Hubel, Lessmann, & Waldmann, 2008). Competitive inhibitor discovery using chemical probes provides an alternative strategy that is applicable *in vitro* with purified protein but also *in situ* with live cells (Chandrasekar, Hong, & van der Hoorn, 2017; Niphakis & Cravatt, 2014; Prothiwa, Englmaier, & Böttcher, 2018; Zuhl et al., 2012). Hereby, an activity-based chemical probe with a reporter tag binds covalently into the active site of the target enzyme and allows to monitor its activity via fluorescence labeling (Fig. 1A). Proteins are then separated by polyacrylamide gel electrophoresis and the fluorescence of the probe-labeled target protein is recorded. Potent enzyme inhibitors prevent binding of the active site-directed probe and can be detected by reduction of the fluorescence signal. This strategy is particularly well suited for enzymes with a nucleophilic active site residue which consequently can be targeted with activity-based electrophilic probes. There is a great diversity of literature-known electrophilic warheads that have been established as protein reactive components of chemical probes (Böttcher, Pitscheider, & Sieber, 2010). Commonly employed electrophilic warheads are, for example, fluorophosphonates, sulfonates, α-chloroacetamides, epoxides, and Michael acceptors, some of which even can be integral parts of the structures of natural products (Fonovic & Bogyo, 2008; Gersch, Kreuzer, & Sieber, 2012). Their well-tuned electrophilicity in combination with general ligand properties of the probe's structure provide the selectivity and reactivity required to address only nucleophilic cysteine, serine, and threonine residues in an enzyme's active site (Fig. 1B). The reporter tag of a chemical probe can be either directly a fluorescence dye or a functional group for bioorthogonal conjugation with a corresponding fluorescent

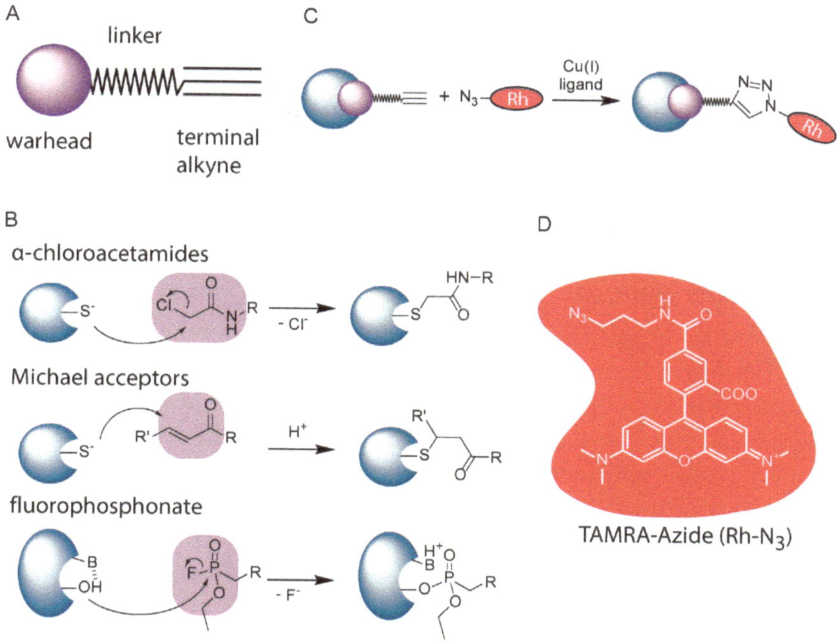

Fig. 1 Activity-based probes. (A) Design of a chemical probe with an electrophilic warhead conferring protein reactivity and a terminal alkyne reporter tag. (B) Selection of electrophilic warheads that covalently modify nucleophilic active site cysteine and serine residues. (C) Copper (I) catalyzed alkyne-azide click chemistry allows to append a fluorescence marker to the probe-labeled proteins. (D) Example of a fluorophore azide used for click chemistry.

marker (Willems, Overkleeft, & van Kasteren, 2014). Since steric demanding fluorophores may limit access of the probe to the active site, probe designs with a terminal alkyne tag or an azide group are often preferred (Fig. 1). Following protein binding, these can be conjugated under denaturing conditions by the Huisgen cycloaddition with a fluorophore equipped with an azide or alkyne group, respectively (Fig. 1C) (Rostovtsev, Green, Fokin, & Sharpless, 2002; Tornoe, Christensen, & Meldal, 2002). Testing chemical probes in native proteomes will typically result in the labeling of multiple protein targets and activity-based proteomics (ABPP) experiments are needed to identify the corresponding targets by mass spectrometry (Cravatt, Wright, & Kozarich, 2008). An alternative strategy is thus to overproduce the enzyme of interest and test a focused probe library with different electrophilic warheads and structural diversity for labeling selectivity. Once an appropriate probe has been identified that displays efficient and selective labeling of the active site of the target enzyme, the probe can be used in competitive screening experiments. In this chapter, we will showcase the application of inhibitor discovery by competitive profiling on the example of the quinolone biosynthesis enzyme PqsD.

2. Targeting quinolone biosynthesis

2.1 Biosynthesis of quinolones by *Pseudomonas aeruginosa*

The human pathogen *Pseudomonas aeruginosa* and some other species of the genus *Pseudomonas* and *Burkholderia* produce a great diversity of 2-alkyl-4 (1*H*)-quinolones (AQs) and their derivatives (Heeb et al., 2011; Lepine, Milot, Deziel, He, & Rahme, 2004). Some of these metabolites like the Pseudomonas Quinolone Signal (PQS, or 2-heptyl-3-hydroxy-4(1*H*)-quinolone) and its biosynthetic precursor 2-heptyl-4(1*H*)-quinolone (HHQ) are important quorum sensing signals that allow to coordinate bacterial virulence and various types of population behavior of *P. aeruginosa* (Dubern & Diggle, 2008; Huse & Whiteley, 2011). In addition, PQS can also directly modulate the human immune response (Lin, Cheng, Wang, & Shen, 2018) and has narrow spectrum antibiotic activities against gram-negative bacteria (Szamosvári, Schuhmacher, Hauck, & Böttcher, 2019). The related class of 2-alkyl-4(1*H*)-quinolone *N*-oxides (AQNOs) is not involved in signaling but mediates interspecies competition by inhibiting growth of other microbes (Szamosvári & Böttcher, 2017, 2018). Inhibitors of quinolone biosynthesis pathways are thus not only interesting from an anti-virulence perspective but also as chemical tools to dissect their roles in signaling and microbial interactions. In *P. aeruginosa* the divergent biosynthesis of all quinolones is executed by the proteins encoded by the *pqsABCDE* gene cluster, *pqsH* and *pqsL* (Fig. 2). The key metabolite hereby is anthranilic acid which can be provided via the chorismic acid and the kynurenine pathways (Farrow & Pesci, 2007). Anthranilic acid is activated by PqsA to anthraniloyl-CoA, which is subsequently transacylated to the active site cysteine residue of PqsD, where the compound undergoes a decarboxylative Claisen-like condensation with malonyl-CoA (Fig. 2) (Hutter et al., 2014). After hydrolysis of the CoA-thioester via PqsE (Drees & Fetzner, 2015), the labile precursor 2′-aminobenzoylacetate (2-ABA) is formed, which is condensed and cyclized with activated fatty acids of different chain length via the PqsBC complex to 2-alkyl-4(1*H*)-quinolones (AQs) (Drees et al., 2016). Alternatively, monooxygenation of 2-ABA by PqsL generates

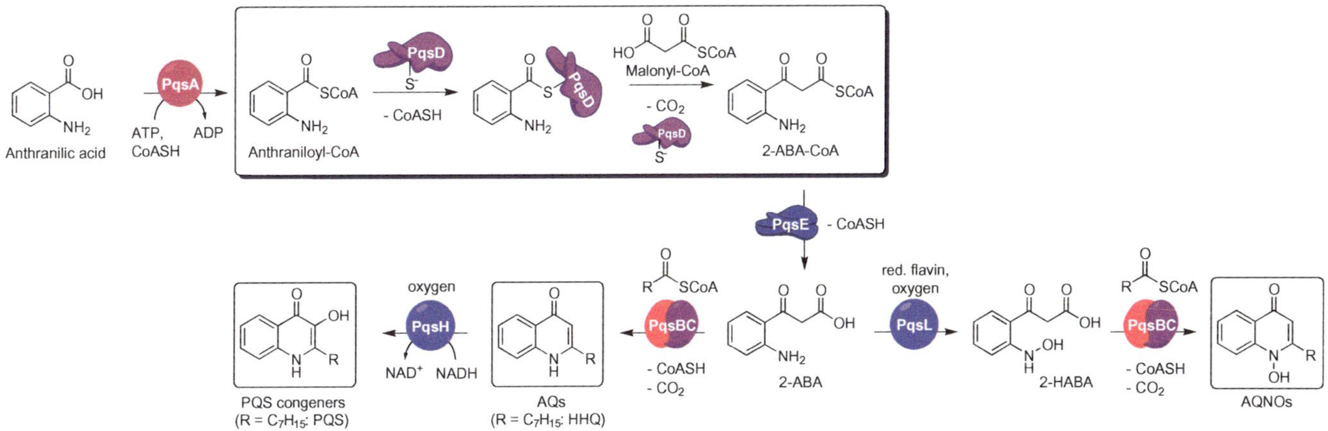

Fig. 2 Scheme of the divergent quinolone biosynthesis pathway of *P. aeruginosa* leading to quorum sensing signals PQS and HHQ and their congeners as well as the antimicrobial AQNOs. Mechanistic details are presented according to the most recent understanding. PqsD with a nucleophilic active site cysteine forms a covalent intermediate with the substrate and plays a central role in the entire pathway.

2-hydroxylaminobenzoylacetate (2-HABA), which also is a substrate of PqsBC and leads to the corresponding 2-alkyl-4(1*H*)-quinolone *N*-oxides (AQNOs) (Drees et al., 2018). Finally, AQs can be further oxidized by PqsH to yield congeners of the main signal PQS (Schertzer, Brown, & Whiteley, 2010). In this divergent biosynthetic route, PqsD plays a central role in generating the common precursor of all quinolone derivatives and has hence been in the focus of medicinal chemistry approaches for inhibitor design (Fig. 2) (Schütz & Empting, 2018). The covalent enzyme-substrate intermediate formed by attack of the nucleophilic active site cysteine on anthraniloyl-CoA makes this enzyme particularly interesting for targeting by electrophilic chemical probes and an ideal example for inhibitor discovery by competitive labeling.

2.2 Labeling of PqsD by chemical probes

We thus synthesized a small library of electrophilic probes comprising α-chloroacetamide and Michael acceptor warheads and a terminal alkyne that allowed to append a fluorescent dye via bioorthogonal Cu(I)-catalyzed alkyne-azide cycloaddition (click chemistry). Since enzymes of interest are frequently only produced in low levels in the native species or not produced constitutively, it is advisable to use a suitable expression system to overproduce the target enzyme. A further advantage of this strategy is that efficient labeling can be achieved even when binding affinity or target specificity of a probe are too low for target labeling in the native proteome. Indeed, none of the probes labeled PqsD in live cells of *P. aeruginosa*. Thus, PqsD-overproducing cells of *Escherichia coli* were used as model system. In the example presented here, the *pqsD* gene of *P. aeruginosa* PAO1 with a sequence encoding a N-terminal Strep-tag was cloned into an expression vector and recombinant PqsD was produced in *E. coli*. The only requirement for applicability of a probe in competitive profiling is its selectivity for active site labeling which can be tested using an active site mutant. The active site nucleophile of wild type PqsD is cysteine C112. In order to exclude unspecific binding of the probe to amino acids in the periphery of the enzyme, we constructed a C112A mutant where the nucleophilic cysteine was replaced by alanine using site-directed mutagenesis. Wild type and mutant PqsD were purified using an affinity column and labeling was tested for different probes (Prothiwa, Szamosvari, Glasmacher, & Böttcher, 2016). After incubation with the probe, click chemistry was performed with fluorescent tetramethylrhodamine (TAMRA) azide followed by

SDS-polyacrylamide gel electrophoresis (SDS-PAGE). Fluorescence scanning revealed that the most efficient labeling of PqsD was achieved with α-chloroacetamide probes of which probe *CA* was used in all further experiments (Prothiwa et al., 2016). Mutant PqsD C112A was not labeled, suggesting selectivity for the nucleophilic active site cysteine. This discrimination of one of the in total six cysteine residues in PqsD by the probe was further confirmed by mass spectrometry (Prothiwa et al., 2016). We describe below the synthesis of this simple probe (*CA* probe), which can be achieved by the reaction of 1-amino-3-butyne with chloroacetyl chloride in high yields. This reaction is robust and can be achieved even in laboratories that are not specifically equipped for organic synthesis.

2.3 Competitive profiling strategy

Since the probe *CA* targeted only the nucleophilic active site cysteine of PqsD, it was suitable for competitive labeling experiments (Fig. 3). *In vitro* competition experiments with purified PqsD and a small library of quinolones-, chromenone- and thiochromenone-derivatives resulted in sets of active competitors that prevented labeling by the probe (Prothiwa et al., 2016). While several compounds displayed promising *in vitro* activity, these compounds did not block PqsD labeling in live recombinant *E. coli* cells suggesting insufficient cell permeability. This underscores the need for live cell assays. Adding the *CA* probe to live cells of PqsD producing *E. coli* led to surprising selectivity for PqsD in the background of the entire *E. coli* proteome with labeling of PqsD down to submicromolar concentrations. Competitive screening was performed using a small library of covalently binding synthetic as well as commercially available α-chloroketones and α-chloroacetamides (Fig. 3). Hereby, those compounds mimicking the anthraniloyl moiety of native substrate anthraniloyl-CoA were particularly potent inhibitors that blocked probe labeling already in the low micromolar range. In contrast, compounds that were structurally unrelated to the anthraniloyl moiety or contained bulky residues did not affect labeling even at concentrations as high as 50 μM suggesting that these compounds were unable to bind into the active site of PqsD (Prothiwa et al., 2018).

Potent inhibitors from the initial screening at 50 μM were then investigated over a range of different concentrations. Fluorescence imaging allowed to quantify the labeling intensities of *CA* probe in presence of corresponding competitive inhibitors. The relative fluorescence intensities of the gel bands were used to determine live cells IC_{50} values. Finally, potent

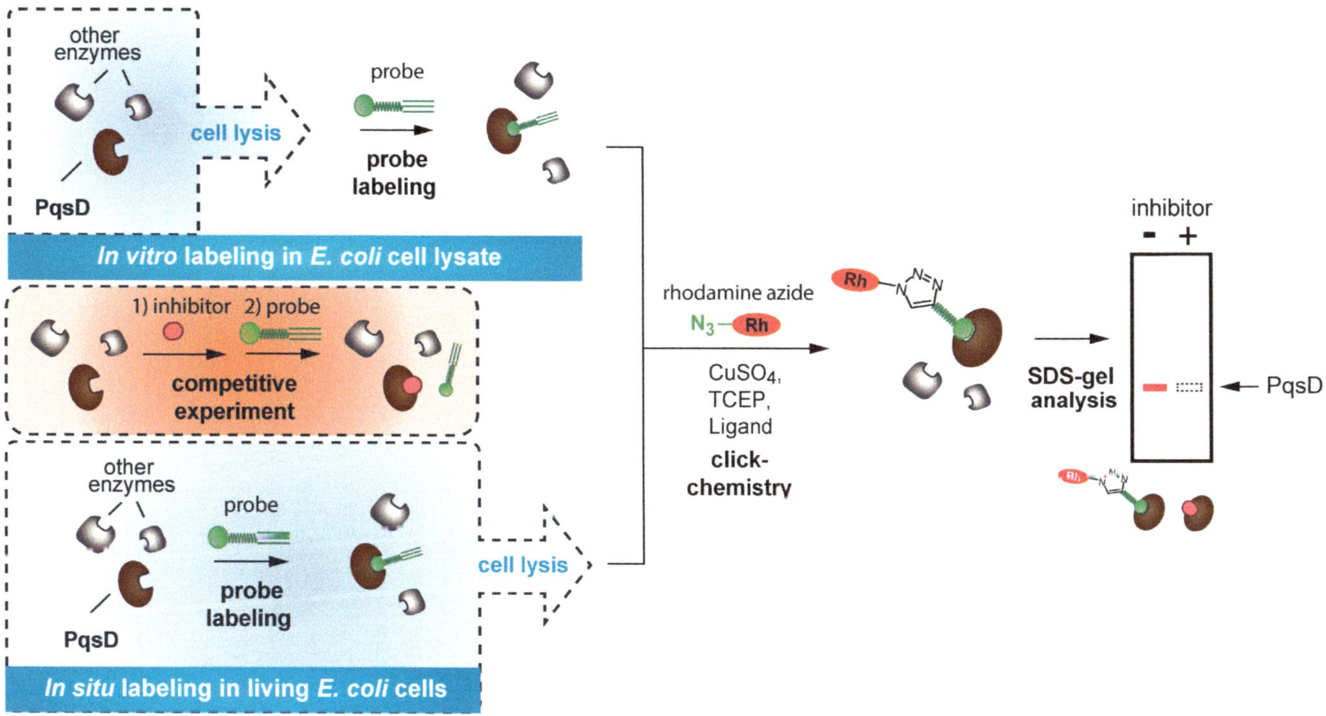

Fig. 3 General scheme for *in vitro* and *in situ* competitive labeling method.

competitive inhibitors with activity in PqsD-overproducing *E. coli* were also tested in *P. aeruginosa*. Hereby, 2′-amino-α-chloroacetophenone was one of the most active inhibitors which displayed activity in *P. aeruginosa* and blocked in concentration dependence the biosynthesis of all classes of native quinolone-derived metabolites such as PQS as well as different AQ and AQNO congeners (Prothiwa et al., 2018). The successful discovery of potent PqsD inhibitors highlights the power of competitive profiling with chemical probes. Particularly live cell profiling has major advantages and allows to assess efficacy of an inhibitor directly in a cell-based system accounting for pharmacokinetic properties of a compound such as cell permeability and stability in a cell.

3. Protocol

3.1 Materials

3.1.1 Equipment and supplies

- Spot Prep II Liquid Chromatography system (Armen Instrument, France) for reversed phase high performance liquid chromatograph (RP-HPLC)
- HPLC column Reprosil 100 C18, 10 μM, 250 × 20 mm (Dr. Maisch GmbH)
- 600 MHz NMR (nuclear magnetic resonance) instrument
- LC-HRMS instrument
- Rotary evaporator
- Magnetic stirrer
- Thermocycler instrument for PCR (e.g., peqSTAR thermal cycler (PEQLAB))
- Cell disruptor (e.g., Branson 450 Digital Sonifier with microtip)
- Microcentrifuge (cooled to 4 °C)
- Disposable 1.5 and 2 mL microcentrifuge tubes
- Disposable 15 mL culture tubes
- Shaking incubator (temperature set to 37 °C)
- PerfectBlue™ Twin S system for small gels (10 × 10 cm) or other SDS-PAGE unit with power supply
- Instrument for in-gel fluorescence scanning (e.g., Fusion-FX7 Advanced of Vilber Lourmat)
- Puregene Yeast/Bacteria Kit B (Qiagen)
- Appropriate polymerase, primer and buffers for PCR (e.g., Phusion High-Fidelity PCR kit (ThermoFisher scientific, Ulm, Germany))

3.1.2 Reagents and solvents

- 1-Amino-3-butyne (CAS No.: 14044-63-4)
- Triethylamine (CAS No.: 121-44-8)
- Chloroacetyl chloride (CAS No.: 79-04-9)
- Tetramethylethylenediamine (TEMED) (Thermo Scientific pierce) (CAS No.: 110-18-9)
- Dichloromethane (dry)
- Acetonitrile (HPLC-grade)
- Isopropanol (HPLC-grade)
- 10% HCl solution
- Brine
- Magnesium sulfate (anhydrous)
- 7-(Chloroacetyl)indole (test inhibitor, Sigma Aldrich)
- Pierce™ Coomassie Plus (Bradford) Assay Reagent (ThermoFisher Scientific)
- Rotiphorese® Gel 30

3.1.3 Buffers and solutions

- Carbenicillin 1 mg/mL (1000 ×) stock solution in sterile H_2O, stored at −20 °C
- Kanamycin 25 mg/mL (1000 ×) stock solution in sterile H_2O, stored at −20 °C
- Lysogenic broth-Lennox (LB) medium and LB agar
- Phosphate-buffered saline (PBS)
- 0.650 mM 5-carboxytetramethylrhodamine azide (5-TAMRA azide; CAS No.: 1192590-89-8) working stock in DMSO, prepared from a 193 mM stock in MeOH
- 2 mM tris[(1-benzyl-1*H*-1,2,3-triazol-4-yl)methyl]amine (TBTA; CAS No.: 510758-28-8) ligand stock in *tert*-butanol/DMSO (8:2 v/v)
- 50 mM tris(2-carboxyethyl)phosphine (TCEP) hydrochloride solution in water
- 50 mM $CuSO_4$ stock solution in water
- 2 × SDS-loading buffer (63 mM Tris-HCl, 10% (v/v) glycerol, 2% (w/v) SDS, 0.25% (w/v) bromophenol blue and 5% (v/v) β-mercaptoethanol in H_2O_{deion})
- 1 × SDS running buffer (24.8 mM Tris, 191.8 mM glycin and 3.5 mM SDS in H_2O_{deion}, pH 8.3)
- SDS solution (10% in H_2O_{deion})
- SDS solution (21% in H_2O_{deion})

- Coomassie staining solution (25% (w/v) Coomassie Brilliant Blue R250, 9.2% (v/v) acetic acid (conc.) and 45.5% ethanol (abs.) in H_2O_{deion})
- Coomassie destainer (10% acetic acid (conc.) and 20% ethanol (abs.) in H_2O_{deion})
- Separation gel buffer (3.0 M Tris in H_2O_{deion}, pH 8.8)
- Stacking gel buffer (0.5 M Tris in H_2O_{deion}, pH 6.8)
- Ammonium peroxodisulfate (APS) solution (10% w/v)

3.1.4 Bacterial cells, expression vectors and enzymes

- *Pseudomonas aeruginosa* PAO1 (DSM 22644)
- Chemically competent One Shot® TOP10 *E. coli* cells (Invitrogen)
- Chemically competent *E. coli* BL21 (DE3) cells (ThermoFisher Scientific)
- Gateway pDONR 201 donor vector (ThermoFisher Scientific)
- pDEST destination vector encoding streptavidin tag II and ampicillin resistance; or similar Gateway pDEST destination vector
- Gateway LR Clonase II Enzyme Mix (ThermoFisher Scientific)
- Gateway BP Clonase II Enyzme Mix (ThermoFisher Scientific)
- *Dpn*I endonuclease (NEW ENGLAND BioLabs® Inc.)
- T4 DNA ligase (NEW ENGLAND BioLabs® Inc.)

3.2 Synthesis of α-chloroacetamide probe

A reaction scheme is given in Fig. 4.

1. Add 2.5 mL dichloromethane (DCM), 1-amino-3-butyne (1 mmol, 82 μL, 1 eq) and triethylamine (1.1 mmol, 153 μL, 1.1 eq) into a dried round-bottom flask under argon or nitrogen atmosphere
2. Cool the solution under stirring to 0 °C (ice bath)
3. Add chloroacetyl chloride (1.2 mmol, 95 μL, 1.2 eq) and stir for 60 min at 0 °C
4. Warm the reaction mixture to room temperature and stir for additional 2 h

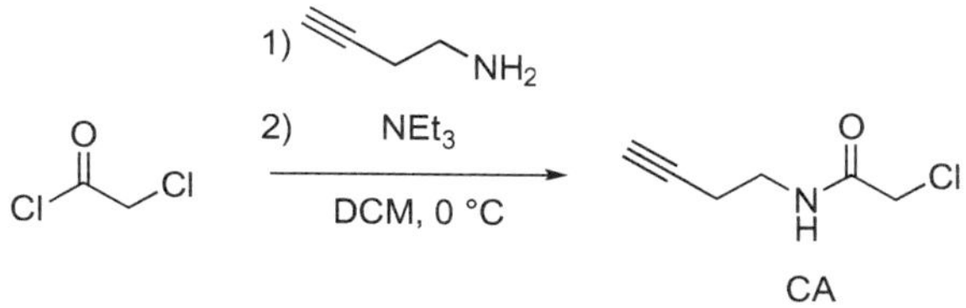

Fig. 4 Synthetic scheme for the preparation of α-chloroacetamide (*CA*) as activity-based probe.

5. Add 5 mL water to stop the reaction and extract it with DCM (3 × 5 mL)
6. Wash the combined DCM layers with 10% HCl solution and brine
7. Dry the organic layer over anhydrous magnesium sulfate
8. Remove the solvent under reduced pressure using a rotary evaporator
9. Perform preparative HPLC for purification (gradient: 5–95% acetonitrile in MilliQ water in 35 min. Flow-rate: 15 mL/min) to yield the product as white solid (0.8 mmol, 126 mg, 86%)
10. Identify and characterize product purity by NMR and LC-HRMS, see Prothiwa et al. (2016)
11. Keep product (white solid) at −20 °C for long time storage
12. Prepare a 100 mM stock solution in DMSO and store up to 3 months at −20 °C

    Tip: DMSO contains residues of water that impact probe stability, thus use aliquoted stock solutions to prevent freeze-thaw cycles; prepare new stock solutions after 3 months.

3.3 Overproduction of PqsD in *E. coli*

1. Grow *Pseudomonas aeruginosa* PAO1 in LB medium overnight at 37 °C (shaking at 180 rpm) and extract the genome
2. Use appropriate primers and any cloning workflow (e.g., Invitrogen™ Gateway® Technology (ThermoFisher Scientific)) to generate following expression vectors:
   - **(i)** Construct an expression vector coding for the wild type enzyme PqsD of *P. aeruginosa* PAO1 (for Gateway technology generate pDEST/*pqsDwt*, which contains additionally an N-terminal Strep-tag for affinity purification)
   - **(ii)** Perform site-directed mutagenesis to construct an expression vector coding for the Cys112Ala mutant of the enzyme PqsD (for Gateway technology generate pDEST/*pqsDmt*, which contains additionally an N-terminal Strep-tag for affinity purification)

   Tip: for mutagenesis use site-directed mutagenesis:
   - Replicate the expression vector containing the sequence coding for wild type PqsD with a forward primer enclosing the desired mutation and a reverse primer with a phosphorylated 5′ end
   - Digest template plasmid by *Dpn*I endonuclease
   - Ligate plasmid by T4 DNA ligase
3. Use any standard transformation protocol for transforming the constructed plasmids in competent *E. coli* cells

4. Prepare cryo-stock (8.5 parts glycerol/1.5 parts cells; frozen at −80 °C) of *E. coli* cells containing the expression vector for wild type or mutant PqsD
5. Proceed with instructions in Section 3.4.1

3.4 Labeling of PqsD

3.4.1 Preparation of cells for competitive labeling experiments

1. Prepare a cryo-stock of *E. coli* cells containing the expression vector coding for wild type or mutant PqsD (preparation see Section 3.3)
2. Inoculate cells with a pipette tip from a cryo-stock into 5 mL LB medium supplemented with 100 μg/mL carbenicillin and incubate overnight at 37 °C and shaking at 180 rpm
3. On the next day, inoculate culture 1:50 in LB media containing 100 μg/mL carbenicillin
4. Incubate at 37 °C and shaking at 180 rpm until $OD_{600} = 0.5$
5. Induce expression using 0.2 μg/mL anhydrotetracycline
6. Incubate at 37 °C and shaking at 180 rpm for 120 min or until the protein band corresponding to PqsD is visible after SDS-PAGE (preparation see Section 3.7) and Coomassie staining

   Tip: vary protein expression time to find an optimal protein concentration for labeling.
7. Transfer 1–2 mL of culture into a 2 mL microcentrifuge tube for each sample
8. Centrifuge each sample at 4500*g*, 5 min, 4 °C and discard supernatant
9. Gently resuspend cell pellet of each sample in 1 mL PBS by pipetting (= washing step)
10. Centrifuge at 4500*g*, 5 min, 4 °C and discard supernatant
11. Gently resuspend cell pellet of each sample in 100 μL PBS by pipetting
12. For *in vitro* competitive profiling follow instructions in Section 3.4.2
13. For live cell competitive profiling follow instructions in Section 3.4.3

3.4.2 In vitro competitive profiling

1. Prepare 100 μL of cell suspension containing *E. coli* cells recombinantly overexpressing PqsD for each sample according to Section 3.4.1
2. Disrupt cells by any cell lysis method, e.g. sonication according to Section 3.5
3. *Inhibitor treatment*: add 1 μL of a potential PqsD inhibitor stock in DMSO in an appropriate concentration (e.g., 1 mM, 10 mM, 50 mM stock concentration) to 100 μL of cell lysate and gently mix by pipetting up and down

4. Incubate for 30 min at 37 °C and shaking at 180 rpm
5. *Competitive labeling*: add 1 μL of probe *CA* stock in DMSO in an appropriate concentration (e.g., 100 mM stock concentration) to each sample and gently mix by pipetting up and down (for preparation of probe *CA* follow procedure described in Section 3.2)
6. Incubate for 60 min at 37 °C and shaking at 180 rpm

   Tip: try different experimental conditions: vary inhibitor concentrations, activity-based probe concentration; perform each condition in triplicates.
7. Perform click chemistry according to instructions in Section 3.6

3.4.3 Live cell competitive profiling

1. Prepare 100 μL of cell suspension containing *E. coli* cells recombinantly overexpressing PqsD for each sample according to Section 3.4.1
2. *Inhibitor treatment*: add 1 μL of a potential PqsD inhibitor stock in DMSO (use commercially available 7-(chloroacetyl)indole as positive control) in an appropriate concentration (e.g., 1 mM, 10 mM, 50 mM) to each sample and gently mix by pipetting up and down
3. Incubate for 60 min at 37 °C and shaking at 180 rpm
4. *Competitive labeling*: add 1 μL of probe *CA* stock in DMSO in an appropriate concentration (e.g., 100 mM stock concentration) to each sample and gently mix by pipetting up and down (for preparation of probe *CA* follow procedure described in Section 3.2)

   Tip: try different experimental conditions: vary inhibitor concentration, activity-based probe concentration; perform each condition in triplicates for high reproducibility
5. Incubate for 60 min at 37 °C and shaking at 180 rpm
6. Centrifuge each sample at 4500*g*, 5 min, 4 °C and discard supernatant
7. Gently resuspend cell pellet of each sample in 1 mL PBS by pipetting up and down (=washing step)
8. Centrifuge at 4500*g*, 5 min, 4 °C and discard supernatant
9. Gently resuspend cell pellet of each sample in 100 μL PBS by pipetting up and down
10. Disrupt cells by any cell lysis method, e.g., sonication according to Section 3.5
11. After cell lysis, perform click chemistry according to instructions in Section 3.6

3.5 Cell lysis by sonication

1. For cell lysis of 100 μL cell suspension by sonication use 10% amplitude, 0.5 s on, 1.0 s off, 10 pulses.

   Note: *critical step!* To ensure consistent protein concentration of all samples after cell lysis, it is important to fix every microcentrifuge tube carefully in an equal position to the microtip, which should *almost touch the bottom of the tube* (Fig. 5). *During sonication foam formation should be strictly avoided.* Perform cell lysis on ice or put microcentrifuge tubes on ice directly after cell lysis
2. Centrifuge at 12000*g*, 10 min, 4 °C
3. Transfer 100 μL of supernatant (=cell lysate) in a new microcentrifuge tube

   Tip: check protein concentration (Bradford assay) to obtain an equal protein content of 1–2 mg/mL for each sample
4. For *in vitro* competitive profiling follow instructions in Section 3.4.2
5. For live cell competitive profiling follow instructions in Section 3.4.3

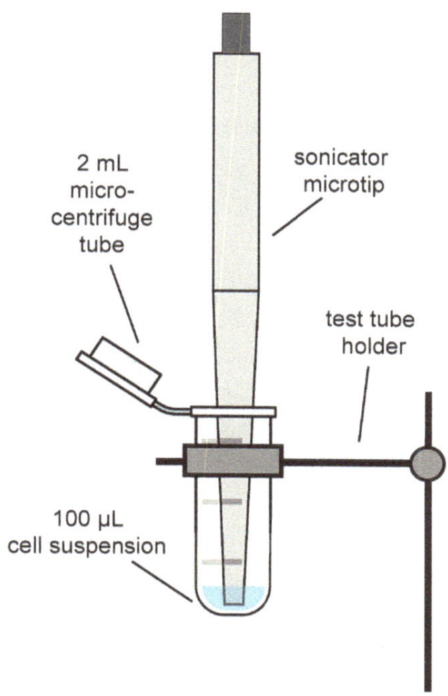

Fig. 5 Experimental set up for cell lysis by sonication.

3.6 Click chemistry

1. Perform competitive profiling experiments according to Section 3.4.2 (*in vitro*) or Section 3.4.3 (in live cells)
2. From each sample, transfer 44.5 μL of cell lysate to a fresh 1.5 mL microcentrifuge tube
3. Add 1 μL 21.5% SDS solution
4. Add 1 μL rhodamine azide stock
5. Add 1.5 μL TBTA ligand stock
6. Add 1 μL TCEP stock solution
7. In the last step add 1 μL $CuSO_4$ stock solution to every centrifugal tube to start the click chemistry reaction and mix immediately by vortexing

   Tip: prepare a master mix of SDS solution, rhodamine azide, TBTA ligand and TCEP for up to 20 samples, finally add $CuSO_4$ to the master mix and immediately add the appropriate amount of master mix to samples by fast pipetting; mix by vortexing
8. Incubate for 60 min at room temperature without shaking
9. Add 2 × SDS-loading buffer to stop the reaction and mix by vortexing

   Tip: lysates can be frozen at this step at −20 °C
10. Proceed with instructions in Section 3.7. For SDS-gel analysis

3.7 SDS-PAGE and in-gel fluorescence scanning

1. Conduct competitive profiling experiments according to Section 3.4.2 (*in vitro*) or Section 3.4.3 (in live cells) followed by click chemistry (Section 3.6)
2. Perform SDS-PAGE according to standard methods to visualize labeling (following descriptions are based on a PeqLab System)

   Note: since acrylamide is toxic, a potential carcinogen and suspected of damaging fertility always wear gloves and goggles, and work under the hood when pouring gels. Also never touch gels without gloves as residues of acrylamide can be present
3. For pouring SDS-gels, assemble the glass plates and spacer into the PeqLab system
4. Prepare 10% separations gels in a beaker containing a stirring bar; for each gel under stirring:
   - **(i)** Add 9 mL separation buffer
   - **(ii)** Add 12 mL Rotiphorese® Gel 30
   - **(iii)** Add 360 μL 10% SDS solution
   - **(iv)** Add 14.4 mL H_2O_{deion}

   (v) Add 15 μL TEMED
   (vi) In the final step add 150 μL APS solution (10% w/v)
   (vii) Immediately after addition of APS solution pour the SDS-gel solution between assembled glass plates to cast the gel
   (viii) Add 100 μL isopropanol on top of each gel to smoothen the surface and avoid bubbles
5. Wait until gel is polymerized
6. Remove isopropanol by pouring into a waste bottle
7. Prepare 3.75% stacking gel in a beaker containing a stirring bar; for each gel under stirring:
   (ix) Add 3 mL stacking buffer
   (x) Add 1.5 mL Rotiphorese® Gel 30
   (xi) Add 120 μL 10% SDS solution
   (xii) Add 7.35 mL H_2O_{deion}
   (xiii) Add 12 μL TEMED
   (xiv) In the final step add 60 μL APS solution (10% w/v)
   (xv) Immediately after addition of APS solution pour the SDS-gel solution on top of the polymerized separation gel to cast the gel
   (xvi) Insert spacer
8. Wait until stacking gel is polymerized

   Note: for storage cover gels in a wet towel, put into air-tight bags and keep at 4 °C
9. For running SDS-gel, assemble the gel into the PeqLab system and fill system with 1 × SDS running buffer
10. Load samples containing denaturing 2 × SDS-loading buffer onto gel (for preparation of samples follow instructions in Section 3.4.2–3.6)
11. Immediately after loading the samples run gel by applying 55 mA
12. After running the gel, visualize labeling by in-gel fluorescence scanning
13. To determine in situ IC_{50} values follow instructions in Section 3.8
14. Perform Coomassie staining to confirm equal protein content and separation in the SDS-gel

3.8 Quantification of probe labeling (IC_{50} values) in live cells

1. Perform competitive profiling experiments in live cells as described in Section 3.4.3

   Note: vary inhibitor concentrations and find an appropriate concentration range for IC_{50} value determination; apply each inhibitor concentration in triplicates (Fig. 6)

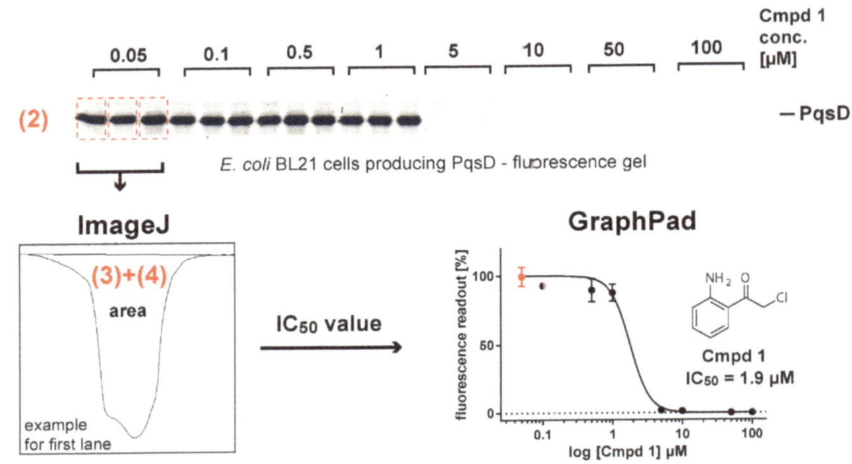

Fig. 6 Example for determination of IC_{50} value of compound **1** against PqsD. Usage of "rectangle tool" (2), "line tool" (3) and "magic wand tool" (4) (ImageJ software) to integrate peak areas for calculation of IC_{50} value.

2. After competitive labeling perform cell lysis (Section 3.5), click chemistry (Section 3.6) and SDS-PAGE (Section 3.7) to visualize labeled proteins via in-gel fluorescence scanning
3. Use ImageJ Software (open source; https://imagej.net/Downloads) to measure integrated fluorescence intensities of protein bands (Fig. 6)
   - **(i)** Open the fluorescence image inside ImageJ
   - **(ii)** Select the "rectangle tool" and draw a box around each gel lane
   - **(iii)** Plot lanes and mark protein peak area using the "line tool"
   - **(iv)** Integrate peak area to obtain fluorescence intensities using the "magic wand tool"
4. Export integrated areas into any statistic software (e.g., GraphPad Prism) and generate a dose-response curve to calculate IC_{50} value (Fig. 6)

4. Summary

Competitive profiling using activity-based chemical probes is a useful strategy for the discovery of enzyme inhibitors especially for enzymes with nucleophilic active site residues. Recent work has demonstrated the utility of this strategy for finding inhibitors of the quinolone biosynthesis enzyme PqsD of the human pathogen *Pseudomonas aeruginosa*. This chapter describes the methods for the synthesis of a chemical probe as well as for performing competitive profiling experiments *in vitro* and in live cells against the enzyme PqsD. We also provide a protocol for the application of probes to determine IC_{50} values of inhibitors in live cells.

Acknowledgment

This work was funded by the Emmy No-ether program of the Deutsche Forschungsgemeinschaft (DFG), EU FP7 Marie Curie Zukunftskolleg Incoming Fellowship Program—University of Konstanz Grant no. 291784, Fonds der Chemischen Industrie (FCI), SFB969, and Konstanz Research School Chemical Biology (KoRS-CB). M.P. acknowledges support by a Zeiss PhD fellowship.

References

Böttcher, T., Pitscheider, M., & Sieber, S. A. (2010). Natural products and their biological targets: Proteomic and metabolomic labeling strategies. *Angewandte Chemie (International Ed. in English)*, *49*(15), 2680–2698. https://doi.org/10.1002/anie.200905352.

Chandrasekar, B., Hong, T. N., & van der Hoorn, R. A. (2017). Inhibitor discovery by convolution ABPP. *Methods in Molecular Biology*, *1491*, 47–56. https://doi.org/10.1007/978-1-4939-6439-0_4.

Cravatt, B. F., Wright, A. T., & Kozarich, J. W. (2008). Activity-based protein profiling: From enzyme chemistry to proteomic chemistry. *Annual Review of Biochemistry*, *77*, 383–414. https://doi.org/10.1146/annurev.biochem.75.101304.124125.

Drees, S. L., Ernst, S., Belviso, B. D., Jagmann, N., Hennecke, U., & Fetzner, S. (2018). PqsL uses reduced flavin to produce 2-hydroxylaminobenzoylacetate, a preferred PqsBC substrate in alkyl quinolone biosynthesis in Pseudomonas aeruginosa. *The Journal of Biological Chemistry*, *293*(24), 9345–9357. https://doi.org/10.1074/jbc.RA117.000789.

Drees, S. L., & Fetzner, S. (2015). PqsE of *Pseudomonas aeruginosa* acts as pathway-specific thioesterase in the biosynthesis of alkylquinolone signaling molecules. *Chemistry & Biology*, *22*(5), 611–618. https://doi.org/10.1016/j.chembiol.2015.04.012.

Drees, S. L., Li, C., Prasetya, F., Saleem, M., Dreveny, I., Williams, P., et al. (2016). PqsBC, a condensing enzyme in the biosynthesis of the *Pseudomonas aeruginosa* quinolone signal: Crystal structure, inhibition, and reaction mechanism. *The Journal of Biological Chemistry*, *291*(13), 6610–6624. https://doi.org/10.1074/jbc.M115.708453.

Dubern, J. F., & Diggle, S. P. (2008). Quorum sensing by 2-alkyl-4-quinolones in *Pseudomonas aeruginosa* and other bacterial species. *Molecular BioSystems*, *4*(9), 882–888. https://doi.org/10.1039/b803796p.

Farrow, J. M., 3rd, & Pesci, E. C. (2007). Two distinct pathways supply anthranilate as a precursor of the Pseudomonas quinolone signal. *Journal of Bacteriology*, *189*(9), 3425–3433. https://doi.org/10.1128/JB.00209-07.

Fonovic, M., & Bogyo, M. (2008). Activity-based probes as a tool for functional proteomic analysis of proteases. *Expert Review of Proteomics*, *5*(5), 721–730. https://doi.org/10.1586/14789450.5.5.721.

Gersch, M., Kreuzer, J., & Sieber, S. A. (2012). Electrophilic natural products and their biological targets. *Natural Product Reports*, *29*(6), 659–682. https://doi.org/10.1039/c2np20012k.

Heeb, S., Fletcher, M. P., Chhabra, S. R., Diggle, S. P., Williams, P., & Camara, M. (2011). Quinolones: From antibiotics to autoinducers. *FEMS Microbiology Reviews*, *35*(2), 247–274. https://doi.org/10.1111/j.1574-6976.2010.00247.x.

Hubel, K., Lessmann, T., & Waldmann, H. (2008). Chemical biology—Identification of small molecule modulators of cellular activity by natural product inspired synthesis. *Chemical Society Reviews*, *37*(7), 1361–1374. https://doi.org/10.1039/b704729k.

Huse, H., & Whiteley, M. (2011). 4-Quinolones: Smart phones of the microbial world. *Chemical Reviews*, *111*(1), 152–159. https://doi.org/10.1021/cr100063u.

Hutter, M. C., Brengel, C., Negri, M., Henn, C., Zimmer, C., Hartmann, R. W., et al. (2014). Mechanistic details for anthraniloyl transfer in PqsD: The initial step in HHQ biosynthesis. *Journal of Molecular Modeling*, *20*(6), 2255. https://doi.org/10.1007/s00894-014-2255-z.

Lepine, F., Milot, S., Deziel, E., He, J., & Rahme, L. G. (2004). Electrospray/mass spectrometric identification and analysis of 4-hydroxy-2-alkylquinolines (HAQs) produced by *Pseudomonas aeruginosa*. *Journal of the American Society for Mass Spectrometry*, *15*(6), 862–869. https://doi.org/10.1016/j.jasms.2004.02.012.

Lin, J., Cheng, J., Wang, Y., & Shen, X. (2018). The pseudomonas quinolone signal (PQS): Not just for quorum sensing anymore. *Frontiers in Cellular and Infection Microbiology*, *8*, 230. https://doi.org/10.3389/fcimb.2018.00230.

Niphakis, M. J., & Cravatt, B. F. (2014). Enzyme inhibitor discovery by activity-based protein profiling. *Annual Review of Biochemistry*, *83*, 341–377. https://doi.org/10.1146/annurev-biochem-060713-035708.

Prothiwa, M., Englmaier, F., & Böttcher, T. (2018). Competitive live-cell profiling strategy for discovering inhibitors of the quinolone biosynthesis of Pseudomonas aeruginosa. *Journal of the American Chemical Society*, *140*(43), 14019–14023. https://doi.org/10.1021/jacs.8b07629.

Prothiwa, M., Szamosvari, D., Glasmacher, S., & Böttcher, T. (2016). Chemical probes for competitive profiling of the quorum sensing signal synthase PqsD of *Pseudomonas aeruginosa*. *Beilstein Journal of Organic Chemistry*, *12*, 2784–2792. https://doi.org/10.3762/bjoc.12.277.

Rostovtsev, V. V., Green, L. G., Fokin, V. V., & Sharpless, K. B. (2002). A stepwise Huisgen cycloaddition process: Copper(I)-catalyzed regioselective "ligation" of azides and terminal alkynes. *Angewandte Chemie (International Ed. in English)*, *41*(14), 2596–2599. https://doi.org/10.1002/1521-3773(20020715)41:14<2596::AID-ANIE2596>3.0.CO;2-4.

Schertzer, J. W., Brown, S. A., & Whiteley, M. (2010). Oxygen levels rapidly modulate *Pseudomonas aeruginosa* social behaviours via substrate limitation of PqsH. *Molecular Microbiology*, *77*(6), 1527–1538. https://doi.org/10.1111/j.1365-2958.2010.07303.x.

Schütz, C., & Empting, M. (2018). Targeting the pseudomonas quinolone signal quorum sensing system for the discovery of novel anti-infective pathoblockers. *Beilstein Journal of Organic Chemistry*, *14*, 2627–2645. https://doi.org/10.3762/bjoc.14.241.

Szamosvári, D., & Böttcher, T. (2017). An unsaturated quinolone N-oxide of *Pseudomonas aeruginosa* modulates growth and virulence of *Staphylococcus aureus*. *Angewandte Chemie (International Ed. in English)*, *56*(25), 7271–7275. https://doi.org/10.1002/anie.201702944.

Szamosvári, D., & Böttcher, T. (2018). 4-Quinolone N-oxides as bacterial weapons. *Synlett*, *29*, 542–547.

Szamosvári, D., Schuhmacher, T., Hauck, C. R., & Böttcher, T. (2019). A thiochromenone antibiotic derived from the pseudomonas quinolone signal selectively targets the Gram-negative pathogen Moraxella catarrhalis. *Chemical Science*, *10*(27), 6624–6628. https://doi.org/10.1039/c9sc01090d.

Tornoe, C. W., Christensen, C., & Meldal, M. (2002). Peptidotriazoles on solid phase: [1,2,3]-triazoles by regiospecific copper(i)-catalyzed 1,3-dipolar cycloadditions of terminal alkynes to azides. *Journal of Organic Chemistry*, *67*(9), 3057–3064. https://doi.org/10.1021/jo011148j.

Wijdeven, R. H., Neefjes, J., & Ovaa, H. (2014). How chemistry supports cell biology: The chemical toolbox at your service. *Trends in Cell Biology*, *24*(12), 751–760. https://doi.org/10.1016/j.tcb.2014.07.002.

Willems, L. I., Overkleeft, H. S., & van Kasteren, S. I. (2014). Current developments in activity-based protein profiling. *Bioconjugate Chemistry*, *25*(7), 1181–1191. https://doi.org/10.1021/bc500208y.

Zuhl, A. M., Mohr, J. T., Bachovchin, D. A., Niessen, S., Hsu, K. L., Berlin, J. M., et al. (2012). Competitive activity-based protein profiling identifies aza-beta-lactams as a versatile chemotype for serine hydrolase inhibition. *Journal of the American Chemical Society*, *134*(11), 5068–5071. https://doi.org/10.1021/ja300799t.

CHAPTER FIVE

The NMR tube bioreactor

Alexandra V. Chatzikonstantinou[a,b], Antonis D. Tsiailanis[b], Ioannis P. Gerothanassis[b], Haralambos Stamatis[a], Enrico Ravera[c,d], Marco Fragai[c,d], Claudio Luchinat[c,d], Giacomo Parigi[c,d], Andreas G. Tzakos[b,*]

[a]Laboratory of Biotechnology, Department of Biological Applications and Technology, University of Ioannina, Ioannina, Greece
[b]Department of Chemistry, Section of Organic Chemistry and Biochemistry, University of Ioannina, Ioannina, Greece
[c]Magnetic Resonance Center (CERM), Department of Chemistry "Ugo Schiff", University of Florence, Sesto Fiorentino, Italy
[d]Consorzio Interuniversitario Risonanze Magnetiche di Metallo Proteine, University of Florence, Sesto Fiorentino, Italy
*Corresponding author: e-mail address: atzakos@uoi.gr

Contents

Methods in Enzymology, Volume 633
ISSN 0076-6879
https://doi.org/10.1016/bs.mie.2019.10.032

Abstract

Enzymes are pliable systems and core cellular components allowing the performance of several processes. They can also be utilized as "green" synthetic factories to generate bioactive therapeutic, diagnostic or theranostic compounds. Methods to enable the mapping of enzyme substrates as well as the understanding of the interactions of the formed products with target proteins could be of importance. This chapter describes the utilization of the "NMR tube bioreactor" method. This method, carried out inside an NMR tube, can allow for the prediction of compounds that are able to serve as potential enzyme substrates, and also explicit the regioselectivity of the enzymatic reactions. Furthermore, it enables the real time monitoring of multiple-biotransformation products in the NMR tube without the need of fractionation or isolation of each individual component. Finally, it allows for the screening of the resulting biotransformation products as ligands for protein targets.

Abbreviations

AcCN acetonitrile
BSA bovine serum albumin
CaLB *Pseudozyma (Candida) antarctica*
Cyt-c cytochrome c iso-1 C107T
D_2O deuterium oxide
DMSO dimethyl sulfoxide
HMBC heteronuclear multiple bond correlation
HSQC heteronuclear single quantum coherence
NMR nuclear magnetic resonance
PBS phosphate-buffered saline
STD saturation transfer difference

1. Introduction

The biochemical reactions in a cell are constantly tuned, intertwined and calibrated in a coordinated fashion so as to build up and break down cellular components and ultimately fulfill all cellular functions. To achieve this fragile equilibrium cells systematize the reactions in enzyme-powered paths. Single enzymatic steps can be linked up and shape numerous metabolic circuits. However, enzymes do not use a single substrate but an array of different compounds leading to chemical diversity. This catalytic promiscuity has often remained elusive and uncharted in the absence of efficient tools to decompose this information. Charting the substrate diversity of an enzyme is of immense importance. For instance, the different metabolites in case of drugs that can emerge from specific enzymes should be identified since they can be harmful for the human health. The recognized enzyme promiscuity has also been utilized to synthesize different compounds with "green" processes. Numerous research groups utilized this enzyme promiscuity to generate bioactive compounds via biocatalysis (Choi, Han, & Kim, 2015; Patel, 2018; Sun, Zhang, Ang, & Zhao, 2018). Another currently active trend in drug discovery is the development of drug precursors (prodrugs) that can be activated to yield the active component in a specific cell microenvironment. For example, in oncology, researchers exploit enzymes that are selectively overexpressed in cancer cells (Giang, Boland, & Poon, 2014; Mahato, Tai, & Cheng, 2011) to build up cancer responsive prodrugs. Thus, in a drug discovery pipeline it is very important to develop tools to rapidly validate such enzyme responsive prodrugs in a simulated cellular environment. Furthermore, there is a quest to discover cellular responsive chemosensors so as to chart cellular functions, like, for instance, the pH microenvironment in cells (Cardone, Casavola, & Reshkin, 2005), the intracellular redox potential (Hegedűs et al., 2018), and the cell metabolic activity (3-(4,5-dimethylthiazol-2-yl)-2,5-diphenyltetrazolium bromide, MTT assay) (Van Meerloo, Kaspers, & Cloos, 2011). Along these lines there is a clear need to establish tools and processes to foster the use of enzymes as systems to generate new compounds, as well as to monitor the products or metabolites. Such tools could enable the charting of enzyme promiscuity, discovery of substrates of uncharacterized enzymes, assist in the development of prodrugs activatable by intracellular enzymes and synthesize bioactive therapeutic, diagnostic or theranostic compounds. Although biotransformation can be implicated in all these processes, several challenges exist to exploit them in full, since the products can be chemically complex and could also constitute complex mixtures; thus the isolation and characterization could be challenging.

In this chapter, we describe how to set up the "NMR tube bioreactor" (Chatzikonstantinou et al., 2018). This is a three step process that is completely carried out inside an NMR tube (Fig. 1). The NMR tube bioreactor could allow for predicting potential enzyme substrates and the regioselectivity of the enzymatic reaction, for charting the formed prodrugs in real time and for evaluating the resulting products as interactors of proteins of interest.

As illustrated in Fig. 1, in the first step, one can quickly predict in an NMR tube: (1) the capacity of a compound or a mixture of compounds

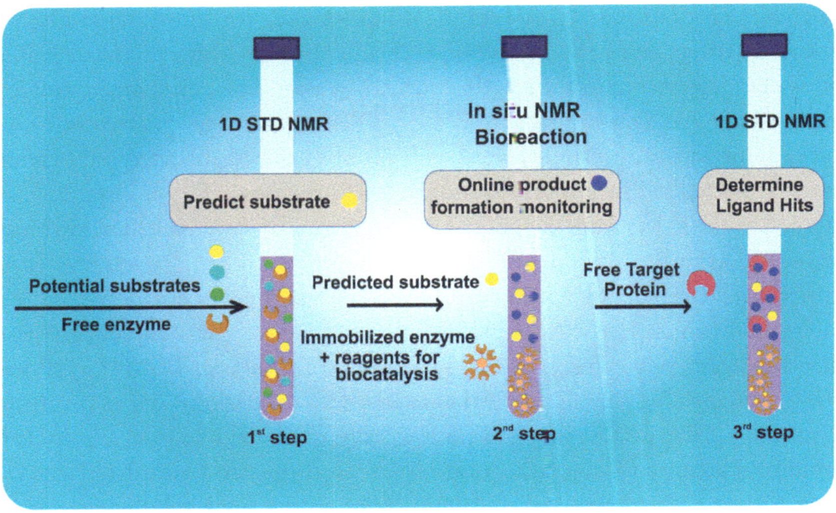

Fig. 1 The great advantages of the use of the NMR tube bioreactor: (i) use of low volumes of aqueous solvent, (ii) elimination of all purification steps of a classical synthetic procedure, (iii) use of mild reaction conditions without protection/de-protection steps, and (iv) immobilization of the enzyme allowing for the recycle of the catalysts for several reactions. The simultaneous screening of several compounds allows a reduction of the experiments that should be performed.

to serve as enzyme substrate(s) and (2) the potential regioselectivity of the biotransformations. Then, in the second step one can monitor in real time the formation of enzymatically generated products as well as optimize the enzymatic reaction yield by rapidly screening different conditions. Of importance, a real time monitoring of the reaction product(s) is conducted in this step without any prior fractionation or purification. In the third step, the biotransformed products are evaluated as ligands in a protein-ligand screening process.

In this chapter, we will illustrate in detail the required steps for the successful application of the NMR tube bioreactor. To illustrate the application of the first two steps, we selected, as putative enzymatic substrates, flavonoids due to their inherent scaffold complexity. As a template enzyme, we use lipase B from *Pseudozyma (Candida) antarctica* (CaLB) for either ester synthesis or alcoholysis. Furthermore, to evaluate the applicability of the process, we utilize the enzyme cytochrome *c* iso-1 C107T (Cyt-c) for the oxidation of a substrate. For the third step, that is ligand-protein interaction screening on the enzymatically derived products, we illustrate two examples using the prototypical protein carrier BSA and the intrinsically disordered protein

α-synuclein. The two main approaches, ligand-based screening and protein-based screening, will be discussed. Through the analyzed protocols the user will be able to understand how to incorporate the process in a drug discovery pipeline as well as in drug metabolite screening.

2. Prediction of the capacity of a compound to serve as an enzyme substrate and the regioselectivity of the biotransformation

Prediction of the potential of a compound to serve as a substrate for an enzyme while at the same time predicting the regioselectivity of the biotransformation is clearly of great importance. Saturation Transfer Difference NMR spectroscopy (STD NMR) is an excellent tool for exploring ligand-protein interactions that offers numerous advantages over other conventional methods. It does not only provide information on a potential interaction of a ligand with a protein but it can also unveil the binding epitope of a ligand. Thus, it can be utilized to identify the binding interface of a putative ligand to a protein. In the case of enzymes it can provide crucial information on the regioselectivity of an enzymatic reaction. Furthermore, this technique offers rapid screening of a library of molecules while at the same time allows for the use of low concentrations of proteins, due to its high sensitivity (Mayer & Meyer, 2001; Pellecchia et al., 2008). Complementing this ligand-based NMR technique with protein-based NMR techniques, i.e., ^{15}N HSQC chemical shift perturbation mapping, in screening ligand-protein interactions, a global picture on the studied system can be obtained in a short time.

In the first step of the NMR tube bioreactor, one has to apply STD NMR in the solution containing the putative substrate with the enzyme but in the absence of components needed for the reaction to occur. Thus, recording the STD NMR spectra of this solution the protons of the ligand implicated in the interaction will be investigated and the regioselectivity of an enzymatic reaction can be monitored.

In order to validate that the selectivity profile of a biotransformation can be predicted, two enzymes, lipase from *Pseudozyma (Candida) antarctica* (CaLB), a hydrolase, and cytochrome *c* iso-1 C107T (Cyt-c), a peroxidase, have been selected. CaLB is a versatile hydrolase which can catalyze reactions such as acetylation (Kazlauskas & Bornscheuer, 1998), alcoholysis (Kyriakou et al., 2018), as well as ester synthesis (Papadopoulou et al., 2013). Recently, our group demonstrated that CaLB can transform the flavonoid quercetin to ester analogues, thus this group of compounds could serve as efficacious substrates for this enzyme (Kyriakou et al., 2012). As alternative biocatalyst Cyt-c was used, which is a key protein for the initialization of cell apoptosis, and a radical scavenger. Cyt-c is one of the best characterized proteins and presents high peroxidase activity in vitro. Cyt-c is used extensively in biocatalysis and nanomedicine, helping in understanding the molecular mechanisms in vivo (Patila et al., 2018).

STD NMR experiments are performed to chart the interactions between different flavonoids like quercetin and naringenin (or rutin, luteolin-7-*O*-glucoside and naringin) and free CaLB in 65% (v/v) phosphate-buffered saline (PBS) in D_2O and 35% (v/v) AcCN-d_3. Under these conditions no enzymatic reaction can take place. The recorded STD NMR experiments confirm a specific interaction of quercetin with CaLB. More specifically, the protons 5′H and 6′H of ring B of quercetin are found to interact with the enzyme. This pinpoints that the main interaction site of the flavonoid with the enzyme is located on ring B of the putative substrate (Fig. 2). Through this experiment it is evident that quercetin interacts with the enzyme, and also that this interaction is mediated with a specific region of this compound. Thus, it can be predicted that quercetin can serve as a substrate for the specific enzyme, and also that the regioselectivity of the reaction could take place in ring B that is susceptible for the interaction. Indeed, according to literature (Kyriakou et al., 2012), ring B of quercetin is regioselectively biotransformed by CaLB. STD NMR experiments are also performed to investigate the interactions between naringenin and CaLB. Naringenin has similar structure to quercetin; however, it lacks the OH at position 3 and a double bond in the ring C. STD NMR experiments demonstrate that all aromatic protons of naringenin interact with CaLB. This result could be explained by the high flexibility of the naringenin scaffold due to the absence of the double bond between carbon C2 and C3, thus enabling a better conformational adaptation to the enzyme binding site (Fig. 2). STD NMR experiments are also performed for glycosylated flavonoids such as luteolin-7-*O*-glucoside, naringenin and rutin. A specific interaction with CaLB is observed for all the studied compounds (Fig. 2). Interestingly, several possible regioselectively acetylated locations can be predicted (Figs. 2 and 4).

To further prove the applicability of this technique, the transformation of pyrogallol to purpurogallin (Ono & Goto, 2006), a compound with high biological activity (Caruana et al., 2011; Inamori et al., 1997), catalyzed by Cyt-c is explored. STD NMR demonstrates a specific interaction between pyrogallol and Cyt-c in aqueous solution (Fig. 2). In the absence

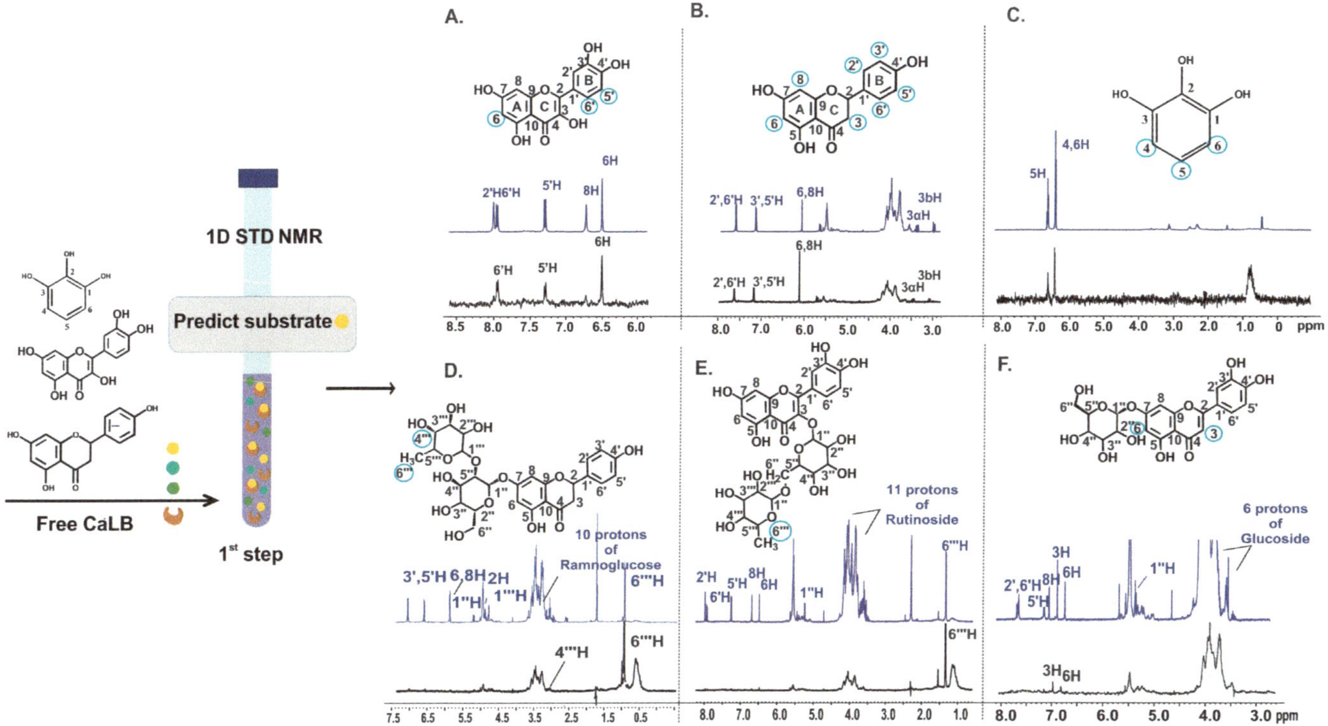

Fig. 2 Selected region of the STD NMR spectra of (A) quercetin, (B) naringenin, (D) naringin, (E) rutin, (F) luteolin-7-*O*-glucoside with free CaLB in 65% PBS 10 mM, pD = 7.0 in D_2O and 35% AcCN-d_3 at 298 K and (C) pyrogallol with free Cyt-c in 20 mM sodium phosphate buffer at pD = 6.8 in D_2O at 298 K. The reference ^1H NMR spectra are colored in blue and the difference ^1H NMR spectra are colored in black. The protons of the enzymatic substrates identified to interact with the free enzyme are highlighted by a blue circle. A schematic representation of the first step applied in a single compound is shown on the left.

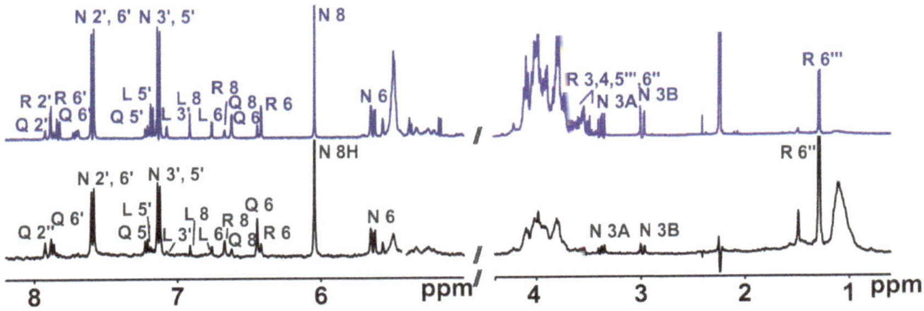

Fig. 3 Selected region of the STD NMR spectrum of a mixture of quercetin (Q), naringenin (N), rutin (R) and luteolin-7-*O*-glucoside (L) in the presence of free CaLB in 65% PBS 10 mM pD = 7 in D_2O and 35% AcCN-d_3. The reference ^{1}H NMR spectra are colored in blue and the difference ^{1}H NMR spectra are colored in black.

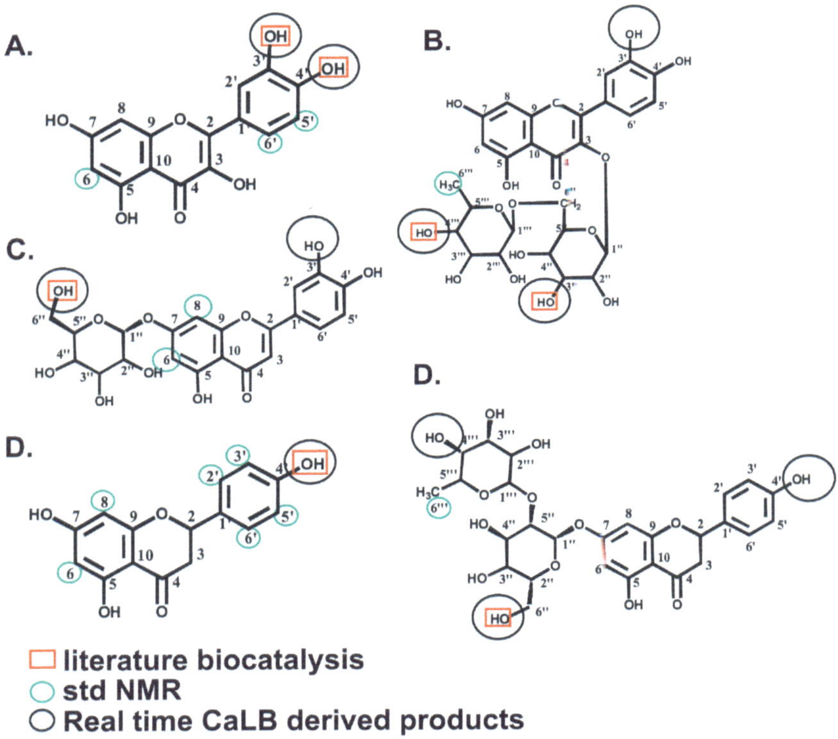

Fig. 4 The protons of the enzymatic substrates (A) quercetin, (B) rutin, (C) luteolin-7-*O*-glucoside, (D) naringenin and (E) naringin identified through STD NMR to interact with free CaLB are highlighted on the 2D structures by blue circles, the positions reported in the literature to be biotransformed are labeled by red squares, whereas the real time CaLB derived products are labeled by black circles.

of the electron acceptor, in the positions where oxidation occurs, purpurogallin is produced, indicating the potential of this method to be applicable in numerous classes of substrates and enzymes.

The potential of this method to be used for screening multiple putative substrates in the same NMR tube is demonstrated in a mixture of four glycosylated or non-glycosylated flavone and flavonones (quercetin, naringenin, rutin and luteolin-7-*O*-glucoside) in the presence of free CaLB. In these 4 substrates 23 OH groups exist that could be possible locations for biotransformation. However, biotransformation from CaLB takes place in only six OH groups according to literature (Chebil, Humeau, Falcimaigne, Engasser, & Ghoul, 2006; De Oliveira et al., 2010; Mellou et al., 2005; Viskupicova, Sturdik, & Ondrejovic, 2009) (Fig. 4). In the first step of the NMR tube bioreactor a specific interaction with CaLB can be observed for all the tested compounds, suggesting that all can act as putative substrates (Fig. 3). We then explored whether the mapped interacting parts of the tested compounds are in proximity to the location where regioselective transformation is performed with the specific enzyme. Indeed, as it can be seen in Fig. 4, the OH groups subject to biocatalysis are largely surrounded by CaLB interacting protons as determined by STD NMR.

These results imply that it may be possible to predict the substrate efficacy and also to propose the regioselective positions where the biocatalysis could occur even in a mixture of different compounds and before performing the actual biotransformation.

2.1 Equipment, materials, and methodology

2.1.1 General safety

Nitrile gloves, protective mask.

2.1.2 Equipment

1. 5 mm NMR tube
2. Micropipettes (2–20, 20–200, 100–1000 μL)
3. Pipette tips
4. NMR spectrometer (Bruker AV-500 spectrometer equipped with a TXI cryoprobe, Bruker BioSpin, Rheinsetten, Germany)

2.1.3 Chemicals and materials

1. Lipase from *Pseudozyma (Candida) antarctica* (CaLB) was from c-LEcta GmbH, Perlickstrase.
2. Flavonoids (quercetin, rutin, luteolin-7-*O*-glucoside, naringenin) were purchased from Extrasynthese.

3. C107T yeast iso-1-cytochrome c (Cyt-c) expressed and purified as described in Barker et al. (2001) and was purchased from Giotto Biotech (https://www.giottobiotech.com).
4. Pyrogallol was from Sigma-Aldrich (St. Louis, MO, USA).
5. Deuterium oxide (D_2O) (99.9%) and Acetonitrile-d_3 (AcCN-d_3) from Deutero (Kastellaun, Germany).
6. Phosphate-buffered saline (PBS) 10 mM pD = 7 in D_2O.
7. Sodium phosphate buffer 20 mM, pD = 6.8, in D_2O.

2.2 Screening of potential substrates for CaLB through STD NMR

2.2.1 Methodology

2.2.1.1 Duration

1 h.

2.2.1.2 Protocol

1. 2 mM of quercetin (or naringenin, naringin, rutin and luteolin-7-O-glucoside) dissolved in 500 μL solvent mixture of 65% (v/v) phosphate-buffered saline (PBS) 10 mM pD = 7.0 in D_2O, and 35% (v/v) AcCN-d_3 and transferred in 5 mm NMR tube
2. 2.2 mg (~0.02 mM protein) or 4.4 mg (~0.04 mM protein), in the case of the mixture of four flavonoids, of CaLB finally added to the mixture in the NMR tube.
3. After loading the sample into the Bruker AV-500 spectrometer the sample is locked, tuned and matched according to manufacturer's guidelines. STD NMR experiments are recorded at 298 K on a Bruker AV-500 spectrometer equipped with a TXI cryoprobe. The on-resonance irradiation of the protein is performed at 1.03 ppm. Off-resonance irradiation is conducted at 40 ppm outside the ^1H NMR resonances region of the protein. Each Gaussian pulse required 50 ms separated by a delay of 1 ms. Saturation time was set at 2 s and the experimental time was ~1 h (128 scans). The water suppression was conducted by using an excitation sculpting pulse sequence.

2.3 Screening of potential substrates for Cyt-c through STD NMR

2.3.1 Methodology

2.3.1.1 Duration

1 h

2.3.1.2 Protocol

1. 5 mM of pyrogallol are dissolved in 500 μL of 20 mM sodium phosphate buffer 20 mM, pD = 6.8, in D_2O, and transferred in 5 mm NMR tube.
2. 0.05 mM of C107T yeast iso-1-cytochrome c are added in the NMR tube.
3. After loading the sample into the Bruker AV-500 spectrometer the sample is locked, tuned and matched according to manufacturer's guidelines. STD NMR experiments are recorded at 298 K on a Bruker AV-500 spectrometer equipped with a TXI cryoprobe. The on-resonance irradiation of the protein is performed at 0.88 ppm. Off-resonance irradiation is conducted at 40 ppm outside the ^{1}H NMR resonances region of the protein. Each Gaussian pulse required 50 ms separated by a delay of 1 ms. Saturation time was set at 2 s and the experimental time was ~1 h (128 scans). The water suppression was conducted by using an excitation sculpting pulse sequence.
   TIP 1: To enhance the solubility of the flavonoids (especially in case of the mixture) the solvent should be heated to 60 °C for 5–10 min before the addition of the biocatalyst.
   TIP 2: For high resolution NMR spectra store the deuterated NMR solvents under nitrogen.

3. Real time monitoring of biotransformation products in the NMR tube bioreactor

In the first step of the process, the compounds that could serve as potential substrates for a target enzyme are screened with STD NMR and also the regioselectivity of the enzymatic reaction can be determined. In the second step the actual biotransformation reaction on the selected putative substrates is monitored inside the NMR tube in the fully assembled enzymatic reaction.

In a conventional way of monitoring biotransformation products, a tedious process is followed where the products have to be isolated from the reaction mixture, fractionated, purified, separated and analyzed. To speed up the specific process, in the NMR tube bioreactor the biocatalytic reactions are performed directly in the NMR tube. On line monitoring of the biotransformation of natural products by in situ ^{1}H NMR is a valuable tool to rapidly screen different conditions, so as to optimize the enzymatic reaction yield, or to gain data about kinetics and mechanisms of biotransformations (Chatzikonstantinou et al., 2018). Thus, in the NMR tube bioreactor procedure, the enzymatic reaction is performed in the NMR tube and the formation of the products can be identified, in real time, by 1D and 2D NMR experiments.

Before moving to the assembly of the enzymatic reaction in the NMR tube bioreactor, one can first conduct, for validation purposes, a conventional enzymatic reaction by taking advantage of the results obtained in the first step of the process. After determining that quercetin, naringenin, rutin, luteolin-7-*O*-glucoside and naringin could act as substrates of CaLB, the conventional enzymatic acetylation of the specific flavonoids can be assembled. In an eppendorf tube, 16 mM of each flavonoid and 250 μL of vinyl acetate are added in 0.5 mL of acetonitrile, in the presence of 60 mg mL^{-1} of CaLB. Then the reaction mixture is incubated in a thermomixer at 750 rpm at 333 K for 3 days (Kyriakou et al., 2012). Finally, the biotransformed products are isolated from the reaction mixture, fractionated, purified, separated and analyzed.

In the real time NMR tube bioreactor process the biotransformation of the targeted substrates is recorded rapidly by NMR. To record high resolution 1D and 2D NMR spectra of the biotransformation products and avoid the presence of NMR signals from the targeted enzyme, immobilization of the enzyme can be beneficial. Actually, the enzyme is immobilized and confined at the bottom of the NMR tube and thus is located outside the active volume of the NMR coil. Further, the NMR signals of the enzyme are severely broadened, therefore in situ monitoring of the reaction products can be achieved avoiding at the same time the interference from protein signals. To achieve high resolution in situ 1D and 2D NMR spectra, the conditions of the biotransformation as well as the conditions of the NMR experiments should be optimized. Fig. 5 illustrates the optimization process recording ^{1}H NMR spectra of the in situ esterification of quercetin catalyzed by CaLB. To achieve high quality NMR spectra with sharp peaks the concentration of the immobilized enzyme (optimum concentration: 30 mg mL^{-1}) and the amount of the acetyl donor (optimum amount: 125 μL) reduced while the NMR experiments on a medium field NMR spectrometer are performed at low temperature (282 K). As it can be seen in Fig. 5 after the optimization the line-widths of the originally broad resonances of C-5 OH groups reduced by a factor of two (from 8 Hz in Fig. 5A to 4 Hz in Fig. 5D).

After the preliminary optimization, the biotransformation of quercetin, naringenin and pyrogallol are performed under the optimum conditions and high resolution in situ 1D ^{1}H NMR and 2D ^{1}H-^{13}C HSQC, 2D ^{1}H-^{13}C

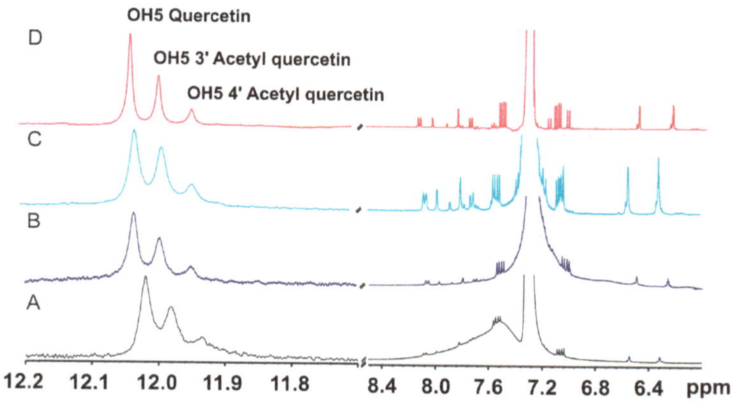

Fig. 5 ^{1}H NMR spectra of in situ esterification of quercetin catalyzed by CaLB. (A) On Bruker AV-400 spectrometer, at 333 K, with 30 mg CaLB and 250 μL vinyl acetate. (B) On Bruker AV-400 spectrometer, at 298 K, with 30 mg CaLB and 250 μL vinyl acetate. (C) On Bruker AV-400 spectrometer, at 298 K, with 15 mg CaLB and 125 μL vinyl acetate. (D) On Bruker AV-500 spectrometer, at 282 K, with 15 mg CaLB and 125 μL vinyl acetate.

HMBC spectra are recorded. In the case of Cyt-c the enzyme is first immobilized on celite in order to perform in situ monitoring of the biotransformation of pyrogallol catalyzed by Cyt-c. The analysis of the spectra as a function of time allows for monitoring in real time the formation of the products as well as the concurrent consumption of the substrates (Figs. 6 and 7) (Chatzikonstantinou et al., 2018).

The NMR tube bioreactor has the potential to operate also as a high throughput process. To probe this potential the method has been applied in a mixture of five different flavonoids as putative enzymatic substrates. In this way, the formation of eight different products can be recorded in real time. It is worth mentioning that although 13 different compounds are present in the NMR tube bioreactor (5 substrates and 8 products), high resolution NMR spectra can be recorded in the presence of the targeted enzyme (Fig. 7). Thus, it is evident that through this process numerous potential enzymatic substrates, for a targeted enzyme, can be investigated and the NMR tube bioreactor can be utilized to expand the chemical space in a biologically relevant way.

Through this process the regioselective biotransformation on the targeted substrates (quercetin and naringenin) can be rapidly determined in accordance also to previous work (Kyriakou et al., 2012). Furthermore, the high potential of the NMR tube bioreactor is evident since it enables the determination of biotransformation products that were overlooked in the past. Specifically,

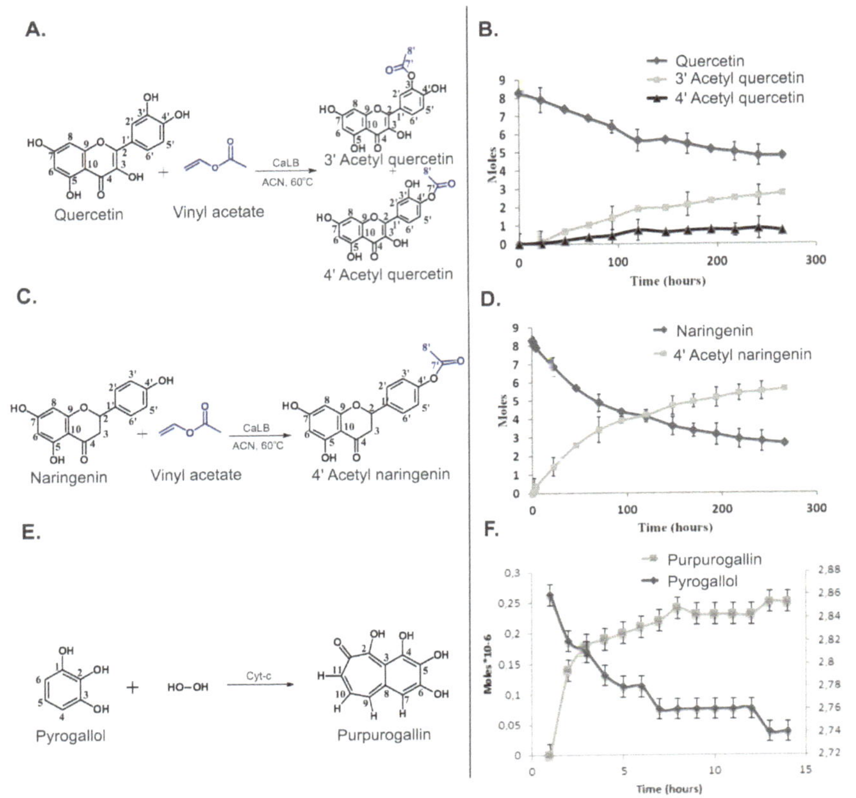

Fig. 6 (A and C) The acetylation of quercetin and naringenin catalyzed by CaLB. (E) The oxidation of pyrogallol catalyzed by Cyt-c. (B, D and F) The real time monitoring of the biotransformation of quercetin, naringenin and pyrogallol, respectively.

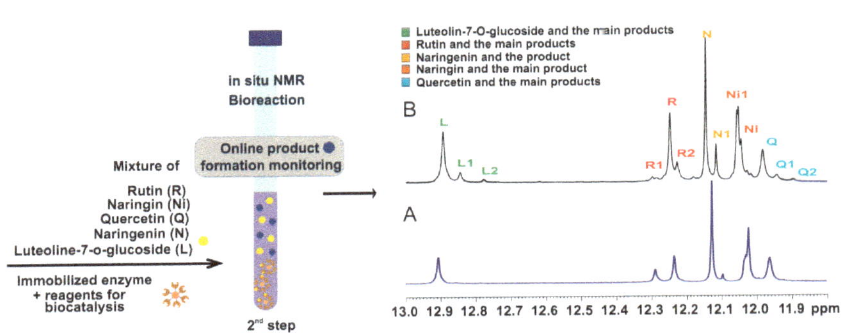

Fig. 7 Selected region (OH5 protons) of the ^{1}H NMR spectra of the CaLB mediated biotransformation of a mixture of five flavonoids (quercetin (Q), naringenin (N), rutin (R) and luteolin-7-*O*-glucoside (L), naringin (Ni)) (A), at 240 h and (B), at 48 h, in AcCN-d_3 at 282 K. A schematic representation of the second step applied in a mixture of five compound is shown on the left.

although specific CaLB mediated biotransformation products of the glycosylated flavonoids such as rutin, luteolin-7-*O*-glucoside and naringin have been reported in the literature, the NMR tube bioreactor allowed the identification of new acetylated products of rutin, luteolin-7-*O*-glucoside and naringin (Fig. 8). In the case of rutin it is known that the OH groups accessible for CaLB biotransformation are the C4′ and C3′ OH (Chebil et al., 2006; De Oliveira et al., 2010; Viskupicova et al., 2009). However, in the present experiment one more OH group is proved to be susceptible to CaLB mediated biotransformation (C3′ OH), leading to the synthesis of diacetyl and triacetyl rutin as indicated in Fig. 8. Similar results can be also found for the case of naringin. The primary C6′ OH group is known to be accessible to CaLB (Chebil et al., 2006; Mellou et al., 2005; Viskupicova et al., 2009). Through the NMR tube bioreactor two more OH groups of naringin, that have not been identified formerly to be biotransformed, are found to be accessible to CaLB biocatalysis (C4″ and C4′ OH) leading to the formation of diacetyl and triacetyl naringin (Fig. 8). The same results are also found for the case of luteolin-7-*O*-glucoside. The primary C6′ OH group is known to be accessible to CaLB biotransformation (Chebil et al., 2006; Viskupicova et al., 2009), but herein the C3″ and C3 OH groups are also found to be susceptible for biotransformation leading to the formation of diacetyl and triacetyl luteolin-7-*O*-glucoside (Fig. 8).

3.1 Characterizations of all the substrates and the biotransformed products through 1D ^{1}H NMR and 2D ^{1}H-^{13}C HSQC, 2D ^{1}H-^{13}C HMBC

The NMR characterization of all the substrates and the biotransformed products through 1D ^{1}H NMR and 2D ^{1}H-^{13}C HSQC, 2D ^{1}H-^{13}C HMBC can be found below:

Quercetin: ^{1}H NMR (500 MHz, DMSO-d_6): δ 12.54 (s, 1H), 10.85 (s, 1H), 9.66 (s, 1H), 9.44 (s, 1H), 9.37 (s, 1H), 7.71 (d, J=2.2 Hz, 1H), 7.57 (dd, J=2.23/6.28 1H), 6.92 (d, J=8.5 Hz, 1H), 6.44 (d, J=2.08 Hz, 1H), 6.22 (d, J=2.08 Hz, 1H); ^{13}C NMR (500 MHz, DMSO-d_6): δ 165.0, 161.9, 157.4, 148.7, 147.8, 146.2, 136.8, 123.1, 120.6, 116.2, 115.7, 103.9, 98.8, 94.0;

3′acetyl-quercetin: ^{1}H NMR (500 MHz, DMSO-d_6): δ 12.47 (s, 1H), 10.89 (s, 1H), 10.48 (s, 1H), 9.66 (s, 1H), 7.99 (dd, J=2.21/6.44 Hz, 1H), 7.91 (d, J=2.2 Hz 1H), 7.12 (d, J=8.5 Hz, 1H), 6.50 (d, J=2.08 Hz, 1H), 6.23 (d, J=2.08 Hz, 1H), 2.33(s, 3H); ^{13}C NMR (500 MHz, DMSO-d_6):

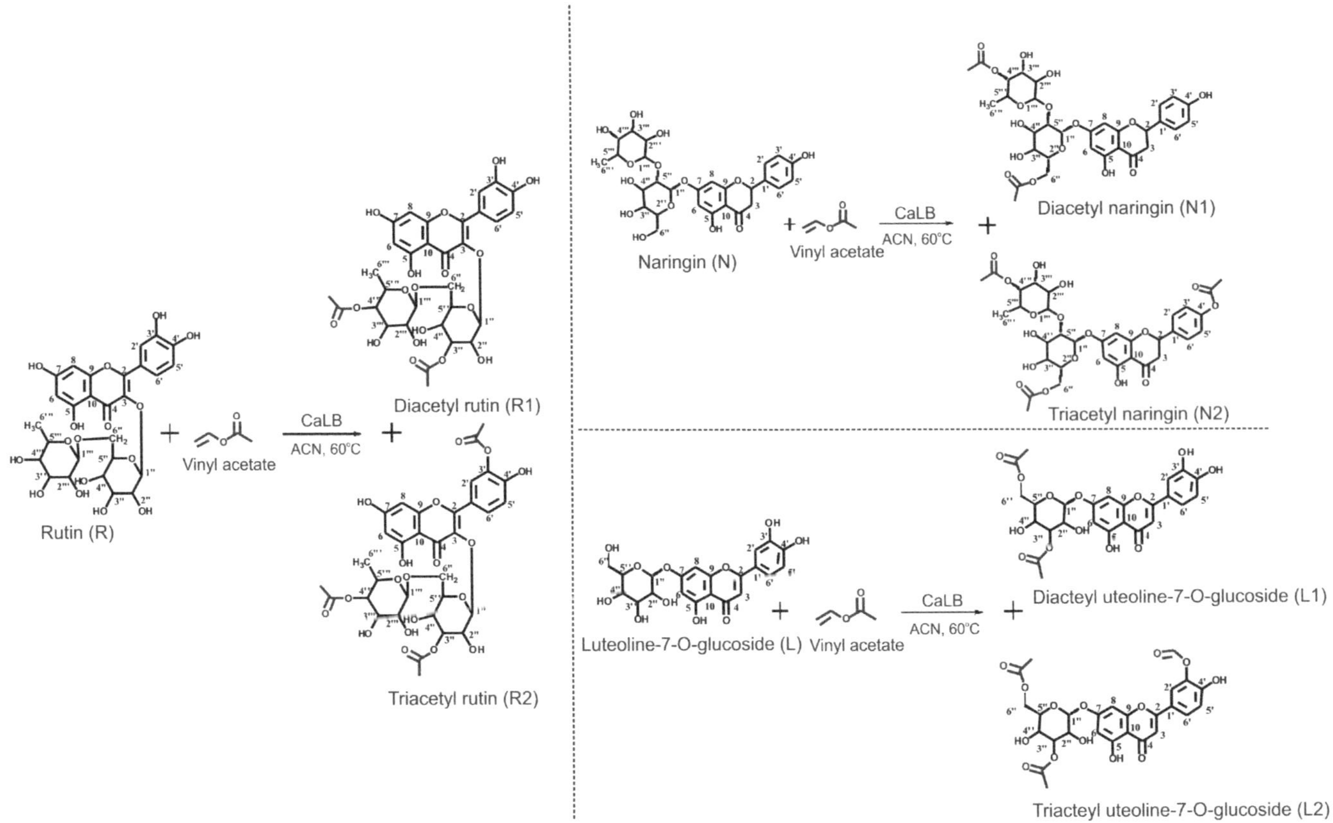

Fig. 8 CaLB mediated acetylation of (A) rutin, (B) naringin and (C) luteolin-7-*O*-glucoside.

δ 177.2, 169.9, 165.2, 161.9, 157.2, 152.1, 146.7, 139.2, 137.2, 127.2, 123.3, 123.2, 117.6, 98.9, 94.2, 21.4;

4′acetyl-quercetin: ^{1}H NMR (500 MHz, DMSO-d_6): δ 12.41 (s, 1H), 10.94 (s, 1H), 10.09 (s, 1H), 9.78 (s, 1H), 7.83 (dd, J=2.12/6.37 Hz, 1H), 7.64 (d, J=8.46 Hz 1H), 7.22 (d, J=8.5 Hz, 1H), 6.47 (d, J=2.08 Hz, 1H), 6.25 (d, J=2.08 Hz, 1H), 2.32 (s, 3H); ^{13}C NMR (500 MHz, DMSO-d_6): δ 177.4, 169.8, 165.4, 161.9, 157.4, 150.0, 146.4, 140.6, 138.2, 130.3, 124.0, 119.4, 116.7, 104.2, 98.9, 94.1, 21.3;

Naringenin: ^{1}H NMR (500 MHz, DMSO-d_6): δ 12.14 (s, 1H), 10.78 (s, 1H), 9.58 (s, 1H), 7.31 (d, J=8.57 Hz, 2H), 6.8 (d, J=8.66 Hz, 2H), 5.88 (tt, J=2.2/2.04 Hz, 2H), 5.42 (dd, J=3.05/12.88, 1H), 3.17 (dd, J=13.03/18.3 Hz, 1H), 3.25 (dd, J=13.04/17.2 Hz, 1H); ^{13}C NMR (500 MHz, DMSO-d_6): δ 196.9, 167.3, 164.2, 163.2, 158.6, 129.2, 115.9, 102.6, 96.6, 95.9, 78.9, 42.48;

4′ acetyl-naringenin: ^{1}H NMR (500 MHz, DMSO-d_6): δ 12.84 (s, 1H), 12.12 (s, 1H), 7.56 (d, J=8.62 Hz, 2H), 7.31 (d, J=8.58 Hz, 2H), 5.93 (dd, J=11.05/2.14 Hz, 2H), 5.9 (s, 1H), 5.6 (dd, J=12.78/3.08 Hz 1H), 2.83 (dd, J=17.17/3.12 Hz 1H), 2.28 (s, 3H); ^{13}C NMR (500 MHz, DMSO-d_6): δ 196.9, 169.8, 167.3, 164.2, 163.4, 150.9, 136.6, 128.4, 122.3, 102.6, 96.6, 95.9, 78.2, 42.6, 21.43;

Pyrogallol: ^{1}H NMR (500 MHz, D_2O): δ 6.73 (t, J=8.18 Hz, 1H), 6.53 (d, J=8.21 Hz, 2H); ^{13}C NMR (500 MHz, D_2O): δ 146.1, 133, 107, 118.3;

Purpurogallin: ^{1}H NMR (500 MHz, D_2O): δ 7.36 (d, J=11.3 Hz, 1H), 7.13 (d, J=9.42 Hz, 1H) 6.87 (s, 1H), 6.78 (dd, J=9.73/11.73 Hz, 1H); ^{13}C NMR (500 MHz, D_2O): δ 146.6, 153.2, 136.2, 128.1, 128.4, 126.6, 124.3, 106.3;

Rutin (R): ^{1}H NMR (500 MHz, DMSO-d_6): δ 12.63 (s, 1H), 10.87 (s, 1H), 9.69 (s, 1H), 9.2 (s, 1H), 7.57 (s, 1H), 7.49 (dd, J=2.2/10.4 Hz, 1H), 6.87 (d, J=8.0 Hz, 1H), 6.43 (d, J=2.0 Hz 1H), 6.23 (d, J=2.0 Hz, 1H), 5.35 (d, J=1.3 Hz, 1H), 4.4 (d, J=7.3 Hz, 1H), 3.74 (d, J=10.5 Hz, 1H), 3.42 (m, 1H), 3.32 (d, J=3.08 Hz, 1H), 3.3 (d, J=2.9 Hz, 1H), 3.32 (d, J=1.7 Hz, 3H), 3.26 (m, 1H), 3.2 (m, 1H), 3.1 (s, 1H), 3.09 (m, 1H), 1.03 (d, J=6.15 Hz, 3H); ^{13}C NMR (500 MHz, DMSO-d_6): δ 178.0, 164.9, 162.0, 157.6, 157.3, 149.3, 145.6, 134.2, 122.2, 116.8, 115.9, 104.8, 101.9, 101.5, 99.28, 94.17, 77.1, 77.0, 72.6, 71.3, 71.1, 70.8, 69.8, 67.8, 67.6, 18.42;

Diacetyl rutin (R1): ^{1}H NMR (500 MHz, DMSO-d_6): δ=12.62 (s, 1H), 10.88 (s, 1H), 9.7 (s, 1H), 9.25 (s, 1H), 7.55 (s, 1H), 7.54 (s, 1H), 6.88 (d, J=8.2 Hz, 1H), 6.43 (d, J=2.0 Hz, 1H), 6.23 (d, J=2.2, 1H),

5.58 (d, $J=7.72$, 1H,), 4.87 (s, 1H), 4.65 (s, 1H), 4.49 (s, 1H), 3.71 (1H), 3.57 (s, 1H), 3.52 (s, 1H), 3.47 (s, 1H), 3.46 (s, 1H), 3.45 (s, 1H), 3.41 (s, 1H), 2.09 (s, 3H), 1.94 (s, 3H), 0.78 (s, 3H); ^{13}C NMR (500 MHz, DMSO-d_6): $\delta=170.77$, 170.51, 164.99, 164.87, 162.1, 157.17, 157.15, 149.13, 145.69, 133.75, 122.31, 116.95, 116.14, 104.55, 101.27, 101.16, 99.51, 94.31, 78.15, 75.86, 74.31, 72.64, 71.04, 68.79, 68.21, 67.12, 66.58, 21.91, 21.64, 17.76;

Triacetyl rutin (R2): ^{1}H NMR (500 MHz, DMSO-d_6): $\delta=12.54$ (s, 1H), 10.93 (s, 1H), 10.46 (s, 1H), 7.89 (dd, $J=2.2/10.4$ Hz, 1H), 7.84 (s, 1H), 7.07 (d, $J=8.7$ Hz, 1H), 6.46 (d, $J=2.0$ Hz, 1H), 6.25 (d, $J=2.1$ Hz, 1H), 5.58 (s, 1H), 4.87 (s, 1H), 4.65 (s, 1H), 4.48 (s, 1H), 3.57 (s, 1H), 3.5 (s, 1H), 3.46 (s, 1H), 3.45 (s, 1H), 3.37 (s, 1H), 3.71 (s, 1H), 3.33 (s, 1H), 2.32 (s, 3H), 2.09 (s, 3H), 1.94 (s, 3H), 0.78 (s, 3H); ^{13}C NMR (500 MHz, DMSO-d_6): $\delta=170.52$, 170.09, 169.49, 165.09, 164.91, 162.05, 157.29, 156.4, 152.61, 138.45, 122.78, 124.97, 117.39, 104.77, 101.27, 101.20, 99.54, 94.52, 78.19, 75.88, 74.33, 72.79, 71.06, 68.8, 68.42, 67.13, 66.59, 21.85, 21.65, 21.43, 17.68;

Naringin (Ni): ^{1}H NMR (500 MHz, DMSO-d_6): δ 12.09 (s, 1H), 9.65 (s, 1H), 7.63 (d, 2H), 7.49 (dd, $J=6.30/2.20$ Hz, 1H), 6.83 (d, 2H), 6.15 (dd, $J=2.23$ Hz 1H), 6.12 (dd, $J=2.12$ Hz, 1H), 5.53 (m, 1H), 5.34 (s, 1H), 5.13 (m, 1H), 5.17 (m, 1H), 4.75 (s, 1H), 4.69 (s, 1H), 4.61 (s, 1H), 4.51 (s, 1H), 3.73 (m, 1H), 3.71 (m, 1H), 3.68 (m, 1H), 3.48 (m, 1H), 3.46 (m, 2H), 3.39 (m, 1H), 3.35 (m, 1H), 3.23 (m, 1H), 3.2 (m, 2H), 2.76 (m, 1H), 1.19 (d, $J=6.68$ Hz, 3H); ^{13}C NMR (500 MHz, DMSO-d_6): δ 198.2, 165.6, 163.9, 163.7, 158.6, 129.3, 115.9, 104.2, 101.1, 116.8, 97.95, 97.0, 95.83, 79.39, 77.8, 77.7, 76.9, 72.53, 71.3, 71.2, 70.40, 68.9, 61.03, 42.7, 18.78;

Diacetyl naringin (Ni1): ^{1}H NMR (500 MHz, DMSO-d_6): $\delta=12.07$ (1H, OH 5), 9.63 (1H, OH 4′), 7.35 (1H, H 2′), 7.35 (1H, H6′), 6.83 (1H, H 5′), 6.83 (1H, H 3′), 6.18 (1H, H 8), 6.12 (1H, H6), 5.56 (1H, H2), 5.23 (1H, H 1″), 5.22 (1H, H 1‴), 4.82 (1H, H 4‴), 4.31 (1H, H6″α), 4.08 (1H, H 6″β), 4.01 (1H, H5″), 3.76 (1H, H2″), 3.76 (1H, H2‴), 3.56 (1H, H 3‴), 3.54 (1H, H 3″), 3.52 (1H, H 5″), 3.42 (1H, H 3α), 3.22 (1H, H 4″), 2.79 (1H, H 3β), 2.01 (3H, H 8″), 2.01 (1H, H 8‴), 1.1 (3H, H 6‴) ppm. ^{13}C NMR (500 MHz, DMSO-d_6): $\delta=198.17$, 170.7, 170.7, 165.11, 163.11, 158.65, 129.1, 120.91, 115.82, 115.82, 104.4, 100.3 C1, 97.9 C1, 96.96, 95.81, 79.42, 77.51, 76.28, 74.68, 74.36, 70.52, 70.52, 68.57, 66.3, 63.85, 42.71, 21.54, 21.54, 18.31;

Triacetyl naringin (Ni2): ^{1}H NMR (500 MHz, DMSO-d_6): $\delta=12.04$ (1H, OH 5), 7.6 (1H, H 2′), 7.6 (1H, H6′), 7.24 (1H, H 5′),

7.24 (1H, H 3′), 6.23 (1H, H 8), 6.13 (1H, H6), 5.71 (1H, H2), 5.23 (1H, H 1″), 5.22 (1H, H 1‴), 4.82 (1H, H 4‴), 4.31 (1H, H6″α), 4.08 (1H, H 6″β), 4.01 (1H, H 5‴), 3.76 (1H, H2″), 3.76 (1H, H2‴), 3.56 (1H, H 3‴), 3.54 (1H, H 3″), 3.52 (1H, H 5″), 3.41 (1H, H 3α), 3.22 (1H, H 4″), 2.89 (1H, H 3β), 2.32 (3H, H 8′) 2.01 (3H, H 8″), 2.01 (1H, H 8‴), 1.1 (3H, H 6‴); ^{13}C NMR (500 MHz, DMSO-d_6): δ = 197.84, 170.7, 170.7, 170.1, 165.38163.9, 151,57, 151.53, 137.0, 128.65, 120.91, 122.6, 122.6, 104.2, 100.3, 97.9′, 97.01, 95.97, 78.89, 77.51, 76.28, 76.68, 74.36, 70.52, 70.57, 68.57, 66.3, 63.85, 42.73, 21.61, 21.54, 21.54, 18.31;

Luteolin-7-O-glycoside (L): ^{1}H NMR (500 MHz, DMSO-d_6): δ = 13.03 (s, 1H), 10.03 (s, 1H), 9.41 (s, 1H), 7.49 (dd, J = 8.21/2.4 Hz, 1H), 7.46 (d, J = 2.3 Hz, 1H), 6.94 (d, J = 1H), 6.82 (d, J = 2.2 Hz, 1H), 6.78 (s, 1H), 6.48 (d, J = 2.2 Hz, 1H), 5.12 (d, J = 7.5 Hz, 1H), 3.75 (d, J = 10.8 Hz, 1H), 3.51 (d, J = 10.8 Hz, 1H), 3.48 (m, 1H), 3.3 (m, 1H), 3.2 (t, J = 9 Hz, 1H); ^{13}C NMR (500 MHz, DMSO-d_6): δ = 182.89, 165.34, 163.78, 162.03, 157.8, 150.69, 146.66, 122.3, 119.84, 116.68, 114.23, 106.22, 103.83, 100.55, 100.29, 95.37, 77.82, 77.05, 73.79, 70.23, 74.17, 61.26;

Diacetyl luteolin-7-O-glycoside (L1): ^{1}H NMR (500 MHz, DMSO-d_6): δ = 13.05 (s, 1H), 10.05 (s, 1H′), 9.42 (s, 1H), 7.48 (dd, J = 8.27/2.16 Hz, 1H), 7.46 (s, 1H), 6.94 (dd, J = 8.40/3.3Hz, 1H), 6.82 (d, J = 2.25 Hz, 1H), 6.8 (s, 1H), 6.5 (dd, J = 3.09/2.3 Hz, 1H), 5.77 (s, 1H), 5.61 (s, 1H), 5.33 (t, J = 8.3/7 Hz,), 4.95 (m, 1H), 4.35 (m, 1H), 4.14 (m, 1H), 3.94 (t, J = 10.0/7.9 Hz, 1H), 3.5 (m, 1H), 3.44 (m, 1H), 2.09 (d, J = 7 Hz, 3H), 2.06 (dd, J = 8.5/2.25 Hz, 3H); ^{13}C NMR (500 MHz, DMSO-d_6): δ = 182.64, 170.47, 170.29, 165.2, 163.14, 161.9, 157.5, 150.57, 146.58, 122.04, 119.92, 116.64, 114.26, 106.32, 103.89, 100.19, 99.75, 95.57, 77.71, 74.17, 71.6, 68.35, 63.78, 21.61, 21.3;

Triacetyl luteolin-7-O-glycoside (L2): ^{1}H NMR (500 MHz, DMSO-d_6): δ = 12.96 (s, 1H), 10.74 (s, 1H), 7.91 (dd, J = 8.4/1.40 Hz, 1H), 7.86 (t, J = 2.85/2.48 Hz, 1H), 7.12 (dd, J = 8.8/5.23 Hz, 1H), 6.96 (d, J = 1 Hz, 1H), 6.76 (d, J = 2.2 Hz, 1H), 6.45 (dd, J = 6.45/2.26 Hz, 1H), 5.62 (s, 1H), 5.55 (s, 1H), 5.46 (t, 1H), 4.86 (t, 1H), 4.4 (m, 1H), 3.89 (m, 1H), 3.33 (m, 2H), 2.33 (d, J = 1.2 Hz, 3H), 2.08 (d, J = 7 Hz, 3H), 2.05 (dd, 1 H); ^{13}C NMR (500 MHz, DMSO-d_6, 25 OC): δ = 182.78, 170.87, 170.01, 169.3, 164.13, 162.6, 161.76, 157.6, 153.83, 139.47, 126.43, 122.49, 118.11, 106.31, 104.51, 100.00, 97.83′, 95.65, 74.7, 74.5, 74.19, 70.54, 63.87, 21.5, 21.5, 21.2.

3.1.1 *General safety*

Nitrile gloves, protective mask.

3.1.2 *Equipment*

1. 5 mm NMR tube
2. Pipette (2–20, 20–200, 100–1000 μL)
3. Pipette tips
4. Thermoshaker
5. NMR spectrometer (Bruker AV-500 spectrometer equipped with a TXI cryoprobe, Bruker BioSpin, Rheinstetten, Germany)

3.1.3 *Chemicals and materials*

1. Quercetin, rutin, luteolin-7-*O*-glucoside, naringenin and naringin were from Extrasynthese.
2. Pyrogallol was from Sigma-Aldrich (St. Louis, MO, USA).
3. Deuterium oxide (D_2O) (99.9%) and Acetonitrile-d_3 (AcCN-d_3) were from Deutero (Kastellaun, Germany).
4. Lipase from *Pseudozyma (Candida) antarctica* (CaLB) immobilized on acrylic resin (Novozyme 435) was kindly provided from Novozyme.
5. C107T yeast iso-1-cytochrome c (Cyt-c) expressed and purified as described in Barker et al. (2001) and was purchased from Giotto Biotech (https://www.giottobiotech.com).
6. Vinyl acetate was from Sigma-Aldrich (St. Louis, MO, USA).
7. Sodium phosphate buffer 20 mM, pD = 6.8, in D_2O.
8. Phosphate-buffered saline (PBS) 10 mM, pD = 6.8 in D_2O.

3.2 Real time monitoring and characterization of biotransformation products from CaLB in the NMR tube bioreactor

3.2.1 *Methodology*

3.2.1.1 Duration

For the real time monitoring of the formation of the biotransformation products ~15 min.

For characterization of the biotransformation products ~8 h (depending on the concentration of the studied compounds).

3.2.1.2 Protocol

1. 16 mM of flavonoid (naringenin or quercetin, rutin, luteolin-7-*O*-glucoside and naringin) (or 16 mM of quercetin, and naringenin, rutin, luteolin-7-*O*-glucoside and naringin in case of the mixture) dissolved in 500 μL AcCN-d_3 in a 1.5 mL eppendorf.

2. 125 μL of the acyl donor vinyl acetate are added and the mixture is transferred in a 5 mm NMR tube.
3. 15 mg of CaLB Novozyme 435 are added to the mixture in the NMR tube.
4. The NMR tube bioreactor is incubated in a thermomixer at 60 °C under shaking.
5. Every 24 h the progress of the reaction is monitored by ^{1}H NMR, ~15 min (64 scans) (or 2D ^{1}H-^{13}C HSQC, 2D ^{1}H-^{13}C HMBC NMR spectra are recorded when needed. The number of scans depends on the concentration of the studied compounds). NMR experiments were recorded at 282 K on a Bruker AV-500 spectrometer equipped with a TXI cryoprobe.

3.3 Immobilization of Cyt-c on celite through absorbance

3.3.1 Methodology

3.3.1.1 Duration

35 min

3.3.1.2 Protocol

6. 180 μL of cytochrome *c* solution is added in 18 mg celite and mixed gently.
7. The mixture is incubated in ice for 30 min with intermediate mixing.
8. The sample is washed with 800 μL sodium phosphate buffer pH 6.8, 20 mM.
9. The sample is centrifuged for 5 min at 12,000 rpm, at 4 °C.
10. The supernatant is discarded.
11. The sediment is the immobilized Cyt-c on celite.

3.4 Real time monitoring of the biotransformation of pyrogallol catalyzed by Cyt-c in the NMR tube bioreactor

3.4.1 Methodology

3.4.1.1 Duration

For the real time monitoring of the formation of the biotransformation products ~15 min.

For characterization of the biotransformation products ~8 h (depending on the concentration of the studied compounds).

3.4.1.2 Protocol

1. 5 mM of pyrogallol are dissolved in 500 μL solvent (70% v/v buffer sodium phosphate 20 mM pD = 6.8 in D_2O and 30% AcCN-d_3) in an 1.5 mL eppendorf.

2. Then 5 mM of H_2O_2 are added and the mixture is transferred to a 5 mm NMR tube.
3. 18 mg of Cyt-c immobilized on celite (64.9 μg of protein) are added to the NMR tube.
4. The NMR tube bioreactor is incubated in a thermoshaker at 25 °C under shaking 600 rpm.
5. Every hour the progress of the reaction is monitored by ^{1}H NMR ~15 min (64 scans) (or 2D ^{1}H-^{13}C HSQC, 2D ^{1}H-^{13}C HMBC NMR spectra are recorded when assignment of the final products is needed. The number of scans depends on the concentration of the studied compounds). NMR experiments are recorded at 282 K on a Bruker AV-500 spectrometer equipped with a TXI cryoprobe.
   TIP 1: To enhance the solubility of the flavonoids (especially in the case of mixtures) in AcCN-d_3, the solvent should be heated to 60 °C for 5–10 min before use and if needed the flavonoids should be incubated in the solvent at 60 °C for 5–10 min before the addition of the biocatalyst.
   TIP 2: After the addition of the immobilized CaLB or Cyt-c make sure that the biocatalyst is confined at the bottom of the NMR tube.
   TIP 3: Dehydrate the acyl donor vinyl acetate before use with 4 Å molecular sieves.
   TIP 4: For high resolution NMR spectra store the deuterated NMR solvents under nitrogen.
   TIP 5: For the characterization of the biotransformed products, when the reaction is completed, the immobilized biocatalyst can be removed through filtration. If in the reaction products there are chemical groups that are susceptible to chemical exchange and thus can lead to severe line broadening in the NMR spectra, it is advised to exchange the aqueous buffer with DMSO-d_6. Recording in this solvent the 2D ^{1}H-^{13}C HSQC, and 2D ^{1}H-^{13}C HMBC NMR spectra of the crude enzymatic products could assist in obtaining high resolution NMR spectra.

4. Screening of the real time generated biotransformation products as ligands for protein targets

In the last (third) step the NMR tube bioreactor can offer the possibility to screen in real time the biotransformed products as ligands for protein targets of interest. This is of importance since it can provide, for example, answers to potential questions whereas in a cellular environment a compound can be biotransformed by a specific enzyme and whether the resulting product can interact with a specific target protein. This can be achieved either by recording information directly on the ligand (ligand-based screening) or recording the effect of the ligand to the protein (protein-based screening). As mentioned above, enzymes present substrate pliability and activity. The lipase CaLB can operate in two different ways, either to acetylate substrates or, if the buffer solution are altered, can lead to alcoholysis. To illustrate the potential of the method to monitor the pliability of the enzyme as also to map the interaction of the formed products with target proteins we present in the following part how it is possible to monitor the interactions of the products resulted by the acetylation of different flavonoids and the alcoholysis of penta-acetyl quercetin with two different target proteins. Also the ligand-based screening and the protein-based screening approaches are presented.

4.1 Ligand-based screening

Bovine serum albumin (BSA) serves as a transport carrier for drugs, and it is widely used as a test model to study the interactions between albumin and potent pharmaceutical agents (Sułkowska, 2002). Furthermore, BSA has been used because of its ability to interact with a large number of aromatic and hydrophobic molecules in the bloodstream (Goodman, 1958; Ravera et al., 2014). The interactions between CaLB mediated acetylation products and BSA can be explored via STD NMR. Since the biocatalyst is confined in the bottom of the NMR tube, the target protein (i.e., BSA in this case) can be directly titrated in the NMR tube that contains the immobilized biocatalyst and the biotransformation products. If compatibility issues exist, concerning the buffer for the enzymatic reaction (second step) and the buffer for the target protein (in the third step), then a simple method can be applied to change the buffer: The immobilized enzyme is easily removed through filtration, the original solvent is evaporated by vacuum and the new buffer of the target protein is introduced to the NMR tube. This process is illustrated for the case of CaLB utilized to acetylate different flavonoids and the resulted products have been screened for their interaction with BSA. After following the buffer exchange in the reaction mixture of the CaLB mediated acetylation of naringenin, a STD NMR spectrum has been recorded with BSA. The results illustrate that both the parent compound (naringenin) and the in situ biotransformed product bind on BSA (Fig. 9). It is interesting to note that the biotransformation product binds stronger to

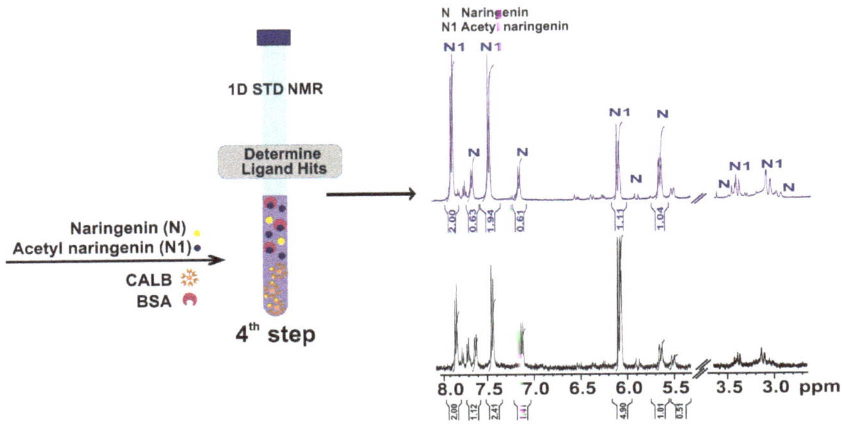

Fig. 9 Selected region of the STD NMR spectrum of the crude CaLB mediated biotransformation (acetylation) of naringenin (N) and the formed product (N_1) with BSA in 65% PBS in D_2O, 35% AcCN-d_3. The reference ^1H NMR spectra is colored in blue and the difference ^1H NMR spectra is colored in black.

the target protein than the parent compound, as can be seen from their enhanced STD signal (Fig. 9).

This method can be also applied in the case of mixtures of substrates. Herein, this method was employed to the mixture of three different flavonoids, rutin (R), naringenin (N) and quercetin (Q), and their acetyl derivatives (denoted as PR, PQ and PN, respectively) derived from the CaLB mediated reaction. STD NMR demonstrates the interactions between BSA and the parent molecules, as well as with their derivatives (Fig. 10).

4.2 Protein-based screening

The application of the methodology on a different pharmaceutical target, an intrinsically disordered protein (α-synuclein), as also for monitoring the effect of the ligand on the protein, is presented for the regioselective alcoholysis, catalyzed by CaLB, of penta-acetyl quercetin.

In the second step, penta-acetyl quercetin was used as a substrate for the regioselective alcoholysis catalyzed by CaLB. Three acetylated products were formed from the biotransformation of penta-acetyl quercetin that have high antiproliferative activity in three breast cancer cell lines (Kyriakou et al., 2018). The mixture of these products was tested for potential interactions with α-synuclein. α-Synuclein was studied due to the fact that it is associated with numerous neurological diseases including Alzheimer (Herrera et al., 2008; Oh et al., 2014) and identification of potential binders is of

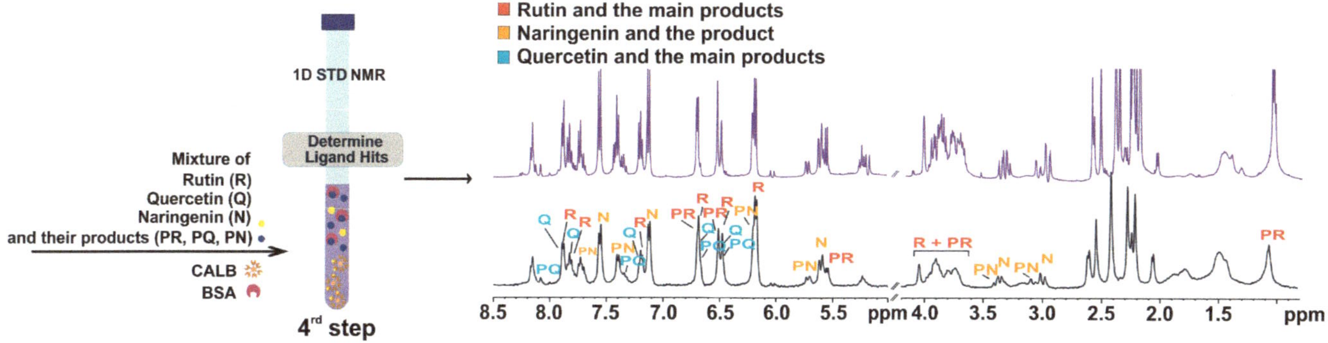

Fig. 10 Selected region of the STD NMR spectrum of quercetin (Q), naringenin (N), rutin (R) and their acetylated derivatives (PQ, PN, PR) with BSA in 65% PBS in D_2O, 35% AcCN-d_3. With blue is illustrated the reference ^{1}H NMR spectrum and with black the difference ^{1}H NMR spectrum.

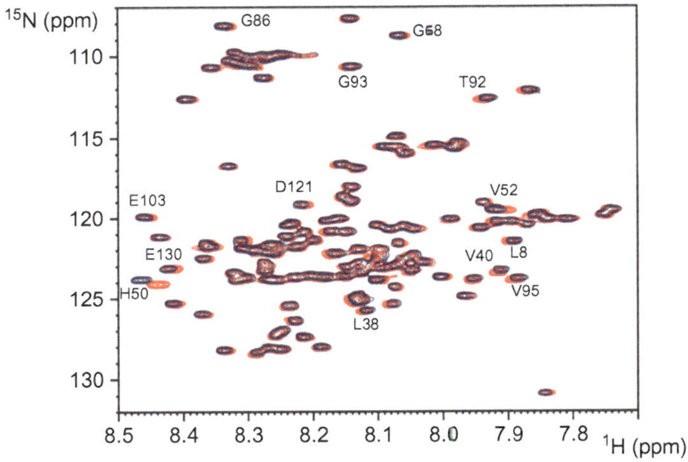

Fig. 11 2D ^1H–^{15}N SOFAST HMQC spectra of α-synuclein (at 283 K) in the presence of the products of the alcoholysis of penta-acetyl quercetin at the molar ratio of 1:6 (α-synuclein:products) (red spectrum), and of the protein alone (black spectrum).

importance. To monitor the interactions of the in situ generated ligands with the target protein α-synuclein, the protein was labeled with ^{15}N. After titration experiments, several chemical shifts that are associated with interactions of the products were observed in 2D ^1H–^{15}N SOFAST HMQC NMR spectra of α-synuclein (Fig. 11).

4.2.1 General safety

Nitrile gloves, protective mask.

4.2.2 Equipment

1. 5 mm NMR tube
2. Pipette (2–20, 20–200, 100–1000 μL)
3. Pipette tips
4. Thermoshaker
5. NMR spectrometer (Bruker AV-500 MHz spectrometer equipped with a TXI cryoprobe/Bruker Avance III 950 MHz spectrometer, equipped with a TCI, Bruker BioSpin, Rheinstetten, Germany)

4.2.3 Chemicals and materials

1. Lipase from *Pseudozyma (Candida) antarctica* (CaLB) immobilized on acrylic resin (Novozyme 435) was kindly provided by Novozyme

2. Flavonoids (quercetin, rutin, naringenin) were purchased from Extrasynthese
3. Penta-acetyl quercetin was synthesized according to Kyriakou et al. (2018)
4. *tert*-Butyl methyl ether was from Merck
5. Bovine Serum Albumin (BSA) was from Sigma-Aldrich (St. Louis, MO, USA)
6. α-Synuclein expressed and purified as described in reference Bertini et al. (2007) and purchased from Giotto Biotech (https://www.giottobiotech.com)
7. Deuterium oxide (D_2O) (99.9%) was from Deutero (Kastellaun, Germany)
8. Acetonitrile-d_3 (AcCN-d_3) was from Deutero (Kastellaun, Germany)
9. Phosphate-buffered saline (PBS) 10 mM pD = 7.0 in D_2O

4.3 Screening of potential interactions of biotransformed products with BSA through STD NMR

4.3.1 Methodology

4.3.1.1 Duration

1 h

4.3.1.2 Protocol

1. 16 mM of flavonoid naringenin or quercetin (or 16 mM of quercetin, and naringenin, and rutin in case of the mixture) are dissolved in 500 μL AcCN-d_3 in a 1.5 mL eppendorf.
2. 125 μL of the acyl donor vinyl acetate are added and the mixture is transferred in a 5 mm NMR tube.
3. 15 mg of CaLB Novozyme 435 are added in the NMR tube.
4. The NMR tube bioreactor is incubated in a thermoshaker at 60 °C at 600 rpm and the reaction is monitored on line through 1D ^1H NMR.
5. After the consumption of the substrates, if compatibility issues exist, the immobilized enzyme can be separated through filtration and the solvent removed in vacuum.
6. 0.1 mM of BSA is added in 500 μL of solvent mixture of 65% (v/v) phosphate-buffered saline (PBS) 10 mM pD = 7.0 in D_2O, and 35% v/v AcCN-d_3,
7. After loading the sample into the 500 MHz NMR spectrometer the sample is locked, tuned and matched according to manufacturer's guidelines. STD NMR experiments are recorded at 298 K on a Bruker AV-500

spectrometer equipped with a TXI cryoprobe. The on-resonance irradiation of the protein is performed at 0.76 ppm. Off-resonance irradiation is conducted at 40 ppm outside the ^1H NMR resonances region of the protein. Each Gaussian pulse required 50 ms separated by a delay of 1 ms. Saturation time was set at 2 s and the experimental time was ~1 h (128 scans). The water suppression was conducted by using an excitation sculpting pulse sequence.

4.4 Screening of potential interactions of biotransformed products with α-synuclein with 2D ^1H–^{15}N SOFAST HMQC NMR

4.4.1 Methodology

4.4.1.1 Duration

10 min

4.4.1.2 Protocol

1. 5 mg (0.001 mmol) of penta-acetyl quercetin and 1.3 μL of n-butanol (0.014 mmol) are dissolved in 125 μL *tert*-butyl methyl ether and the mixture is transferred in a 5 mm NMR tube.
2. 7.15 mg of CaLB Novozyme 435 are added to the mixture in the NMR tube.
3. The NMR tube bioreactor is incubated in a thermomixer at 40 °C, at 600 rpm and the reaction is monitored on line through 1D ^1H NMR.
4. After the consumption of the substrate the immobilized enzyme, if compatibility issues exist with the buffer that has to been used with the target protein, is removed through filtration and the solvent is removed in vacuum.
5. 2D ^1H–^{15}N SOFAST HMQC spectra of α-synuclein in the absence of the products of the alcoholysis of penta-acetyl quercetin and in its presence at the molar ratio of 1:6 are recorded on a 22.3 T magnet, with a Bruker Avance III spectrometer operating at 950 MHz 1H Larmor frequency, equipped with a cryogenically cooled probehead. The experiments are acquired with 2048 points in the direct dimension and 128 in the indirect dimension, with 16 scans (~10 min). The interscan delay is set to 0.3 s with 90° excitation pulse.

   TIP 1: To enhance the solubility of the flavonoids (especially in the case of the mixtures) the solvent should be heated to 60 °C for 5–10 min before the addition of the biocatalyst.

TIP 2: For high resolution NMR spectra store the deuterated NMR solvents under nitrogen.

TIP 3: α-Synuclein should be purified from protein aggregates before the use in the NMR experiments.

5. Conclusions

Herein, we described the steps that should be followed in the NMR tube bioreactor process. The process enables for the rapid determination of substrates for specific enzymes, for recording in real time the product formation, as also for conducting ligand-protein interaction screening of the parent compounds and resulted products with protein targets.

Acknowledgments

This research has been financed by the European Union (European Social Fund—ESF) and Greek national funds through the Operational Program "Education and Lifelong Learning" of the National Strategic Reference Framework (NSRF)—Research Funding Program: ARISTEIA II [grant number: 5199], by Greek national funds through the operational program "THESSALY-MAINLAND GREECE AND EPIRUS-2007 to 2013" of the National Strategic Reference Framework (NSRF 2007–13) and Greek State Scholarship Foundation (IKY) (MIS 5000432); A.V.C. gratefully acknowledges the IKY Foundation for the financial support. The PhD research was implemented with an IKY scholarship funded by the Operational Program Human Resources Development, Education, and Lifelong Learning, 2014–20, with MIS 5000432, and co-financed by the European Social Fund (ESF) and the Greek National Funds; C.L., G.P., M.F. and E.R. acknowledge the Instruct-ERIC Core Center CERM/CIRMMP; A patent regarding the experimental data described herein has been filed (number 1009100).

The authors thank colleagues and collaborators named in joint publications for their valuable contributions.

This research has been partially financially supported by the General Secretariat for Research and Technology (GSRT) and the Hellenic Foundation for Research and Innovation (HFRI) (Code: 991) under the 1st Call for proposals for research projects for the support of Faculty members and Researchers working in the Greek Universities and Research Centers and the procurement of strategic research equipment.

References

Barker, P. D., Bertini, I., Del Conte, R., Ferguson, S. J., Hajieva, P., Tomlinson, E., et al. (2001). A further clue to understanding the mobility of mitochondrial yeast cytochrome c. *European Journal of Biochemistry*, *268*(16), 4468–4476. https://doi.org/10.1046/j.1432-1327.2001.02369.x.

Bertini, I., Gupta, Y. K., Luchinat, C., Parigi, G., Peana, M., Sgheri, L., et al. (2007). Paramagnetism-based NMR restraints provide maximum allowed probabilities for the different conformations of partially independent protein domains. *Journal of the American Chemical Society*, *129*(42), 12786–12794. https://doi.org/10.1021/ja0726613.

Cardone, R. A., Casavola, V., & Reshkin, S. J. (2005). The role of disturbed pH dynamics and the NA+/H+ exchanger in metastasis. *Nature Reviews Cancer*, *5*(10), 786–795. https://doi.org/10.1038/nrc1713.

Caruana, M., Högen, T., Levin, J., Hillmer, A., Giese, A., & Vassallo, N. (2011). Inhibition and disaggregation of α-synuclein oligomers by natural polyphenolic compounds. *FEBS Letters*, *585*(8), 1113–1120. https://doi.org/10.1016/j.febslet.2011.03.046.

Chatzikonstantinou, A. V., Chatziathanasiadou, M. V., Ravera, E., Fragai, M., Parigi, G., Gerothanassis, I. P., et al. (2018). Enriching the biological space of natural products and charting drug metabolites, through real time biotransformation monitoring: The NMR tube bioreactor. *Biochimica et Biophysica Acta—General Subjects*, *1862*(1), 1–8. https://doi.org/10.1016/j.bbagen.2017.09.021.

Chebil, L., Humeau, C., Falcimaigne, A., Engasser, J.-M., & Ghoul, M. (2006). Enzymatic acylation of flavonoids. *Process Biochemistry*, *41*(11), 2237–2251. https://doi.org/10.1016/j.procbio.2006.05.027.

Choi, J. M., Han, S. S., & Kim, H. S. (2015). Industrial applications of enzyme biocatalysis: Current status and future aspects. *Biotechnology Advances*, *33*(7), 1443–1454. https://doi.org/10.1016/j.biotechadv.2015.02.014.

De Oliveira, E. B., Humeau, C., Maia, E. R., Chebil, L., Ronat, E., Monard, G., et al. (2010). An approach based on Density Functional Theory (DFT) calculations to assess the *Candida antarctica* lipase B selectivity in rutin, isoquercitrin and quercetin acetylation. *Journal of Molecular Catalysis B: Enzymatic*, *66*(3–4), 325–331. https://doi.org/10.1016/j.molcatb.2010.06.009.

Giang, I., Boland, E. L., & Poon, G. M. K. (2014). Prodrug applications for targeted cancer therapy. *The AAPS Journal*, *16*(5), 899–913. https://doi.org/10.1208/s12248-014-9638-z.

Goodman, D. (1958). The interaction of human serum albumin with long-chain fatty acid anions. *Journal of the American Chemical Society*, *80*(3), 3892–3898.

Hegedűs, C., Kovács, K., Polgár, Z., Regdon, Z., Szabó, É., Robaszkiewicz, A., et al. (2018). Redox control of cancer cell destruction. *Redox Biology*, *16*, 59–74. https://doi.org/10.1016/j.redox.2018.01.015.

Herrera, F. E., Chesi, A., Paleologou, K. E., Schmid, A., Munoz, A., Vendruscolo, M., et al. (2008). Inhibition of alpha-synuclein fibrillization by dopamine is mediated by interactions with five C-terminal residues and with E83 in the NAC region. *PLoS One*, *3*(10), e3394, https://doi.org/10.1371/journal.pone.0003394.

Inamori, Y., Muro, C., Sajima, E., Katagiri, M., Okamoto, Y., Tanaka, H., et al. (1997). Biological activity of purpurogallin. *Bioscience, Biotechnology, and Biochemistry*, *61*(5), 890–892. https://doi.org/10.1271/bbb.61.890.

Kazlauskas, R. J., & Bornscheuer, U. T. (1998). Biotransformations with lipases. In H. J. Rehm & G. Reed (Eds.), *Biotechnology Set* (pp. 37–191). https://doi.org/10.1002/9783527620999.ch3h.

Kyriakou, E., Kostagianni, A. D., Kellici, T. F., Giannopoulou, E., Siatis, K. E., Sayyad, N., et al. (2018). Three regioselectively acylated flavonoid aglycone derivatives in equimolar yield at one blow. *ChemistrySelect*, *3*, 5207–5211. https://doi.org/10.1002/slct.201703002. Communications.

Kyriakou, E., Primikyri, A., Charisiadis, P., Katsoura, M., Gerothanassis, I. P., Stamatis, H., et al. (2012). Unexpected enzyme-catalyzed regioselective acylation of flavonoid aglycones and rapid product screening. *Organic & Biomolecular Chemistry*, *10*(9), 1739–1742. https://doi.org/10.1039/c2ob06784f.

Mahato, R., Tai, W., & Cheng, K. (2011). Prodrugs for improving tumor targetability and efficiency. *Advanced Drug Delivery Reviews*, *63*(8), 659–670. https://doi.org/10.1016/j.addr.2011.02.002.

Mayer, M., & Meyer, B. (2001). Group epitope mapping by saturation transfer difference NMR to identify segments of a ligand in direct contact with a protein receptor. *Journal of the American Chemical Society*, *123*(25), 6108–6117. https://doi.org/10.1021/ja0100120.

Mellou, F., et al. (2005). Biocatalytic preparation of acylated derivatives of flavonoid glycosides enhances their antioxidant and antimicrobial activity. *Journal of Biotechnology*, *116*(3), 295–304. https://doi.org/10.1016/j.jbiotec.2004.12.002.

Oh, M., Lee, J. H., Wang, W., Lee, H. S., Lee, W. S., Burlak, C., et al. (2014). Potential pharmacological chaperones targeting cancer-associated MCL-1 and Parkinson disease-associated a-synuclein. *Proceedings of the National Academy of Sciences of the United States of America*, *111*(30), 11007–11012. https://doi.org/10.1073/pnas.1320556111.

Ono, T., & Goto, M. (2006). Peroxidative catalytic behavior of cytochrome c solubilized in reverse micelles. *Biochemical Engineering Journal*, *28*, 156–160. https://doi.org/10.1016/j.bej.2005.10.005.

Papadopoulou, A. A., Katsoura, M. H., Chatzikonstantinou, A., Kyriakou, E., Polydera, A. C., Tzakos, A. G., et al. (2013). Enzymatic hybridization of α-lipoic acid with bioactive compounds in ionic solvents. *Bioresource Technology*, *136*, 41–48. https://doi.org/10.1016/j.biortech.2013.02.067.

Patel, R. N. (2018). Biocatalysis for synthesis of pharmaceuticals. *Bioorganic and Medicinal Chemistry*, *26*, 1252–1274. https://doi.org/10.1016/j.bmc.2017.05.023. May.

Patila, M., Diamanti, E. K., Bergouni, D., Polydera, A. C., Gournis, D., & Stamatis, H. (2018). Preparation and biochemical characterisation of nanoconjugates of functionalized carbon nanotubes and cytochrome c. *Nanomedicine Research Journal*, *3*(1), 10–18. https://doi.org/10.22034/nmrj.2018.01.002.

Pellecchia, M., Bertini, I., Cowburn, D., Dalvit, C., Giralt, E., Jahnke, W., et al. (2008). Perspectives on NMR in drug discovery: A technique comes of age. *Nature Reviews Drug Discovery*, 7, 738–745. https://doi.org/10.1038/nrd2606.

Ravera, E., Corzilius, B., Michaelis, V. K., Luchinat, C., Griffin, R. G., & Bertini, I. (2014). DNP-enhanced MAS NMR of bovine serum albumin sediments and solutions. *Journal of Physical Chemistry B*, *118*(11), 2957–2965. https://doi.org/10.1021/jp500016f.

Sułkowska, A. (2002). Interaction of drugs with bovine and human serum albumin. *Journal of Molecular Structure*, *614*, 227–232. https://doi.org/10.1016/S0022-2860(02)00256-9.

Sun, H., Zhang, H., Ang, E. L., & Zhao, H. (2018). Biocatalysis for the synthesis of pharmaceuticals and pharmaceutical intermediates. *Bioorganic and Medicinal Chemistry*, *26*, 1275–1284. https://doi.org/10.1016/j.bmc.2017.06.043.

Van Meerloo, J., Kaspers, G. J. L., & Cloos, J. (2011). Cancer Cell Culture: Cell senstitivity assays: The MTT assay. *Methods in Molecular Biology*, *731*, 237–245. https://doi.org/10.1007/978-1-61779-080-5.

Viskupicova, J., Sturdik, E., & Ondrejovic, M. (2009). Enzymatic transformation of flavonoids. *Acta Chimica Slovenica*, *2*(1), 88–106. https://doi.org/https://ojs.fchpt.stuba.sk/ojs/index.php/ACS/article/view/71.

CHAPTER SIX

A chemically-controlled system for activating RAS GTPases

Emily M. Dieter[a], Dustin J. Maly[a,b,*]
[a]Department of Chemistry, University of Washington, Seattle, WA, United States
[b]Department of Biochemistry, University of Washington, Seattle, WA, United States
*Corresponding author: e-mail address: djmaly@uw.edu

Contents

Abstract

RAS GTPases are involved in a number of dynamic signaling processes and have been a major focus of research due to the prevalence of activating RAS mutations in cancer. However, despite decades of research, some fundamental aspects of RAS biology are still not well understood. Difficulty in fully defining RAS-driven signaling stems from the overall complexity of downstream pathways and a lack of tools for specifically perturbing RAS function. To better characterize RAS-driven signaling, we recently developed a chemical genetic system for activating endogenous RAS with a small molecule. In this chapter, we describe the use of chemically inducible activator of RAS (CIAR), a single-protein, chemical genetic system that allows the rapid and dose-dependent activation of endogenous RAS. Methods in this chapter also describe the validation of RAS activation with CIAR through the analysis of downstream signaling.

Methods in Enzymology, Volume 633
ISSN 0076-6879
https://doi.org/10.1016/bs.mie.2019.10.033

1. Introduction

GTPases of the RAS superfamily play numerous roles in diverse cellular processes (Karnoub & Weinberg, 2008). While there are over 150 members of the RAS superfamily, 4 RAS proteins (HRAS, NRAS, KRAS4A, and KRAS4B) are the most studied due to the frequency that they are mutated in cancer (Zhou, Der, & Cox, 2016). RAS genes were the first mutated genes to be identified in human cancers, and it is estimated that RAS mutations are found in over 25% of human tumors (Papke & Der, 2017). RAS mutations are implicated in approximately 33% of lung cancers, 51% of colon cancers, and 91% of pancreatic cancers, three of the four most lethal cancers in the United States (Cox, Fesik, Kimmelman, Luo, & Der, 2014; Prior, Lewis, & Mattos, 2012; Simanshu, Nissley, & McCormick, 2017).

In healthy cells, RAS proteins function as molecular "on/off" switches, existing both in GDP-bound (off) and GTP-bound forms (on), and are capable of controlling complex, diverse, and often conflicting processes. Activation of wild-type RAS is achieved through the activity of guanine nucleotide exchange factors (GEFs), which catalyze the exchange of GDP for the more cellularly abundant GTP (Boriack-Sjodin, Margarit, Bar-Sagi, & Kuriyan, 1998; Cherfils & Zeghouf, 2013). GTPase-activating proteins (GAPs) return RAS to the inactive GDP-bound state by catalyzing the hydrolysis of GTP (Hunter et al., 2015). Once activated, the GTP-bound form of RAS mediates downstream signaling events by recruiting cellular effectors. In addition to the well-characterized RAF/MEK/ERK kinase cascade, additional signaling pathways are believed to be activated by the GTP-bound form of RAS (Rajalingam, Schreck, Rapp, & Albert, 2007).

To date, studies of downstream RAS signaling have primarily relied on overexpression of oncogenic RAS mutants or activation of receptor tyrosine kinases with pleiotropic growth factors. While significant advances in our understanding of RAS signaling have been made with these methods, they are not without shortcomings. Overexpressed oncogenic RAS mutants do not provide information on the temporal features of RAS signaling and rely on expression levels not seen in tumors or non-cancerous cells. Furthermore, there is no mechanism for providing a graded RAS signaling output. Growth factor stimulation provides high temporal RAS activation and can be dose-dependently tuned but the parallel activation of multiple signaling pathways obscures the features that are driven by RAS. For these reasons, we and others have developed methods that allow direct activation of endogenous RAS with high temporal precision (Cunningham-Bryant et al., 2019; Rose et al., 2017; Rose, Dieter, Cunningham-Bryant, & Maly, 2018; Toettcher, Weiner, & Lim, 2013).

1.1 Introduction to CIAR

To facilitate a greater understanding of RAS-mediated signaling, we engineered a single-component protein switch that can dose-dependently activate endogenous RAS with high temporal precision in response to a small molecule (Rose et al., 2018, 2017). Our engineered protein switch, which we call chemically inducible activator of RAS (CIAR), consists of the constitutively active catalytic domain of the GEF Son of Sevenless (SOScat), flanked by the anti-apoptotic protein Bcl-xL and a BH3 peptide (Fig. 1A). Bcl-xL is a single-domain protein of limited size (~25 kDa) that

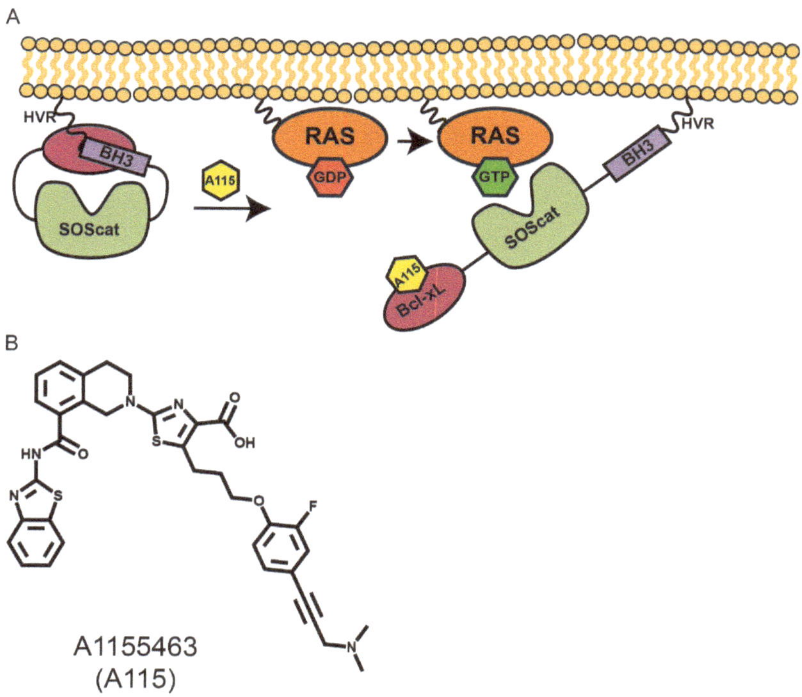

Fig. 1 Schematic of CIAR components and activation. (A) In the closed state, SOScat of CIAR is autoinhibited by the Bcl-xL (magenta)/BH3 interaction. CIAR is localized to membranes due to the inclusion of the HRAS hypervariable domain. Upon addition of A115, which binds to Bcl-xL, the Bcl-xL/BH3 interaction is disrupted, and the autoinhibition is relieved. Since SOScat is already localized to the membrane, it can then activate RAS. (B) Chemical structure of A115.

binds with high affinity to BH3 peptides of less than 30 amino acids (Goreshnik & Maly, 2010). Linkers connecting Bcl-xL and the BH3 peptide to SOScat were computationally optimized such that, in the absence of a small molecule disruptor, the closed Bcl-xL/BH3 complex blocks the active site of SOScat, preventing it from activating RAS. Upon addition of a small molecule inhibitor that targets the BH3 domain-binding site of Bcl-xL, autoinhibition is released and SOScat is capable of catalyzing the exchange of GDP with GTP in RAS.

In cells, we target autoinhibited CIAR to membranes by fusing the hypervariable region (HVR) from HRAS to its C-terminus (Hancock, 2003). This co-localizes CIAR with endogenous RAS and allows RAS activation upon addition of small molecule Bcl-xL inhibitors without requiring relocalization of the CIAR construct. CIAR constructs lacking a membrane-targeting motif are largely incapable of activating RAS in cells. We have found that cells expressing membrane-targeted CIAR show rapid activation of downstream RAS-mediated signaling pathways and that a graded response can be obtained by varying the concentration of Bcl-xL inhibitors used (Rose et al., 2017). The single-protein component architecture of CIAR facilitates studies in diverse cell types and, potentially, allows for its integration with intermolecular systems that use chemically-induced dimerization to activate cell signaling (Foight et al., 2019; Ross, Mehta, & Zhang, 2016; Stanton, Chory, & Crabtree, 2018; Voß, Klewer, & Wu, 2015).

1.2 Small molecule activators of CIAR

The BH3 peptide/Bcl-xL interaction was selected as the autoinhibitory switch for controlling SOScat's activity in CIAR due to the availability of potent small molecules that selectively target Bcl-xL. Bcl-xL is a member of the BCL-2 anti-apoptotic protein family, whose members are partially redundant in their anti-apoptotic function (Cory, Huang, & Adams, 2003; Um, 2015). Small molecules that target a subset of the BCL-2 family generally have a minimal effect on most mammalian cell lines that have not already been primed for apoptosis genetically or with cytotoxic agents (Konopleva et al., 2006; Wei et al., 2012). Since the initial discovery of a small molecule that potently and selectively targets Bcl-xL and Bcl-2, inhibitors that are even more potent and selective for Bcl-xL have been developed (Ashkenazi, Fairbrother, Leverson, & Souers, 2017; Shoemaker et al., 2006). We have tested several Bcl-xL-selective inhibitors and found that they all were able to activate CIAR in cells (Rose et al., 2018). Currently, we recommend using the Bcl-xL-selective inhibitor A-1155463 (A115), which is >800-fold selective for Bcl-xL over the BCL-2 family members Bcl-2 and Mcl-1, resulting in low cytotoxicity in most mammalian cell lines and *in vivo* (Tao et al., 2014) (Fig. 1B).

2. Drug inducible RAS activation

Here, we describe the activation of endogenous RAS using CIAR. In this example, human embryonic kidney 293 (HEK293) cells that inducibly express CIAR were used. We then provide a protocol for measuring downstream RAS activation. Briefly, cells are plated on poly-D-lysine coated plates and expression of CIAR is initiated through the addition of doxycycline (dox). The following day, cells are serum starved. Serum starvation reduces basal signaling, allowing for the observation of elevated RAS-mediated pathway activation. After serum starvation, CIAR is activated with the Bcl-xL inhibitor A115 and the activation of downstream ERK kinase is measured.

2.1 Seeding of CIAR cells

HEK293 cells that stably express CIAR were generated following the manufacturer's instructions using the Flp-In™ 293 T-REx™ cell line from ThermoFisher. These cells contain a single stably integrated Flp Recombination Target (FRT) site and express the tetracycline repressor. CIAR, under tetracycline/dox control, is stably integrated into Flp-In™ 293 T-REx™ cells. Stable integration is achieved upon co-transfection of Flp-In™ 293 T-REx™ cells with a pcDNA™5/FRT expression vector containing CIAR and a Flp recombinase vector (pOG44), followed by hygromycin drug selection. After cells have been selected, they can be used in downstream experiments, and expression of CIAR will be induced at the time of plating.

2.1.1 Equipment

- CO_2 incubator
- Biosafety cabinet
- Centrifuge
- Hemocytometer or automated cell counter

2.1.2 Reagents and buffers

- Flp-In™ 293 T-REx™ cell line (ThermoFisher)
- Dulbecco's phosphate buffered saline (DPBS) (ThermoFisher)
- 0.25% Trypsin-EDTA (ThermoFisher)
- High glucose Dulbecco's Modified Eagle Medium (DMEM) (ThermoFisher)
- Fetal Bovine Serum (FBS) (Millipore Sigma)
- Doxycycline (Millipore Sigma, 1 mg/mL aliquots)
- Poly-D lysine coated 12-well plate (ThermoFisher)

2.1.3 Procedure

This procedure is used to seed and induce expression of CIAR in Flp-In™ 293 T-REx™ cells in a 12-well plate. All amounts and volumes are given on a per well basis, so readers can scale up or down according to plate type.

1. Flp-In™ 293 T-REx™ cells with stably integrated CIAR are maintained in high glucose DMEM supplemented with 10% FBS at 37 °C with 5% CO_2 until cells are ~80% confluent in a T75 flask.
2. Once cells have reached confluence, remove the DMEM, and wash with 3 mL of DPBS. Add 2 mL of 0.25% Trypsin-EDTA at room temperature until cells detach, then add complete DMEM to quench the trypsin.
3. Move cells to a conical tube, then centrifuge for 3 min at 800 rpm to pellet cells. Remove media and resuspend in complete DMEM, making sure to disrupt all cell clumps.
4. Count cells using a hemocytometer or automatic cell counter.
5. Dilute cells to 2.5×10^5 cells/mL. Plate 1 mL of cells on a poly-D lysine coated plate. Add 1 μg/mL doxycycline in complete DMEM to initiate expression of CIAR.
6. Gently rock the plate back and forth to distribute cells and doxycycline evenly throughout the well.
7. Return the cells to the incubator. Incubate cells overnight at 37 °C with 5% CO_2.

2.1.4 Notes

- In subsequent steps, cells will be serum starved. We have found that Flp-In™ 293 T-REx™ cells tend to detach under these conditions. To prevent this, we use pre-coated poly-D lysine plates. Plates can also be manually coated using a poly-D lysine solution (Millipore Sigma) following manufacturer's instructions. For HEK293 cells, a 10 μg/cm^2 poly-D lysine solution works well.

- At each time point, entire wells are collected, so separate plates should be used for each time point to prevent disruption to the cells.

2.2 Serum starvation of CIAR-expressing cells

2.2.1 Equipment

- CO_2 incubator
- Biosafety cabinet

2.2.2 Reagents and buffers

- Dulbecco's phosphate buffered saline (DPBS) (ThermoFisher)
- High glucose Dulbecco's Modified Eagle Medium (DMEM) (ThermoFisher)
- Doxycycline (Millipore Sigma, 1 mg/mL aliquots)

2.2.3 Procedure

This procedure is used to serum starve Flp-In™ 293 T-REx™ cells expressing CIAR in a 12-well plate. All amounts and volumes are given on a per well basis, so readers can scale up or down according to plate type (Fig. 2).

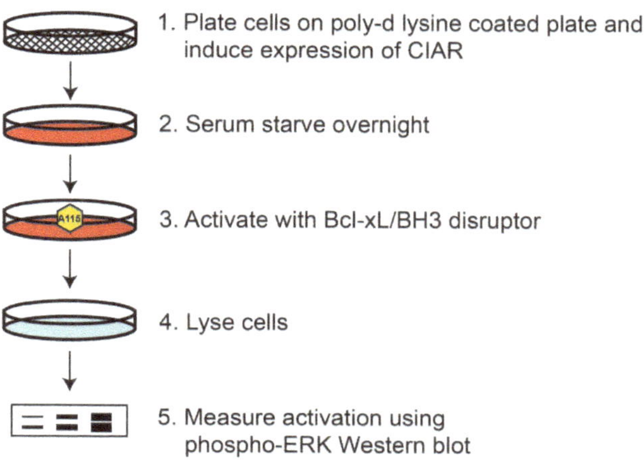

Fig. 2 Design of CIAR activation assay. Briefly, plate cells that have stably incorporated CIAR on poly-D lysine coated plates. Induce expression of CIAR using doxycycline. The next day, wash cells with DPBS, then switch to serum-free media. The following morning add A115, then lyse after the desired amount of time. Activation can be quantified through Western blot and comparing levels of phospho-ERK, a kinase downstream of RAS.

1. Remove cells from step 7 (Section 2.1.3) from the incubator 24 h after plating.
2. Remove the media from the adherent cells. Wash cells twice with 0.5 mL DPBS by gently adding down the side of the well, then gently rock the plate back and forth to cover the entire surface. Remove DPBS from well.
3. Add 1 μg/mL doxycycline in 1 mL high glucose DMEM (no FBS).
4. Return cells to incubator and incubate cells overnight at 37 °C with 5% CO_2.

2.2.4 Notes

- The serum starvation step is used to reduce endogenous signaling. We have found that 16–18 h works best for Flp-In™ 293 T-REx™/HEK293 cells but longer serum starvation times (up to 24 h) can be used.
- Cells should be ~70–80% confluent at the time of serum starvation.

3. Activation of CIAR using Bcl-xL inhibitors

After serum starvation, the small molecule Bcl-xL inhibitor A115 is added. Following treatment for the desired length of time, cells are harvested at the desired time points. Here, treatment time and the concentration of A115 can be varied to provide a graded downstream signaling response.

3.1 Generation of A115 stocks and activation of CIAR cells

3.1.1 Equipment

- CO_2 incubator
- Biosafety cabinet
- Balance

3.1.2 Reagents and buffers

- A-1155463 (A115) (ChemieTek, Catalog #CT-A115)
- Dimethylsulfoxide (DMSO) (ThermoFisher, Catalog #85190)

3.1.3 Procedure

This procedure is used to activate CIAR in Flp-In™ 293 T-REx™ cells in a 12-well plate. All amounts and volumes are given on a per well basis, so readers can scale up or down according to plate type.

1. Weigh out 25 mg of A115.
2. Add 3.540 mL DMSO to A115 to make 10 mM stocks.

3. Pipet solution into 50 μL aliquots. Store aliquots at −20 °C.
4. Dilute A115 aliquot to desired concentration in DMSO. For time course studies, we typically dilute the aliquots to 200 μM, resulting in a final concentration of 200 nM A115 when 1 μL is added to the wells.
5. 16–18 h after switching media, remove the cells from step 4 (Section 2.2.3) from the incubator. Add DMSO or 1 μL of the desired concentration of A115 directly to the media.
6. Rock the plate gently back and forth to distribute the A115, then return the cells to the incubator for the desired length of time.

3.1.4 Notes

- Inhibitors should be aliquoted to avoid multiple freeze-thaw cycles.
- The total percentage of DMSO control or DMSO with A115 in the well should remain under 0.1%.
- *Variations on CIAR activation.* One of the advantages of CIAR is that it serves as a RAS rheostat. By varying the amount of A115 added, a graded output can be obtained. We have observed that RAS signaling increases in a dose-dependent manner (Fig. 3A and B). Additionally, the duration of RAS activation can be varied by lysing cells at different times after initial A115 treatment. We have observed that unlike transient growth factor stimulation, downstream signaling activation in cells expressing CIAR treated with A115 persists until the cells are lysed (Ross et al., 2016) (Fig. 4A and B).

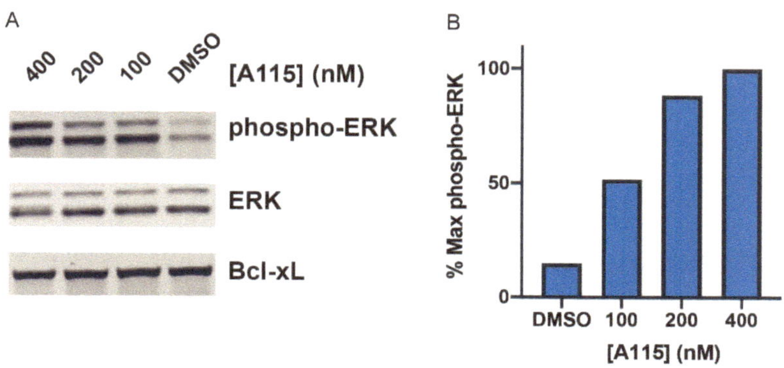

Fig. 3 Dose dependent activation of CIAR. (A) Representative Western blot of cells stably expressing CIAR treated with increasing amounts of A115. As the dose increases, the amount of phospho-ERK increases, while levels of ERK and CIAR (shown by the Bcl-xL antibody) remain the same. (B) Quantified values from Western blot. The means of duplicate values are shown.

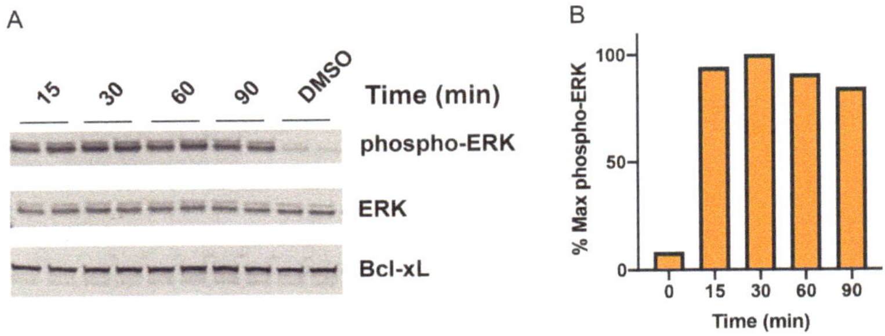

Fig. 4 Time dependent activation of CIAR. (A) Representative blots of cells transfected with CIAR treated with 100 nM A115 for varying lengths of time. (B) Quantified Western blot values. The means of duplicate values are shown.

3.2 Harvest of activated cells and preparation of Western blot samples

3.2.1 Equipment

- Biosafety cabinet
- Vortex
- Tabletop centrifuge
- Heat bath set to 98 °C

3.2.2 Reagents and buffers

- Ice
- Ice cold DPBS (Thermo Fisher)
- Modified RIPA lysis buffer (50 mM Tris pH 8, 150 mM NaCl, 1 mM EDTA, Pierce protease inhibitor (Millipore Sigma), 1% IGEPAL CA630 (Millipore Sigma), 2 mM Na_3VO_4, 30 mM NaF)
- 3 × SDS (240 mM Tris pH 6.8, 30% glycerol, 16% betamercaptoethanol, 6% w/v SDS, 0.03% bromophenol blue in water)
- Cell Scraper (ThermoFisher)

3.2.3 Procedure

This procedure is used to lyse CIAR-activated Flp-In™ 293 T-REx™ cells in a 12-well plate. All amounts and volumes are given on a per well basis, so readers can scale up or down according to plate type.

1. After desired length of A115 treatment time, remove plate from step 6 (Section 3.1.3) from the incubator.
2. Remove the media and place cells on ice. Wash cells once with 0.5 mL ice cold DPBS, then add 100 μL modified RIPA lysis buffer.

3. Detach cells from well by using a cell scraper or by gently pipetting up and down to release cells from plate. Put lysate in a labeled sterile 1.5 mL Eppendorf tube on ice.
4. Vortex cells on medium speed three times for 3 s, then pellet debris by centrifuging in a tabletop centrifuge for 10 min, max speed, 4 °C.
5. Transfer 60 μL supernatant to a clean Eppendorf. Add 30 μL 3 × SDS, then boil the samples for 7 min at 98 °C.

3.2.4 Notes

- Once media has been removed from cells, it is extremely important that cells remain on ice/stay cold until they are boiled in the SDS solution.
- At the time of lysis, cells should be confluent enough that detachment from the plate can be visualized by eye.

4. Examination of RAS-mediated signaling through Western blot

Once cells have been lysed, we often use Western blot analysis of downstream signaling to examine pathway activation and validate CIAR. A variety of antibodies can be used, depending on which downstream signaling events are being analyzed. Here, we use a Bcl-xL antibody to measure CIAR expression levels. Additionally, we look at canonical RAS/MEK/ERK signaling by using ERK and phospho-ERK antibodies to measure the activation state of downstream ERK kinase.

4.1 Western blot of CIAR samples

4.1.1 Equipment

- SDS-PAGE system (Bio-Rad)
- Trans-Blot® Turbo™ system (Bio-Rad)
- Western blot incubation box
- Orbital Shaker

4.1.2 Reagents and buffers

- Precision Plus Protein™ Ladder (Bio-Rad)
- Any kD™ Mini-PROTEAN® TGX™ precast polyacrylamide gel (Bio-Rad)
- NuPAGE™ MOPS SDS Buffer (20 ×) (ThermoFisher), diluted to 1 × with water

- Trans-Blot® Turbo™ Midi Transfer Packs (Bio-Rad)
- Odyssey® Blocking Buffer (Li-Cor)
- TBST (20 mM Tris, 150 mM NaCl, 0.1% Tween® 20, pH 7.6)
- Primary antibodies
  - Anti-Bcl-xL Rabbit mAb (CST #2762)
  - p44/42 MAPK (Erk1/2) (3A7) Mouse mAb (CST #9107)
  - Phospho-p44/42 MAPK (Erk1/2) (Thr2-2/Tyr204) (D13.14.4E) XP Rabbit mAb (CST #4370)
- Secondary antibodies
  - IRDye® 800CW Goat anti-Rabbit IgG Secondary Antibody (Li-Cor)
  - IRDye® 680LT Goat anti-Mouse IgG Secondary Antibody (Li-Cor)

4.1.3 Procedure

1. Remove gel from packaging. Set up SDS-PAGE system and add 1 × MOPS running buffer. Load 2 μL of the protein ladder and 10 μL of the boiled samples from step 5 (Section 3.2.3) on the precast gel. Run the gel at 180 V until the dye front reaches the bottom of the gel, usually about 30 min.
2. Assemble the transfer sandwich. Remove the transfer pack from packaging and place the top stack on the bottom tub of the Trans-Blot® Turbo™ cassette.
3. Once the dye front has reached the bottom of the gel, remove the gel from the gel rig. Separate the plastic casing from the gel and place the gel on the nitrocellulose membrane.
4. Place the top stack from the Trans-Blot® Turbo™ Midi Transfer Packs on top of the gel, then close the cassette cover. Place the cassette in the Trans-Blot® Turbo™ and run the MIDI 7-min program. This will transfer the samples from the gel to the nitrocellulose.
5. Once the program has finished, remove the transfer sandwich from the cassette. Cut the nitrocellulose down to the size of the gel, then remove the gel from the nitrocellulose. Put the nitrocellulose into the Western blot incubation box.
6. Add Odyssey® Blocking Buffer to the incubation block and incubate for 1 h at room temperature with gentle agitation on the orbital shaker.
7. Remove blocking buffer, then add primary antibodies (anti-Bcl-xL antibody (diluted 1:2500), anti-ERK (diluted 1:2500) and anti-phospho-ERK (diluted 1:2500)) in Odyssey® Blocking Buffer and incubate overnight with gentle agitation at 4 °C on an orbital shaker.

8. The next day, remove primary antibody solution, and wash nitrocellulose membrane three times with TBST for 5 min.
9. Add both secondary antibodies (each diluted 1:10,000) in Odyssey® Blocking Buffer. Incubate at room temperature with gentle agitation for 1 h.
10. Wash membrane three times with TBST for 5 min each, then image the membrane on a Li-Cor Odyssey® infrared imager.

4.1.4 Notes

- When assembling the transfer sandwich, use a roller to ensure that there are no bubbles. Bubbles will prevent proper transfer of the samples from the gel to the nitrocellulose.
- Antibodies for other proteins can also be used to study how direct RAS activation affects expression levels or phosphorylation states.

4.2 Quantitation of CIAR Western blot

4.2.1 Equipment

- Image Studio™ Lite software (Li-Cor)

4.2.2 Procedure

1. Load imaged blot into Image Studio™ Lite software.
2. To quantify bands, select a single channel (700 or 800). Under the "Analysis" tab choose "Draw rectangle." Draw a single rectangle over each band of interest.
3. Under the "Shapes" tab, find the "Signal" value. This corresponds to the background-subtracted band of interest that was in the drawn rectangle.

4.2.3 Notes

- Bcl-xL signal is used to confirm that the CIAR switch is being expressed in the cells. The Bcl-xL signal can also be used for normalization based on CIAR expression levels.
- When using ERK and phospho-ERK antibodies, phospho-ERK signal is divided by total ERK signal in order to determine relative phospho-ERK states. This method could be extended to many different phosphorylated proteins where there is a total protein and phospho-specific antibody available.

5. Summary and conclusions

We have developed a single-protein, chemical genetic system, called chemically inducible activator of RAS (CIAR), that directly activates endogenous RAS without requiring the relocalization of any components. CIAR can be introduced through transient transfection or stable integration into diverse cell types, allowing for the examination of the effects of RAS activation on downstream signaling events. The ability to chemically control CIAR facilitates studies of RAS-mediated signaling dynamics and the effects of active RAS dose dependence on the signaling state of cells. In this protocol, we detail how measurement of downstream ERK kinase activation can be used to validate CIAR but downstream signaling analysis can be expanded to look at other RAS-mediated signaling events. Such studies should shed new light on RAS-mediated cell signaling.

Acknowledgments

This research was funded in part by the NIH (R01GM086858 (D.J.M.)) and a National Science Foundation GRFP (E.M.D.).

References

Ashkenazi, A., Fairbrother, W. J., Leverson, J. D., & Souers, A. J. (2017). From basic apoptosis discoveries to advanced selective BCL-2 family inhibitors. *Nature Reviews Drug Discovery*, *16*, 273–284.

Boriack-Sjodin, P. A., Margarit, S. M., Bar-Sagi, D., & Kuriyan, J. (1998). The structural basis of the activation of Ras by Sos. *Nature*, *394*, 337–343.

Cherfils, J., & Zeghouf, M. (2013). Regulation of small GTPases by GEFs, GAPs, and GDIs. *Physiological Reviews*, *93*, 269–309.

Cory, S., Huang, D. C. S., & Adams, J. M. (2003). The Bcl-2 family: Roles in cell survival and oncogenesis. *Oncogene*, *22*(53), 8590–8607.

Cox, A. D., Fesik, S. W., Kimmelman, A. C., Luo, J., & Der, C. J. (2014). Drugging the undruggable RAS: Mission possible? *Nature Reviews Drug Discovery*, *14*, 828–851.

Cunningham-Bryant, D., Dieter, E. M., Foight, G. W., Rose, J. C., Loutey, D. E., & Maly, D. J. (2019). A chemically disrupted proximity system for controlling dynamic cellular processes. *Journal of the American Chemical Society*, *141*(8), 3352–3355.

Foight, G. W., Wang, Z., Wei, C. T., Greisen, P. J., Warner, K. M., Cunningham-Bryant, D., et al. (2019). Multi-input chemical control of protein dimerization for programming graded cellular responses. *Nature Biotechnology*, *37*, 1209–1216. https://doi.org/10.1038/s41587-019-0242-8, PMID: 29200207.

Goreshnik, I., & Maly, D. J. (2010). A small molecule-regulated guanine nucleotide exchange factor. *Journal of the American Chemical Society*, *132*, 938–940.

Hancock, J. F. (2003). Ras proteins: Different signals from different locations. *Nature Reviews Molecular Cell Biology*, *4*(5), 373–385.

Hunter, J. C., Manandhar, A., Carrasco, M. A., Gurbani, D., Gondi, S., & Westover, K. D. (2015). Biochemical and structural analysis of common cancer-associated KRAS mutations. *Molecular Cancer Research, 13*(9), 1325–1335.
Karnoub, A. E., & Weinberg, R. A. (2008). Ras oncogenes: Split personalities. *Nature Reviews Molecular Cell Biology, 9*(7), 517–531.
Konopleva, M., Contractor, R., Tsao, T., Samudio, I., Ruvolo, P. P., Kitada, S., et al. (2006). Mechanisms of apoptosis sensitivity and resistance to the BH3 mimetic ABT-737 in acute myeloid leukemia. *Cancer Cell, 10*(5), 375–388.
Papke, B., & Der, C. J. (2017). Drugging RAS: Know the enemy. *Science, 355*, 1158–1163.
Prior, I. A., Lewis, P. D., & Mattos, C. (2012). A comprehensive survey of Ras mutations in cancer. *Cancer Research, 72*(10), 2457–2467.
Rajalingam, K., Schreck, R., Rapp, U. R., & Albert, Š. (2007). Ras oncogenes and their downstream targets. *Biochimica et Biophysica Acta (BBA)—Molecular Cell Research, 1773*(8), 1177–1195.
Rose, J. C., Dieter, E. M., Cunningham-Bryant, D., & Maly, D. J. (2018). Examining RAS pathway rewiring with a chemically inducible activator of RAS. *Small GTPases, 1*, 1–8.
Rose, J. C., Huang, P. S., Camp, N. D., Ye, J., Leidal, A. M., Goreshnik, I., et al. (2017). A computationally engineered RAS rheostat reveals RAS–ERK signaling dynamics. *Nature Chemical Biology, 13*(1), 119–126.
Ross, B., Mehta, S., & Zhang, J. (2016). Molecular tools for acute spatiotemporal manipulation of signal transduction. *Current Opinion in Chemical Biology, 34*, 135–142.
Shoemaker, A. R., Oleksijew, A., Bauch, J., Belli, B. A., Borre, T., Bruncko, M., et al. (2006). A small-molecule inhibitor of Bcl-xL potentiates the activity of cytotoxic drugs in vitro and in vivo. *Cancer Research, 66*(17), 8731–8739.
Simanshu, D. K., Nissley, D. V., & McCormick, F. (2017). RAS proteins and their regulators in human disease. *Cell, 170*, 17–33.
Stanton, B. Z., Chory, E. J., & Crabtree, G. R. (2018). Chemically induced proximity in biology and medicine. *Science, 359*(6380), eaao5902.
Tao, Z. F., Hasvold, L., Wang, L., Wang, X., Petros, A. M., Park, C. H., et al. (2014). Discovery of a potent and selective BCL-XL inhibitor with in vivo activity. *ACS Medicinal Chemistry Letters, 5*, 1088–1093.
Toettcher, J. E., Weiner, O. D., & Lim, W. A. (2013). Using optogenetics to interrogate the dynamic control of signal transmission by the Ras/Erk module. *Cell, 155*(6), 1422–1434.
Um, H.-D. (2015). Bcl-2 family proteins as regulators of cancer cell invasion and metastasis: A review focusing on mitochondrial respiration and reactive oxygen species. *Oncotarget, 7*, 5193–5203.
Voß, S., Klewer, L., & Wu, Y.-W. (2015). Chemically induced dimerization: Reversible and spatiotemporal control of protein function in cells. *Current Opinion in Chemical Biology, 28*, 194–201.
Wei, G., Margolin, A. A., Haery, L., Brown, E., Cucolo, L., Julian, B., et al. (2012). Chemical genomics identifies small-molecule MCL1 repressors and BCL-xL as a predictor of MCL1 dependency. *Cancer Cell, 21*(4), 547–562.
Zhou, B., Der, C. J., & Cox, A. D. (2016). The role of wild type RAS isoforms in cancer. *Seminars in Cell and Developmental Biology, 58*, 60–69.

CHAPTER SEVEN

Temporal and rheostatic control of genome editing with a chemically-inducible Cas9

Cindy T. Wei[a,b,c], Dustin J. Maly[b,d], Douglas M. Fowler[c,e,*]
[a]Molecular and Cellular Biology Program, University of Washington, Seattle, WA, United States
[b]Department of Chemistry, University of Washington, Seattle, WA, United States
[c]Department of Genome Sciences, University of Washington, Seattle, WA, United States
[d]Department of Biochemistry, University of Washington, Seattle, WA, United States
[e]Department of Bioengineering, University of Washington, Seattle, WA, United States
*Corresponding author: e-mail address: dfowler@uw.edu

Contents

Abstract

Nuclease-mediated DNA cleavage and subsequent repair lie at the heart of genome editing, and the RNA-guided endonuclease Cas9 has emerged as the most widely-used tool for facilitating this process. Extensive biochemical and biophysical efforts have revealed much regarding the structure, mechanism, and cellular properties of Cas9. This has enabled engineering of Cas9 variants with enhanced activity, specificity, and other features. However, we lack a detailed understanding of the kinetics of Cas9-mediated

Methods in Enzymology, Volume 633
ISSN 0076-6879
https://doi.org/10.1016/bs.mie.2019.11.001

DNA cleavage and repair *in vivo*. To study *in vivo* Cas9 cleavage kinetics and activity dose-dependence, we have engineered a chemically-inducible, single-component Cas9, ciCas9. ciCas9 allows for temporal and rheostatic control of Cas9 activity using a small molecule activator, A115. We have also developed a droplet-digital PCR-based assay (DSB-ddPCR) to directly quantify Cas9-mediated double-stranded breaks (DSBs). The methods in this chapter describe the application of ciCas9 and DSB-ddPCR to study the kinetics and dose-dependence of Cas9 editing *in vivo*.

1. Introduction

CRISPR/Cas (clustered regularly interspaced short palindromic repeats) has become a powerful tool for genome editing with broad applications in basic science research and therapeutic development. The *Streptococcus pyogenes* RNA-guided endonuclease, *Sp*Cas9, is the most widely used tool for generating site-specific double-stranded breaks (DSBs) to drive editing. Cas9 is directed to specific loci within the genome by a single-guide RNA (sgRNA). The sgRNA contains a scaffold sequence required for Cas9 binding and a 20 nucleotide sequence that is complementary to a target site (Fig. 1A) (Cong et al., 2013; Jinek et al., 2012; Mali et al., 2013). Once Cas9 is bound to a target site, it generates a DSB in the DNA. This DSB is subsequently repaired by the cell's endogenous DNA repair machinery though error-free homology-directed repair (HDR) or error-prone non-homologous end joining (NHEJ). The ability to target Cas9 to specific loci by having an RNA define the target site of the endonuclease has made it possible to target different loci without laborious protein engineering.

Many studies have been performed *in vitro* to understand the mechanism of Cas9-mediated DNA targeting and cleavage (Chen et al., 2017; Dagdas, Chen, Sternberg, Doudna, & Yildiz, 2017; Nishimasu et al., 2014; Sternberg, LaFrance, Kaplan, & Doudna, 2015; Sternberg, Redding, Jinek, Greene, & Doudna, 2014). Despite this wealth of knowledge, many unknowns remain, especially *in vivo*. For example, few *in vivo* studies have focused on Cas9 kinetics. Furthermore, because *in vivo* studies generally examine editing, rather than the direct products of Cas9 activity, little is known regarding the dynamics of DNA repair following DSB generation. In order to study the kinetics and dose-dependence of Cas9 activity *in vivo*, we have developed two new tools: a chemically-inducible Cas9 (ciCas9) and double-stranded break droplet-digital PCR (DSB-ddPCR). ciCas9 is a Cas9 variant that possesses low basal nuclease activity but can be rapidly activated

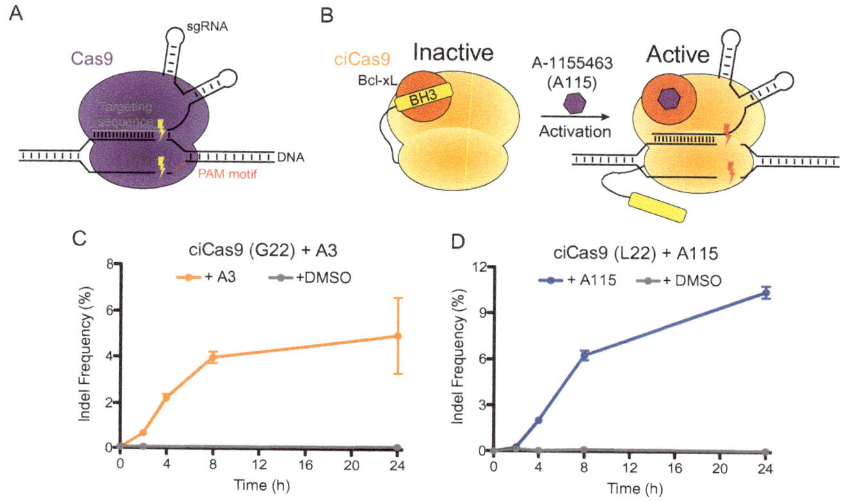

Fig. 1 ciCas9 is a small molecule inducible Cas9. (A) Schematic of WT Cas9. (B) Schematic of ciCas9, which has the REC2 domain of WT Cas9 replaced with the protein Bcl-xL. A BH3 peptide is fused to the C-terminus of the Cas9, which forms an auto-inhibitory interaction with Bcl-xL. The small molecule Bcl-xL inhibitor A115 disrupts the Bcl-xL/BH3 interaction and activates ciCas9. (C) A time course of ciCas9(G22)-mediated indel formation at the AAVS1 locus following the addition of 10 μM A3 or DMSO to HEK293T cells (Rose, Stephany, Valente, et al., 2017). (D) A time course of ciCas9(L22)-mediated indel formation at the AAVS1 locus following the addition of 1 μM A115 or DMSO to HEK293T cells. Data represented as mean ± SEM of three cell culture replicates.

upon addition of small molecule Bcl-xL inhibitors (Fig. 1B). The activity of ciCas9 can be rheostatically tuned by varying the concentration of small molecule added (Rose, Stephany, Valente, et al., 2017; Rose, Stephany, Wei, Fowler, & Maly, 2017). DSB-ddPCR is a droplet-digital PCR-based assay that allows quantification of DSBs formed at a target locus over time (Fig. 2A). By using ciCas9 with DSB-ddPCR and high-throughput sequencing, the *in vivo* kinetics and dose-dependence of Cas9 editing can be explored. Here, we describe methods utilizing ciCas9 and DSB-ddPCR to study the *in vivo* kinetics and dose-dependence of Cas9 editing.

1.1 Introduction to inducible Cas9 systems

To study Cas9 kinetics *in vivo*, a method for temporally controlling Cas9 nuclease activity is required. In addition to our ciCas9 system, a number of chemically-inducible methods that allow temporal control over Cas9 activity are currently available (Gangopadhyay et al., 2019). For example, coarse-grained control of Cas9 activity can be achieved using doxycycline-inducible promoters that regulate Cas9 expression (Dow et al., 2015; González et al., 2014). In addition to transcriptionally-regulated Cas9 systems, several chemically-controlled, post-translational methods for temporal control have been developed. Small molecule-controlled, degradation-based systems have been engineered by creating degron domain fusions to Cas9. The addition of stabilizing small molecules to cells expressing these degradation-based systems allows for accumulation of active Cas9 over time (Kleinjan, Wardrope, Sou, & Rosser, 2017; Maji et al., 2017; Senturk et al., 2017). Control of Cas9 activity through chemically-induced nuclear translocation has also been achieved (Liu et al., 2016; Nguyen et al., 2016; Oakes et al., 2016; Zhao et al., 2018). Notably, the small molecule 4-hydroxytamoxifen both allosterically activates and facilitates the nuclear translocation of one of these Cas9 variants, allosterically regulated Cas9 (arC9) (Oakes et al., 2016). Split-Cas9 that uses chemically-induced dimerization to activate nuclease activity has also been engineered. This split-Cas9 system uses rapamycin-inducible FKBP-FRB heterodimerization domains to reconstitute active Cas9 (Zetsche, Volz, & Zhang, 2015). Finally, a system that facilitates temporal activation of Cas9 by 4-hydroxytamoxifen-mediated removal of an inhibitory intein has been developed (Davis, Pattanayak, Thompson, Zuris, & Liu, 2015). To allow rapid activation of Cas9 activity, we engineered a single-component Cas9 variant, called ciCas9, that is autoinhibited in the basal state but rapidly activated with small molecules.

1.2 Introduction to chemically-inducible Cas9 (ciCas9)

ciCas9 is a single-component, chemically-inducible Cas9 variant that uses an autoinhibitory interaction between the pro-apoptotic protein Bcl-xL and a BH3 peptide to control nuclease activity (Rose, Stephany, Valente, et al., 2017). To ensure that the Bcl-xL/BH3 peptide interaction inhibits Cas9's activity, we replaced the REC2 domain with the anti-apoptotic protein Bcl-xL and appended a BH3 peptide onto the C-terminus. Although the precise molecular details of how the Bcl-xL/BH3 peptide interaction is able to inhibit Cas9's activity are not known, this complex likely physically blocks the ability of Cas9 to bind to DNA target sites. While the Bcl-xL/BH3 peptide interaction efficiently disrupts Cas9's nuclease activity in the basal state, nuclease activity is rapidly restored by a small molecule that selectively targets Bcl-xL.

We used the small molecule A-385358 (A3) to activate our first generation of ciCas9 (ciCas9(G22)). To allow Cas9 activation at a reasonable concentration (<10 μM) of A3, the affinity of the interaction between Bcl-xL and the BH3 peptide was reduced by introducing a glycine at residue 22 within the BH3 peptide. Using A3, we observed rapid DSB and indel formation at the targeted locus, with low basal Cas9 activity in the absence of A3 (Figs. 1C and 2B) (Rose, Stephany, Valente, et al., 2017). The availability of a significantly more potent small molecule, A-1155463 (A115), that selectively targets Bcl-xL allowed us to engineer a second generation of ciCas9 (ciCas9(L22)) with a strengthened interaction between Bcl-xL and the BH3 peptide. Replacing Gly22 in the BH3 peptide with Leu increases the affinity between Bcl-xL and the BH3 peptide by an order of magnitude. As a result, we have found that ciCas9(L22) demonstrates even lower basal Cas9 activity than ciCas9(G22). A115 has low-picomolar affinity for Bcl-xL and can maximally activate ciCas9(G22) or ciCas9(L22) in cells at concentrations significantly below 10 μM (Rose, Stephany, Valente, et al., 2017). The use of ciCas9(L22) with the highly potent A115 allowed us to achieve similar levels of DSBs and indels with minimal basal activity in the absence of small molecule activator at a number of loci (Fig. 1D) (Rose, Stephany, Wei, et al., 2017). We recommend use of the L22 variant for tighter control of Cas9 editing. The methods described in this chapter refer to the use of the ciCas9(L22) variant with activation by A115.

2. Temporal and rheostatic control of genome editing with ciCas9

The ciCas9 system allows for kinetic studies of Cas9 activity through rapid activation of nuclease activity upon addition of the small molecule activator, A115. We commonly introduce ciCas9 into cells using transient transfection along with a sgRNA targeting a locus of choice. The kinetics of Cas9 editing at this locus can then be monitored using a time course experiment. Here, cells are harvested at various time points after A115 addition; DSBs are assessed using DSB-ddPCR and indels are assessed using high-throughput DNA sequencing. We have used this system to study DNA cleavage and indel kinetics at different loci and have found that CRISPR/Cas9 genome editing kinetics can vary at different target sites (Rose, Stephany, Valente, et al., 2017).

This protocol is written based on usage in HEK293T cells, which allow for rapid cell harvesting for temporal precision. We allow around 24h between seeding of cells and transient transfection of ciCas9/sgRNA and then another 24h before A115 addition. In HEK293T cells, this timing allows the cells to reach suitable confluency and stable Cas9 expression following transient transfection. We note that in other cell lines, timing should be adjusted to optimize confluency and expression.

2.1 Seeding cells for *in vivo* kinetic studies of Cas9 activity

2.1.1 *Equipment*

- CO_2 incubator
- Biosafety cabinet
- Centrifuge
- Hemocytometer or automated cell counter

2.1.2 *Reagents*

- Human embryonic kidney 293T cells (ATCC)
- Dulbecco's Modified Eagle Medium (Thermo Fisher Scientific), DMEM
- Fetal Bovine Serum (Millipore Sigma), FBS
- Dulbecco's Phosphate-Buffered Saline (Thermo Fisher Scientific), DPBS
- 0.25% Trypsin-EDTA (Thermo Fisher Scientific)

2.1.3 *Protocol*

(1) HEK293T cells are maintained in DMEM supplemented with 10% FBS (D10) and kept at 37 °C with 5% CO_2 until the cells are 80–90% confluent before seeding for an experiment.

(2) The day of seeding cells, remove the media and wash the cells with DPBS.

(3) Detach the cells from the vessel by adding 0.25% Trypsin-EDTA, just enough to cover the surface area of the cells. Incubate at room temperature until the cells detach. Add D10 to quench the trypsin. Move all the cells into a conical tube.

(4) Centrifuge the cells at $100 \times g$ for 3min and remove the media. Then resuspend the cells in new D10, making sure to disrupt all cell clumps. Count the number of cells using a hemocytometer or an automated cell counter.

(5) Dilute the cells to $2\text{–}2.5 \times 10^5$ cells/mL using new D10.

(6) Add 1 mL of cells into each well of a 12-well plate. This should result in 2–2.5 × 10^5 cells per well. Gently rock the plate to ensure even distribution of cells in each well.
(7) Incubate the plates at 37 °C with 5% CO_2 for 24 h, reaching 40–50% confluency.

2.1.4 Notes

- Entire wells of cells are harvested at each time point, so separate 12-well plates should be used for each time point. This will allow for minimal disruption to the cells that are to be collected at later time points.
- 12-well plates are used in this protocol as this provides enough cells for downstream analysis by high-throughput DNA sequencing and DSB-ddPCR.
- Other cell lines can also be used to study genome editing kinetics with ciCas9. We often use HEK293T cells due to their high transfection efficiency and the ease of harvest at early time points due to the weak adherence of these cells. Cells should be seeded to 40–50% confluency at time of transfection.

2.2 Transient transfection of ciCas9 into cultured human cell lines

2.2.1 Equipment

- CO_2 incubator
- Biosafety cabinet

2.2.2 Reagents

- Turbofectin 8.0 (OriGene)
- Opti-MEM (Thermo Fisher Scientific)

2.2.3 Plasmids

- ciCas9 in mammalian cell expression backbone (e.g., pcDNA5-FRT-TO or pcDNA3.1(+)), Addgene #100551
- sgRNA cloned into Addgene #41824 backbone
- pMAX-GFP transfection control

2.2.4 Protocol

(1) Calculate the volume of transfection mixture required for 100 μL of transfection mix per well. Plasmids should be mixed in a ratio of

450 ng ciCas9, 450 ng sgRNA, and 100 ng pMAX-GFP per well. The remaining volume should be made up with Opti-MEM.

(2) To create the transfection mix, add Opti-MEM to the tube first, followed by adding the plasmids. Mix by inverting the tube 5–6 times.

(3) Add 3 μL of Turbofectin 8.0 reagent per 1 μg of plasmid DNA to the transfection mix. Try not to touch the side of the tube when adding Turbofectin. Invert the tube 5–6 times then incubate the transfection mix with Turbofectin at room temperature for 20 min.

(4) Gently add 100 μL of transfection mix to each well drop-wise and gently rock the plate to mix. Be sure to not pipette the transfection mix onto the cells too quickly, this may result in cells detaching from the plate.

(5) Incubate the plates at 37 °C with 5% CO_2 for 24 h.

2.2.5 Notes

- If DSB-ddPCR will be used to analyze kinetics of DSB formation, half the amount of DNA should be transfected for each well; resulting in 225 ng ciCas9, 225 ng sgRNA, and 50 ng pMAX-GFP.
- Other transfection reagents can also be used to deliver plasmids. We have found that Turbofectin 8.0 works efficiently for delivering large plasmids, such as ciCas9, and also has minimal effects on cell viability.
- We have found that 1 μg of DNA with 3 μL of Turbofectin works well in HEK293, HEK293T, HCT116, and U2OS cell lines for delivering ciCas9 and sgRNA. Transfection has been observed to be most efficient in cells at 40–50% confluency at the time of transfection.
- 1 μg of DNA is suitable for the number of cells plated per well in a 12-well plate, as described in Section 2.1. DNA amounts can be adjusted for the number of cells and well size.

2.3 Activation of ciCas9 to study *in vivo* editing kinetics and dose-dependence

2.3.1 Equipment

- CO_2 incubator
- Biosafety cabinet

2.3.2 Reagents

- A-1155463 dissolved in DMSO at 10 mM stock concentration, aliquoted and stored at −20 °C (ChemieTek, Catalog #: CT-A115), A115
- DMSO (Thermo Fisher Scientific)

2.3.3 Protocol

(1) Dilute the 10 mM stock of A115 to 1 mM using DMSO.

(2) For time course studies, add 1 µL of 1 mM A115 stock to each well where ciCas9 will be activated, reaching a final drug concentration of 1 µM and 0.1% DMSO. For dose-response studies, create serial dilutions of A115 to desired concentrations using DMSO. Add the drug to the side of the well, being sure to not detach any of the cells by forceful pipetting.

(3) Gently rock the plate to mix the drug with the media.

(4) Incubate the plates at 37 °C with 5% CO_2 for the amount of time for each time point.

2.3.4 Notes

- We have observed that 1 µM of A115 can fully activate ciCas9 for editing without affecting cellular viability. In dose-response studies, 1 µM A115 is usually sufficient as the highest dose.
- At any concentration of A115 added, the final DMSO percentage should be 0.1% or less.
- If there are short time points (e.g., under 1 h), it is advised to add A115 to the later time points first to ensure accurate harvesting times without delay for the earlier time points.

2.4 Harvesting of HEK293T cells after ciCas9 activation

2.4.1 Equipment

- Biosafety cabinet
- Centrifuge at 4 °C

2.4.2 Reagents

- Ice cold Dulbecco's Phosphate Buffered Saline (Thermo Fisher Scientific), DPBS

2.4.3 Protocol

(1) Label all 1.5 mL microcentrifuge tubes that the cells will be collected into and pre-chill the tubes on ice for 10 min prior to harvesting cells.

(2) At appropriate time points, move the plates onto ice and immediately remove the growth medium. Wash the cells in each well with 1 mL of ice cold DPBS.

(3) Add 600 μL of ice cold DPBS to each well and pipette up and down to disrupt the cells off the surface of the plate. Move the cells into a pre-chilled 1.5 mL microcentrifuge tube.

(4) Centrifuge the cells at 1500 × *g* for 10 min at 4 °C. Remove DPBS and immediately store cell pellets at −80 °C until isolation of genomic DNA.

2.4.4 Notes

- To ensure temporal precision, all steps must be performed quickly and the cells must be kept cold at all times.
- Cell pellets can be kept at −80 °C indefinitely until isolation of genomic DNA.

2.5 Harvesting of other adherent cell lines after ciCas9 activation

This section describes harvesting of adherent cell lines that are more strongly adherent than HEK293T cells and will require trypsin cleavage to detach cells.

2.5.1 Equipment

- CO_2 incubator
- Biosafety cabinet
- Centrifuge at 4 °C

2.5.2 Reagents

- Dulbecco's Modified Eagle Medium (Thermo Fisher Scientific) supplemented with 10% Fetal Bovine Serum (Millipore Sigma), D10
- Warmed Dulbecco's Phosphate-Buffered Saline (Thermo Fisher Scientific), warmed DPBS
- Ice cold Dulbecco's Phosphate Buffered Saline (Thermo Fisher Scientific), ice cold DPBS
- 0.25% Trypsin-EDTA (Thermo Fisher Scientific)

2.5.3 Protocol

(1) Same as Section 2.4.3, label all 1.5 mL microcentrifuge tubes that the cells will be collected into and pre-chill the tubes on ice for 10 min prior to harvesting cells.

(2) At appropriate time points remove the growth medium and wash the cells in each well with 1 mL of warmed DBPS.

(3) Add 100 μL of 0.25% Trypsin-EDTA and incubate at room temperature until cells detach. Quench trypsin using 1 mL of D10 media.
(4) Resuspend the cells and move into a pre-chilled 1.5 mL microcentrifuge tube and place the tube on ice.
(5) Centrifuge the cells at 1500 × *g* for 10 min at 4 °C. Remove the media.
(6) Resuspend the cell pellet with 600 μL of ice cold DPBS to wash the cells.
(7) Centrifuge the cells again at 1500 × *g* for 10 min at 4 °C. Remove DPBS and immediately store cell pellets at −80 °C until isolation of genomic DNA.

2.5.4 Notes

- The trypsin incubation for detaching cells can occur at 37 °C if the cells are strongly adherent. However, incubation time should be minimized to ensure temporal precision.

3. DSB-ddPCR to measure ciCas9-mediated DSB formation

The outcomes of Cas9 activity *in vivo* are often measured by quantifying the indels formed at the target locus. Quantification of indels by sequencing does not directly measure the outcomes of Cas9 activity as indel formation involves repair of Cas9-mediated DSBs. Cas9 is only involved in the generation of DSBs prior to DNA repair. Thus, to directly measure Cas9 activity, methods for quantifying DSBs are needed. We have developed DSB-ddPCR to directly quantify the frequency of DSBs created by Cas9.

DSB-ddPCR involves using two sets of probes and primers to quantify the percentage of template DNA molecules at a target site that contain DSBs. One set of primers span the sgRNA target locus and the other set form a control template amplicon adjacent to the target locus that does not span the site targeted by the sgRNA (Fig. 2A). Amplification by these two primer pairs are detected using fluorescence-quenched qPCR probes that anneal within the amplicons. Droplets that contain fluorescent signals for both the template and target probes represent uncleaved or repaired DNA targets. Droplets that contain fluorescent signal for only the template probe represent DNA target sites containing DSBs. By counting the number of droplets with these different fluorescent signals, DSB frequencies are quantified.

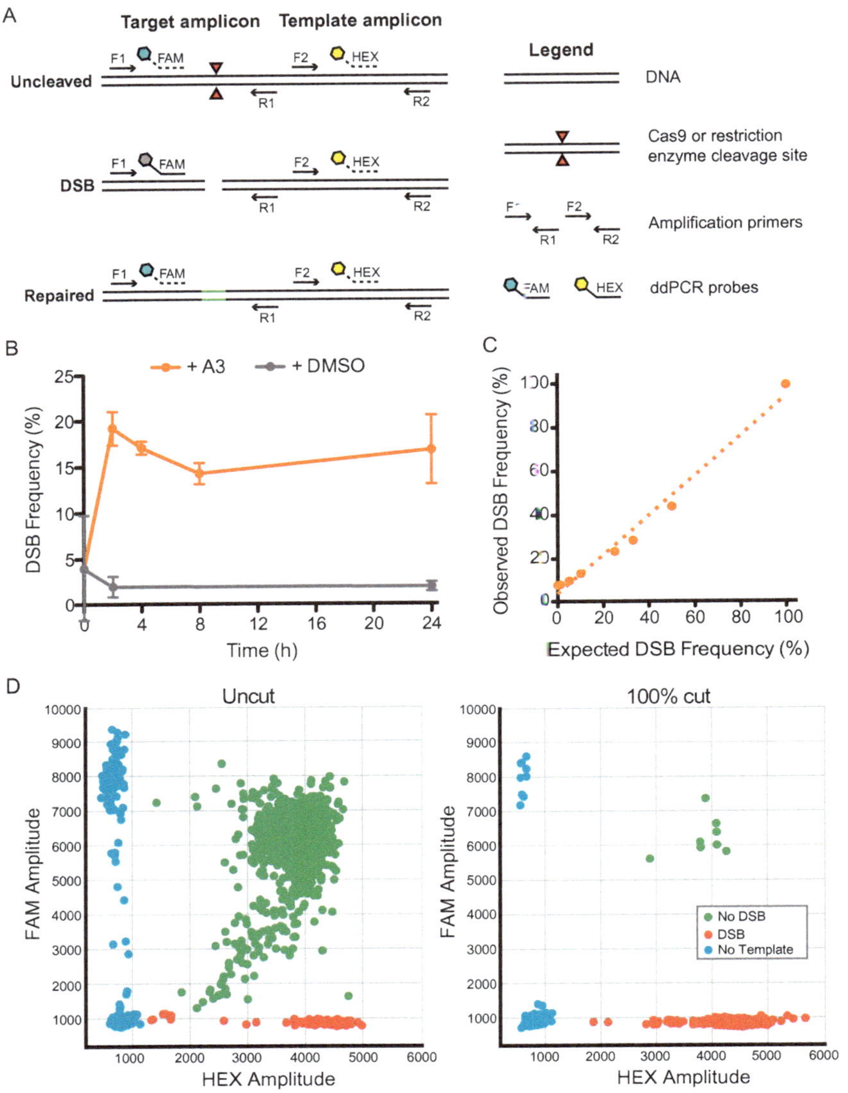

Fig. 2 DSB-ddPCR can be used to measure DSB frequency at target loci. (A) Schematic of DSB-ddPCR, which uses two amplicons to quantify the percentage of DSBs at a target site. One amplicon spans the Cas9 target site, the other amplicon sits 3′ to the target as a template control amplicon. (B) Time course of ciCas9(G22) activated with 10 μM A3 with DSB frequency measured using DSB-ddPCR. ciCas9(G22) was targeted to the AAVS1 locus (Rose, Stephany, Valente, et al., 2017). (C) DSB-ddPCR standard curve at the AAVS1 locus. Untreated DNA was cut with *Bsu*36I restriction enzyme to simulate DSBs. (D) Gating of droplets to form "No DSB" and "DSB" populations to calculate DSB frequency. Example from an uncut DNA sample (left) and 100% cut with *Bsu*36I DNA sample (right). Data represented as mean ± SEM of three cell culture replicates.

The use of DSB-ddPCR coupled with ciCas9-based time course experiments allows for the direct quantification of Cas9-mediated DSBs. Additional insight into DNA repair following DSB generation can be obtained by measuring indel formation using high-throughput sequencing in parallel. Protocols for measuring indels with targeted high-throughput sequencing have been described (Rose, Stephany, Valente, et al., 2017). We have previously coupled these techniques using the original ciCas9 (G22) variant and A3 activation to study the kinetics of various loci (Fig. 2B). Here, we describe methods for the processing and DSB-ddPCR analysis of frozen cell pellets that have been collected from ciCas9 time course experiments. These methods also include guidelines for designing target and template probe and primer sets for DSB-ddPCR and the creation of standard curves for quantifying DSBs.

3.1 Design of primers and probes

3.1.1 Retrieval of genomic DNA sequence using UCSC genome browser

- Use the BLAST-like alignment tool (BLAT) from UCSC Genome Browser (https://genome.ucsc.edu/) to retrieve the genomic DNA sequence flanking the sgRNA and PAM sequence of interest.
- 1000 bp of DNA sequence should be retrieved 5′ and 3′ of the sgRNA target and PAM sequence.
- Use of a DNA sequence visualizer, such as SnapGene, is helpful in identifying the sgRNA target site and visualizing where designed probes and primers will anneal.

3.1.2 Setting parameters in IDT PrimerQuest to design primers and probes

- Amplification primers and double-quenched qPCR probes can be designed and ordered using the Integrated DNA Technologies PrimerQuest tool (https://www.idtdna.com/PrimerQuest/Home/Index). Use the qPCR (2 Primers + Probe) option with the Custom Design Parameters.
- Helpful parameters for designing probes and primers can be found in the Bio-Rad ddPCR experiment guidelines (https://www.bio-rad.com/en-us/applications-technologies/planning-droplet-digital-pcr-experiments?ID=MDV33OKG4).
- General parameters for designing probes and primers for both target and template amplicons can be found in Table 1. These parameters assume 1000 bp of sequence flanking the sgRNA target and PAM sequence has been entered into PrimerQuest.

Table 1 Parameters for IDT PrimerQuest to design probes and primers.

	Input	
Parameter	**Target amplicon (FAM dye)**	**Template amplicon (HEX dye)**
Results		
Results to return	Any	Any
Vary forward primer location	Med	Med
Vary reverse primer location	Med	Med
Reaction conditions		
Monovalent salt (Na^+) (mM)	50	50
Primer DNA conc. (nM)	300	300
Divalent salt (Mg^{++}) (mM)	3	3
Probe DNA conc. (nM)	200	200
dNTP conc. (mM)	0.8	0.8
Primer details		
Primer Tm (°C)	Min: 50, Opt: 62, Max: 65	Min: 50, Opt: 62, Max: 65
Primer GC (%)	Min: 50, Opt: 55, Max: 60	Min: 50, Opt: 55, Max: 60
Primer size (nt)	Min: 17, Opt: 22, Max: 30	Min: 17, Opt: 22, Max: 30
3′ GC clamp (nt)	0	0
Probe details		
Probe Tm (°C)	Min: 64, Opt: 68, Max: 72	Min: 64, Opt: 68, Max: 72
Probe GC (%)	Min: 30, Opt: 50, Max: 80	Min: 30, Opt: 50, Max: 80
Probe size (nt)	Min: 20, Opt: 24, Max: 30	Min: 20, Opt: 24, Max: 30
Amplicon details		
Amplicon size (bp)	Min: 125, Opt: 180, Max: 225	Min: 125, Opt: 180, Max: 225
Regions considered		
Included region	N/A	N/A
Excluded region list	N/A	1–1323 An approximation based on downstream placement of template amplicon (see Section 3.1.3)

Table 1 Parameters for IDT PrimerQuest to design probes and primers—cont'd

	Input	
Parameter	**Target amplicon (FAM dye)**	**Template amplicon (HEX dye)**
Excluded region list—probe	N/A	N/A
Primer focus region		
Target region list	1001–1023	1423–1446
	sgRNA target and PAM location	An approximation based on downstream placement of template amplicon (see Section 3.1.3)
Partial design input		
Forward primer input	N/A	N/A
Probe input	Input probe sequence when designing alternative primers (see Section 3.1.3)	Input probe sequence when designing alternative primers (see Section 3.1.3)
Reverse primer input	N/A	N/A
Design across junctions		
Overlap junction list	N/A	N/A
Min overlap on 3′ end of primer	4	4
Min overlap on 5′ end of primer	7	7
Primer force location settings		
Force forward start position (5′ end)	N/A	N/A
Force forward stop position (3′ end)	N/A	N/A
Force reverse start position (5′ end)	N/A	N/A
Force reverse stop position (3′ end)	N/A	N/A

3.1.3 Considerations for probe and primer selection

- Probes should anneal distal to the sgRNA target and PAM site due to formation of indels after DNA repair.
- Template amplicons can be located anywhere 5′ or 3′ from the target amplicon as long as the amplicons do not overlap. The parameter values suggested for "*Excluded region list*" and "*Target region list*" in Table 1 are based on locating the template amplicon approximately 300bp 3′ of the PAM.
- The target amplicon should contain a restriction enzyme site that is not present between the target and template amplicons. This restriction enzyme site will be used to create the standard curve.
- Probes with multiple primer sets should be selected so many alternative amplicons can be tested for efficiency in detecting cut and uncut DNA when troubleshooting.
- Use of Black Hole quenchers with probes are recommended. The FAM dye is used with the target probe and the HEX dye is used with the template probe.

3.2 Extraction of genomic DNA

3.2.1 Equipment

- Dry block incubator at 56 °C
- Centrifuge

3.2.2 Reagents

- Qiagen DNeasy Blood and Tissue kit
- 96–100% ethanol
- Nuclease free water or elution buffer without EDTA
- Dulbecco's Phosphate-Buffered Saline (Thermo Fisher Scientific), DPBS

3.2.3 Protocol

(1) Thaw frozen cell pellets at room temperature for 3min.
(2) Resuspend cell pellet in 200 μL of DPBS.
(3) Add 20 μL of Proteinase K and then 200 μL buffer AL. Invert tube to mix.
(4) Incubate on a dry block incubator at 56 °C for 1h.
(5) Add 200 μL of 96–100% ethanol. Invert tube to mix.

(6) Transfer mixture into a DNeasy spin column with collection tube. Centrifuge at 8000 × *g* for 1 min. Discard flow through and exchange for a new collection tube.
(7) Add 300 μL of buffer AW1. Centrifuge at 8000 × *g* for 1 min. Discard flow through and exchange for a new collection tube.
(8) Add 300 μL of buffer AW2. Centrifuge at 17,000 × *g* for 1 min. Discard flow through.
(9) Move DNeasy spin column into a clean 1.5 mL microcentrifuge tube.
(10) Add 150 μL of nuclease free water or elution buffer without EDTA. Incubate at room temperature for 1 min. Centrifuge at 8000 × *g* for 1 min. Expected concentration ranges are from 30 to 150 ng/μL.

3.2.4 Notes

- DPBS, Proteinase K, and buffer AL can be made together as a master mix. 400 μL of master mix should be added to each cell pellet followed by quick vortexing to disrupt the cell pellet.
- The extended 1 h incubation at 56 °C ensures Proteinase K digestion of ciCas9 from the target sites that could disrupt DSB-ddPCR analysis.
- Buffer AE from the Qiagen DNeasy Blood and Tissue kit should not be used for elution.
- Extracted genomic DNA can be stored at −20 °C until downstream analysis by DSB-ddPCR or high-throughput sequencing.

3.3 Standard curve sample creation

To create a standard curve, extracted genomic DNA is cut with a restriction enzyme within the target amplicon. Select a restriction enzyme that does not cut between the target and template amplicons or within the template amplicon. DNA from an untreated sample should be used to create the standard curve. Cut DNA is mixed with uncut DNA at different ratios to create the standard curve (Fig. 2C).

3.3.1 Equipment

- Incubators at 37 and 80 °C
- Magnetic microcentrifuge tube rack
- Fume hood

3.3.2 Reagents

- Restriction enzyme with associated buffer
- 70% ethanol

- AMPure XP beads (Beckman Coulter)
- Nuclease free water

3.3.3 Protocol

(1) To create cut DNA, make a 35 μL digest reaction containing: 3.5 μL 10 × buffer, 2 μL restriction enzyme, and 29.5 μL DNA. Restriction enzyme and DNA amounts can be adjusted according to restriction enzyme efficiency.

(2) To create the uncut DNA, make a 70 μL reaction containing: 7 μL 10 × buffer, 63 μL DNA. No restriction enzyme should be added.

(3) Incubate both the cut and uncut reactions separately at 37 °C for 20 min. The length of digestion can be adjusted according to restriction enzyme efficiency.

(4) Add 70 μL of AMPure XP beads to the cut DNA reaction and 140 μL of AMPure XP beads to the uncut DNA reaction. Place reactions on the magnetic microcentrifuge tube rack.

(5) Once all of the beads have settled onto the magnet, aspirate and discard the liquid.

(6) Wash both the cut and uncut DNA reactions with 140 μL of 70% ethanol. Repeat.

(7) Place the open tubes in the fume hood for 10 min to allow any residual ethanol to evaporate.

(8) Add 35 μL of nuclease free water to the cut DNA sample and 70 μL of nuclease free water to the uncut DNA sample.

(9) Mix the cut DNA and uncut DNA samples to create 50%, 33%, 25%, 10%, 5%, and 1% cut DNA samples for the standard curve. A total volume of 12.5 μL of each standard is adequate for several DSB-ddPCR runs.

3.3.4 Notes

- Standard curve samples can be frozen at −20 °C for future use.
- Restriction digest volumes suggested here create enough for five standard curve assays.
- See Section 3.5.2 for calculating and plotting the standard curve after droplet analysis.

3.4 DSB-ddPCR to quantify DSBs at a Cas9/ciCas9 targeted locus

3.4.1 Equipment

- Thermal cycler for 96-well plates
- PX1™ PCR plate sealer (Bio-Rad)

- QX200™ Droplet Generator (Bio-Rad)
- QX200™ Droplet Reader (Bio-Rad)
- DG8™ Cartridge Holder (Bio-Rad)

3.4.2 Consumables

- PCR plate heat seal, foil, pierceable (Bio-Rad)
- ddPCR™ 96-well plates (Bio-Rad)
- DG8™ Cartridges for QX200™ Droplet Generator (Bio-Rad)
- DG8™ Gaskets for QX200™ Droplet Generator (Bio-Rad)

3.4.3 Reagents

- ddPCR™ Supermix for Probes (Bio-Rad)
- Droplet generation oil for probes (Bio-Rad)
- ddPCR™ droplet reader oil (Bio-Rad)
- 10 μM forward and reverse primers mix with both target and template primers
- 5 μM target and template probes mix
- Nuclease free water

3.4.4 Mix for a 1× ddPCR reaction

10 μL 2× ddPCR™ Supermix for Probes
1.8 μL Primers mix (10 μM)
1 μL Probes mix (5 μM)
30–150 ng DNA (acceptable range)
Bring to final volume of 20 μL with nuclease free water

3.4.5 Protocol for generating droplets and PCR cycling

(1) Place a DG8™ cartridge into the DG8™ cartridge holder.
(2) Add 50 μL of droplet generation oil for probes into each oil well in the cartridge.
(3) Transfer samples into the sample wells in the cartridge. Make sure there are no bubbles in the sample at the bottom of the well. If there are less than eight samples, fill the remaining wells with 20 μL of water.
(4) Place a DG8™ gasket onto the cartridge and holder.
(5) Place the cartridge and holder into the QX200™ droplet generator. Close the lid to begin droplet generation.
(6) Pipette droplets carefully from cartridge into a ddPCR™ 96-well plate.
(7) Repeat for all samples.

(8) Seal the plate using a PCR plate pierceable foil heat seal in the PX1™ PCR plate sealer. Make sure to take out the plate support block when heating up the plate sealer to avoid heating samples when placing the plate into the support block.
(9) Load the 96-well plate into a thermal cycler.
(10) Run thermal cycler with program:
    10 min at 95 °C
    40 cycles of:
        30 s at 94 °C
        60 s at 60 °C
    10 min at 98 °C
    Hold at 4 °C

3.4.6 Protocol for loading QX200™ droplet reader

(1) Take the sealed 96-well plate from the thermal cycler and place in the metal plate holder for the QX200™ droplet reader.
(2) Open the QuantaSoft™ software and create a new plate layout template.
(3) Select wells with droplets to be analyzed and input the sample name, experiment type (choose "Rare Event Detection"), master mix used (choose "ddPCR Supermix for Probes"), target name(s), target type(s) (choose "Unknown" for both channels), and fluorescent channels (choose Ch1 for FAM and Ch2 for HEX).
(4) Click "Run" to start the droplet reading process.

3.4.7 Notes

- Probe and primer sets should first be tested using uncut and 100% cut DNA samples to ensure adequate detection. Different primer sets should be tested for a single probe.
- Restriction enzymes that cut distal to both amplicons can be used to improve amplification efficiency. Ideally, selected restriction enzyme cut sites should create a fragment approximately 3000 bp in length and neither enzyme should cut within or between the target and template amplicons. If restriction enzymes are used for this purpose, 0.25 μL of each enzyme should be added to each 1 × ddPCR reaction. No additional incubation steps are required.
- Droplets can be stored at 4 °C for 24 h after PCR cycling. Avoid freezing droplets and storing at 4 °C for longer than 24 h.

- More information about the use and setup of the QX200™ droplet reader can be found on the Bio-Rad website (https://www.bio-rad.com/en-us/product/qx200-droplet-digital-pcr-system?ID=MPOQQE4VY).

3.5 Data analysis and DSB quantification

3.5.1 Processing of raw data and gating of droplets

- Droplets are analyzed using the QuantaSoft™ Analysis Pro software. Open the experiment and select wells and fluorescent channels to be analyzed.
- If there is good separation between positive and negative droplets, the fluorescence thresholds are automatically called and gates are automatically drawn. For probes that do not have good separation between positive and negative droplets, use the uncut and 100% cut samples from the standard curve to draw gates (Fig. 2D).
- Data can be exported as a .csv file for DSB calculations.

3.5.2 Calculation and quantification of DSB frequency

- Raw DSB frequency for all samples and a standard curve can be calculated from the number of droplets using the following equation:

$$\frac{[\mathit{target}-,\ \mathit{template}+]}{([\mathit{target}-,\ \mathit{template}+] + [\mathit{target}+,\ \mathit{template}+])}$$

- The standard curve can be created by plotting raw DSB frequencies to expected DSB frequencies and performing a linear regression.
- Absolute DSB frequencies can then be calculated by fitting the raw DSB frequencies to the standard curve.

4. Summary

We have engineered a rapidly-inducible, single-component Cas9 that can be temporally and rheostatically controlled to study Cas9 editing kinetics and dose-dependence *in vivo*. We have found that ciCas9 rapidly generates DSBs and subsequent indels upon activation with the small molecule A115. In addition, we have developed DSB-ddPCR to directly quantify the outcomes of Cas9 activity by measuring DSB generation *in vivo*. The coupling of these methods with high-throughput sequencing allows the kinetics and dose-dependence of CRISPR/Cas9 genome editing at specific loci of interest to be investigated. DSB-ddPCR can also be readily adapted to measure DSB generation in other contexts.

Acknowledgments

We acknowledge J. Rose and J. Stephany for data previously published in Rose, Stephany, Valente, et al. (2017). This research was funded in part by the NIH (grant no. R01GM086858 (D.J.M.) and grant no. R01GM109110 (D.M.F.)).

References

Chen, J. S., Dagdas, Y. S., Kleinstiver, B. P., Welch, M. M., Sousa, A. A., Harrington, L. B., et al. (2017). Enhanced proofreading governs CRISPR–Cas9 targeting accuracy. *Nature*, *550*(7676), 407–410.

Cong, L., Ran, F. A., Cox, D., Lin, S., Barretto, R., Habib, N., et al. (2013). Multiplex genome engineering using CRISPR/Cas systems. *Science*, *339*(6121), 819–823.

Dagdas, Y. S., Chen, J. S., Sternberg, S. H., Doudna, J. A., & Yildiz, A. (2017). A conformational checkpoint between DNA binding and cleavage by CRISPR-Cas9. *Science Advances*, *3*(8), eaao0027.

Davis, K. M., Pattanayak, V., Thompson, D. B., Zuris, J. A., & Liu, D. R. (2015). Small molecule-triggered Cas9 protein with improved genome-editing specificity. *Nature Chemical Biology*, *11*(5), 316–318.

Dow, L. E., Fisher, J., O'Rourke, K. P., Muley, A., Kastenhuber, E. R., Livshits, G., et al. (2015). Inducible in vivo genome editing with CRISPR-Cas9. *Nature Biotechnology*, *33*(4), 390–394.

Gangopadhyay, S. A., Cox, K. J., Manna, D., Lim, D., Maji, B., Zhou, Q., et al. (2019). Precision control of CRISPR-Cas9 using small molecules and light. *Biochemistry*, *58*(4), 234–244.

González, F., Zhu, Z., Shi, Z.-D., Lelli, K., Verma, N., Li, Q. V., et al. (2014). An iCRISPR platform for rapid, multiplexable, and inducible genome editing in human pluripotent stem cells. *Cell Stem Cell*, *15*(2), 215–226.

Jinek, M., Chylinski, K., Fonfara, I., Hauer, M., Doudna, J. A., & Charpentier, E. (2012). A programmable dual-RNA-guided DNA endonuclease in adaptive bacterial immunity. *Science*, *337*(6096), 816–821.

Kleinjan, D. A., Wardrope, C., Sou, S. N., & Rosser, S. J. (2017). Drug-tunable multidimensional synthetic gene control using inducible degron-tagged dCas9 effectors. *Nature Communications*, *8*(1), 1191.

Liu, K. I., Ramli, M. N. B., Woo, C. W. A., Wang, Y., Zhao, T., Zhang, X., et al. (2016). A chemical-inducible CRISPR–Cas9 system for rapid control of genome editing. *Nature Chemical Biology*, *12*(11), 980–987.

Maji, B., Moore, C. L., Zetsche, B., Volz, S. E., Zhang, F., Shoulders, M. D., et al. (2017). Multidimensional chemical control of CRISPR-Cas9. *Nature Chemical Biology*, *13*(1), 9–11.

Mali, P., Yang, L., Esvelt, K. M., Aach, J., Guell, M., Norville, J. E., et al. (2013). RNA-guided human genome engineering via Cas9. *Science*, *339*(6121), 819–823.

Nguyen, D. P., Miyaoka, Y., Gilbert, L. A., Mayerl, S. J., Lee, B. H., Weissman, J. S., et al. (2016). Ligand-binding domains of nuclear receptors facilitate tight control of split CRISPR activity. *Nature Communications*, 7, 12009.

Nishimasu, H., Ran, F. A., Hsu, P. D., Konermann, S., Shehata, S. I., Dohmae, N., et al. (2014). Crystal structure of Cas9 in complex with guide RNA and target DNA. *Cell*, *156*(5), 935–949.

Oakes, B. L., Nadler, D. C., Flamholz, A., Fellmann, C., Staahl, B. T., Doudna, J. A., et al. (2016). Profiling of engineering hotspots identifies an allosteric CRISPR-Cas9 switch. *Nature Biotechnology*, *34*(6), 646–651.

Rose, J. C., Stephany, J. J., Valente, W. J., Trevillian, B. M., Dang, H. V., Bielas, J. H., et al. (2017). Rapidly inducible Cas9 and DSB-ddPCR to probe editing kinetics. *Nature Methods*, *14*(9), 891–896.

Rose, J. C., Stephany, J. J., Wei, C. T., Fowler, D. M., & Maly, D. J. (2017). Rheostatic control of Cas9-mediated DNA double strand break (DSB) generation and genome editing. *ACS Chemical Biology*, *13*(2), 438–442.

Senturk, S., Shirole, N. H., Nowak, D. G., Corbo, V., Pal, D., Vaughan, A., et al. (2017). Rapid and tunable method to temporally control gene editing based on conditional Cas9 stabilization. *Nature Communications*, *8*, 14370.

Sternberg, S. H., LaFrance, B., Kaplan, M., & Doudna, J. A. (2015). Conformational control of DNA target cleavage by CRISPR–Cas9. *Nature*, *527*(7576), 110–113.

Sternberg, S. H., Redding, S., Jinek, M., Greene, E. C., & Doudna, J. A. (2014). DNA interrogation by the CRISPR RNA-guided endonuclease Cas9. *Nature*, *507*(7490), 62–67.

Zetsche, B., Volz, S. E., & Zhang, F. (2015). A split-Cas9 architecture for inducible genome editing and transcription modulation. *Nature Biotechnology*, *33*(2), 139–142.

Zhao, C., Zhao, Y., Zhang, J., Lu, J., Chen, L., Zhang, Y., et al. (2018). HIT-Cas9: A CRISPR/Cas9 genome-editing device under tight and effective drug control. *Molecular Therapy—Nucleic Acids*, *13*, 208–219.

CHAPTER EIGHT

Synthetic receptors to understand and control cellular functions

Hung-Ju Chang, Jerome Bonnet*
Centre de Biochimie Structurale (CBS), INSERM, CNRS, Univ. Montpellier, Montpellier, France
*Corresponding author: e-mail address: jerome.bonnet@inserm.fr

Contents

Abstract

Cellular functions are controlled by sophisticated signal transduction pathways triggered by receptors responding to myriad environmental stimuli. With the rise of synthetic biology, we can now engineer artificial receptors enabling real-time interrogation and manipulation of cellular signaling, and providing new clues about the design principles of natural sensing systems. In this review, we describe the main classes of synthetic receptors engineered to date, their applications, and highlight recent developments that might improve synthetic receptor design in the future.

Methods in Enzymology, Volume 633
ISSN 0076-6879
https://doi.org/10.1016/bs.mie.2019.11.011

1. Introduction

Synthetic biology aims at designing new biological systems and functions to perform specific tasks of interests. Synthetic biology borrows principles from more mature engineering disciplines like civil engineering or electronics, and try to adapt them to the design of artificial biological systems (Endy, 2005). As an example of such an approach, researchers have aimed at designing modular cellular platforms which function can be programmed by composition of standardized genetic components performing specific functions with a well-characterized behavior (Chen & Smolke, 2011). Engineered cellular devices can be triggered by user-defined signals through various sensor modules responding to environmental stimuli, disease related biomarkers, or synthetic ligands. Multiple signals can be integrated by different signal processing modules such as logic gates, amplifiers, or memory devices (Bonnet, Subsoontorn, & Endy, 2012; Bonnet, Yin, Ortiz, Subsoontorn, & Endy, 2013). Specific cellular functions can be implemented using actuator modules such as fluorescent reporters for biosensors or effector proteins and chemical compounds for therapeutic applications (Bermúdez-Humarán & Langella, 2018; Chang, Voyvodic, Zúñiga, & Bonnet, 2017; Roybal et al., 2016a).

Cellular functions are under the tight control of sophisticated signal transduction pathways (Hill, 1998), and a major challenge for cell biologists is to decipher their molecular logic. Traditional top-down approaches usually deconstruct signaling pathways through tuning one, or at most a few, gene expression levels by knockout or overexpression methods (Ladbury & Arold, 2012). This technique provides a straightforward approach to identify the biological components of a given pathway. However, these approaches do not allow the study of signaling dynamics; in addition, interference or compensation effects from closely related pathways can challenge results interpretation.

Synthetic biology provides brand-new avenues for studying cellular functions by designing and building artificial, altered versions of natural signaling pathways using their elementary components. Through iterative rounds of testing and re-design, critical players controlling cellular functions can be identified and the molecular parameters governing their function and their dynamics can be dissected (Bashor & Collins, 2018). In addition, the interplay among different components can be probed by decomposing pathways into functional modules which can be redesigned and assembled in a combinatorial fashion (Bashor & Collins, 2018; Brophy & Voigt, 2014). Finally, engineered genetic circuits enable direct manipulation of the strength, duration, and on/off state of a particular cellular function. Numerous synthetic tools and approaches have been developed to study signaling pathway at the molecular, system, or multicellular level (Mathur, Xiang, & Smolke, 2017).

In that context, cellular receptors are critical integrating nodes for signaling, and their manipulation can provide tremendous insight into the mechanisms underlying cellular functions. To date, many synthetic receptors triggered by light or synthetic ligands have been used to rewire and study cell proliferation, differentiation, phagocytosis, and chemotaxis (Kim et al., 2016). Moreover, engineering artificial receptors allows researchers to dissect the molecular determinants of their functions. Finally, synthetic receptors are key components for engineering biological devices that can address significant challenges in human healthcare and environmental protection.

One field of application for synthetic receptors is biosensing. Biosensors are biological devices that can detect the presence of target analytes, such as biomarkers or environmental pollutants, and produce a user readable output, like an electronic or a colorimetric signal. Many recombinant biosensors use biomolecules that provide highly specific recognition of ligands biological interests, such as enzymes for blood glucose monitoring or antibodies for antigen identification. Living cells can also be used as biosensors. These whole-cell biosensors (WCB), mostly based on genetically engineered microbes are excellent candidates for the development of scalable and affordable biosensing devices for point-of-care diagnosis due to their ability to detect and process various signals, their cost-effectiveness, and their ease of manipulation and storage (Chang et al., 2017). In recent years, synthetic biology has enabled the design of receptors detecting novel ligands, therefore expanding the sensing capabilities of WCB. In addition, synthetic gene networks can improve the signal processing capabilities of WCB by providing signal amplification, multiplexing logic, and long-term memory (Courbet, Endy, Renard, Molina, & Bonnet, 2015).

Another field of application for synthetic receptors is cellular therapy. Mammalian cells have been extensively engineered for the diagnosis or treatment of various pathologies such as infectious or metabolic diseases, and cancer (Lienert, Lohmueller, Garg, & Silver, 2014; Sedlmayer, Aubel, & Fussenegger, 2018). In this context, synthetic receptors enable therapeutic cells to recognize myriad biomarkers or cell surface antigens.

Synthetic receptors can also be used to improve the precision, biosafety, and performance of therapeutic cells. For example, remote control systems operated by small molecules provide therapeutic treatments with tunable timing, strength, duration and improved biosafety (Bojar, Scheller, Hamri, Xie, & Fussenegger, 2018; Wu, Roybal, Puchner, Onuffer, & Lim, 2015). Engineered microbes can also provide new modalities to treat human diseases, such as metabolic disorders (Isabella et al., 2018), infections (Hwang et al., 2017; Hwang, Koh, Kim, Yew, & Chang, 2016; Mao, Cubillos-Ruiz, Cameron, & Collins, 2018), and obesity (Chen et al., 2014). The simultaneous abilities of microbes to detect signals, deliver therapeutic agents, and their close relationships with human health have generated significant interest in engineering the native human microbiota and nonresident probiotic bacteria as diagnosis and therapeutic platforms (Bober, Beisel, & Nair, 2018). To date, there are different synthetic bacteria strains, such as *E. coli* strain Nissle 1917, *L. lactis*, or *B. ovatus,* that have entered different phases of clinical trials (Bermúdez-Humarán & Langella, 2018). In this context, manipulating the sensing capabilities of these therapeutic microbes is of critical importance for advancing the field of engineered microbial therapeutics.

In this review, we describe the design principles of a few synthetic receptor platforms and highlight recent approaches that could have a critical impact on designing the next-generation of synthetic receptors.

2. Current strategies for engineering synthetic receptors

Here we divide the various approaches for engineering cells with novel sensing capabilities into three main strategies: (i) Plugging a synthetic ligand binding domain to a natural signaling pathway; (ii) Connecting a natural sensing domain to a different natural signaling pathway (i.e., rewiring); (iii) Engineering fully modular receptor platforms with swappable sensing and signaling domains.

2.1 New sensing capabilities for natural pathways through synthetic ligand binding domains

2.1.1 Engineering allosteric receptors

The most intuitive strategy to expand the sensing repertoire of a natural receptor is to engineer its ligand-binding site to accommodate new molecules. However, signal transmission from ligand-binding domains (LBDs) to actuator domains often rely on allosteric mechanisms, and changes in the residues responsible for ligand recognition frequently damage allosteric control (Buskirk & Liu, 2005). One way to address this issue is to identify residues that are critical for signal transmission via high-throughput screening and sequencing. For example, the *E. coli* lac repressor LacI was engineered to respond to new ligands through a combination of computational design, single-site saturation mutagenesis, and random mutagenesis (Fig. 1A) (Taylor et al., 2016). The identified functional variants can respond to novel ligand with a specificity and dynamic range comparable to wild-type LacI to its ligand IPTG. Very interestingly, the mutations responsible for the changes in ligand specificity are not only located in the ligand binding pocket, but also in the dimerization interfaces and DNA binding domains. This method allows the engineering of allosteric receptors detecting novel ligands while providing new insights into allosteric control mechanisms.

2.1.2 Engineering orthogonal ligand/LBD pairs

In addition to engineering ligand-binding pockets to accommodate novel ligands, orthogonal ligand-receptor pairs can also be engineered simultaneously. A group of artificial G-protein coupled receptors (GPCRs), called Receptors Activated Solely by Synthetic Ligands (RASSLs), have been engineered to respond exclusively to synthetic small molecule rather than to their natural ligands (Fig. 1B) (Armbruster, Li, Pausch, Herlitze, & Roth, 2007; Conklin et al., 2008; Coward et al., 1998). This system has been applied for externally controlling different cell functions in mice via systemic drug administration (Campbell & Marchant, 2018; Conklin et al., 2008; Smith et al., 2015). This concept has also been applied to engineering orthogonal recognition through protein-protein interactions: an orthogonal interleukin-2 (IL-2)—IL-2 receptor (IL-2R) pair has been engineered through yeast display technology (Sockolosky et al., 2018). The engineered IL-2/IL-2R can interact only with each other and not with endogenous IL-2/IL-2R. The engineered T cells with orthogonal IL-2 receptors can be triggered specifically by orthogonal IL-2 with limited off-target activity and negligible cell toxicity. These synthetic ligand-receptor pairs provide precise and safe switches for remote control of cell function.

2.1.3 Using synthetic sensing domains

Another strategy for conferring new sensing capabilities is to replace natural LBDs by synthetic ones. Single-chain antibodies represent ideal artificial LBDs because of their high stability, good solubility, versatility (recognize

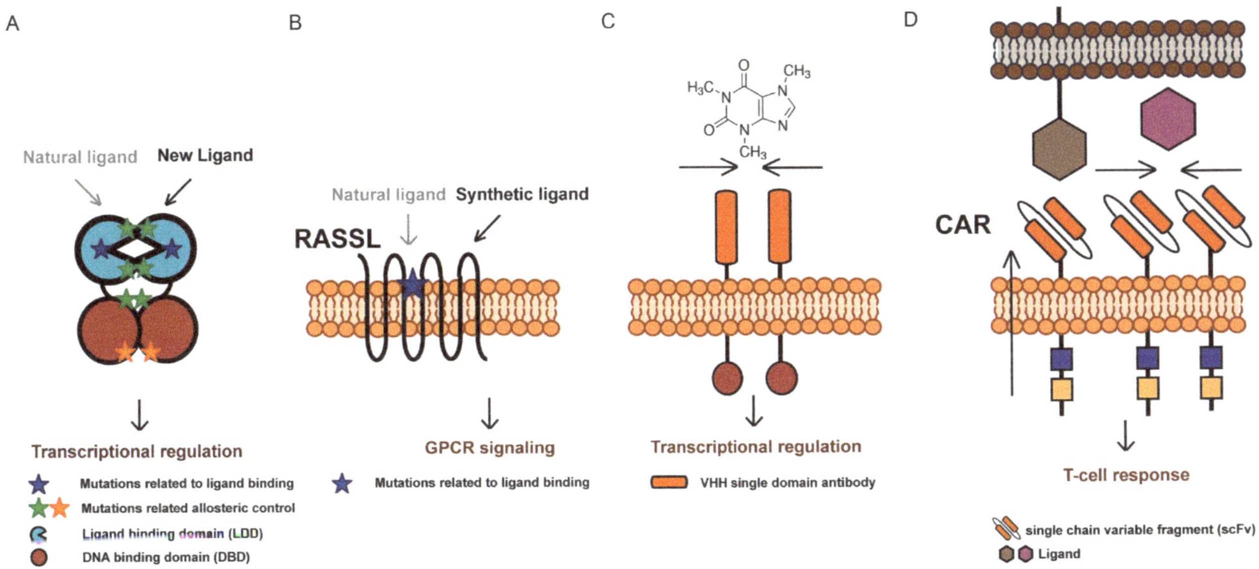

Fig. 1 Strategies to expand the sensing repertoire of natural signaling pathways. (A) Allosteric engineering of transcription factors (TFs). Target specificity of the ligand-binding domain (LBD) of TFs can be altered by mutations in the ligand-binding pockets. However, residues remote from the LBD also need to be mutated to compensate for altered allosteric control. (B) Receptors activated solely by a synthetic ligand (RASSLs). The ligand binding pocket of a GPCR can be engineered to accommodate a novel synthetic ligand, orthogonal to natural GPCR ligands. (C) Using a synthetic ligand-binding domain. The specificity of a synthetic receptor can be changed by fusing it to a novel ligand-sensing module such as caffeine-induced dimerizing VHHs. (D) Chimeric antigen receptor (CAR) T-cells. CARs are composed of an extracellular scFv domain, a transmembrane domain, and an intracellular signaling domain. The activation of CARs can be triggered by a cell surface or a soluble antigen and leads to the activation of downstream T-cell signaling pathways.

proteins or small molecules) and ease to generate new binders by phage-display (Hammers & Stanley, 2014). Recently, a single-domain camelid VHH antibody was connected to different actuator modules such as split-DNA binding domains in bacteria (Fig. 1C) (Chang et al., 2018a) or synthetic transcription factors/receptors in mammalian cells (Bojar et al., 2018), and can control cellular behavior through caffeine-induced dimerization. In addition, high-affinity GFP and mCherry nanobodies were used as extracellular sensing domains to build Synthetic Cytokine Receptors (SyCyRs) which respond to homo- and heterodimeric GFP:mCherry fusion proteins as synthetic ligands to trigger receptor dimerization and activate downstream signaling pathways (Engelowski et al., 2018). The most advanced examples of expanding the sensing repertoire by using an antibody scaffold as sensing module have been provided by Chimeric Antigen Receptors (CARs). CARs are synthetic T-cell receptors that can redirect T cells activity to different cell types through recognition of various cell surface molecules (Sadelain, Brentjens, & Rivière, 2013). CARs use a single-chain variable fragment (scFv) antibody as an extracellular sensing module, connected to an intracellular T-cell signaling domains (Fig. 1D). To date, engineered CAR-T cells can target over 200 cell surface antigens and have been FDA approved to treat leukemia and lymphoma (June, O'Connor, Kawalekar, Ghassemi, & Milone, 2018). Recently, Chang et al. (2018b) have demonstrated that engineered CAR-T cells can also be triggered by diverse soluble ligands, suggesting new applications.

These recent advances highlight the robustness of antibodies as an ideal ligand-sensing domain for different synthetic receptor scaffolds. The binding versatility of antibodies is expected to expand the sensing repertoire to a large variety of ligands.

2.2 Using natural sensing domains and rewiring their output

Rewiring the signal output of natural receptors to artificial signaling pathways can enable direct manipulation of natural sensing capabilities and their use for different applications. The rewiring strategy has been mostly applied to two important receptor families, bacterial two-component systems (TCS) and GPCRs.

2.2.1 Rewiring bacterial two-component systems

Prokaryotic TCSs are ubiquitous among bacteria and sense various kinds of environmental stimuli. They are composed of sensor histidine kinases (HKs) that detect external signals and cognate response regulators (RRs) that modulate downstream gene expression. Two different strategies that have been developed to rewire bacterial two component systems (Fig. 2A):

(I) Bioinformatic analysis have found that the interaction patch between HKs and RRs is located in the HK dimerization and histidine phosphotransfer (DHp) domain (in HK) and the RR receiver (REC) domain (Skerker et al., 2008). Swapping of different DHp domains among different HKs or engineering mutations in DHp domains can switch their specificities toward different RRs without interfering with HK-autophosphorylation activity (Cheng, Morcos, Levine, & Onuchic, 2014; Skerker et al., 2008). Furthermore, orthogonal HKs and RRs pairs can be engineered through analyzing mutational libraries designed based on co-evolution patterns (McClune, Alvarez-Buylla, Voigt, & Laub, 2019). These results indicate that the specificity of HKs is modular and can be re-engineered to connect them to different downstream signaling pathways.

(II) The specificity of TCSs can also be changed by fusing them to synthetic protein–protein interaction modules (Whitaker, Davis, Arkin, & Dueber, 2012); the HK and RR were fused to an SH3-domain and different kind of leucine zippers, respectively. The specificity of fusing HK and RR can be directed by an additional synthetic scaffold composed of an SH3-ligand domain and a corresponding leucine zipper. Rewiring signal transmission pathways can thus be performed by using synthetic scaffolds even without a comprehensive understanding of their inner interaction mechanisms.

2.2.2 Rewiring GPCR signaling: The Tango system

GPCRs constitute a large eukaryotic receptor family detecting many physiological and environmental signals, and are targets of choice for a large number of drugs available on the market (Hauser, Attwood, Rask-Andersen, Schiöth, & Gloriam, 2017). Different protein engineering approaches have been applied to redesign the function or stability of different GPCRs (Keri & Barth, 2018). Besides mutagenesis or directed evolution, fusing GPCRs with functional proteins for rewiring signal output has emerged as a strategy of choice: upon sensing extracellular stimuli, GPCRs with interact and activate different subsets of transmitter proteins, such as arrestins. The synthetic Tango-GPCR system (Fig. 2B) uses the GPCR-arrestin interaction to bring a TEV protease and TEV-cutting site-linked transcription factor into close proximity, and the subsequent proteo-cleavage releases the transcription factor from membrane to enter

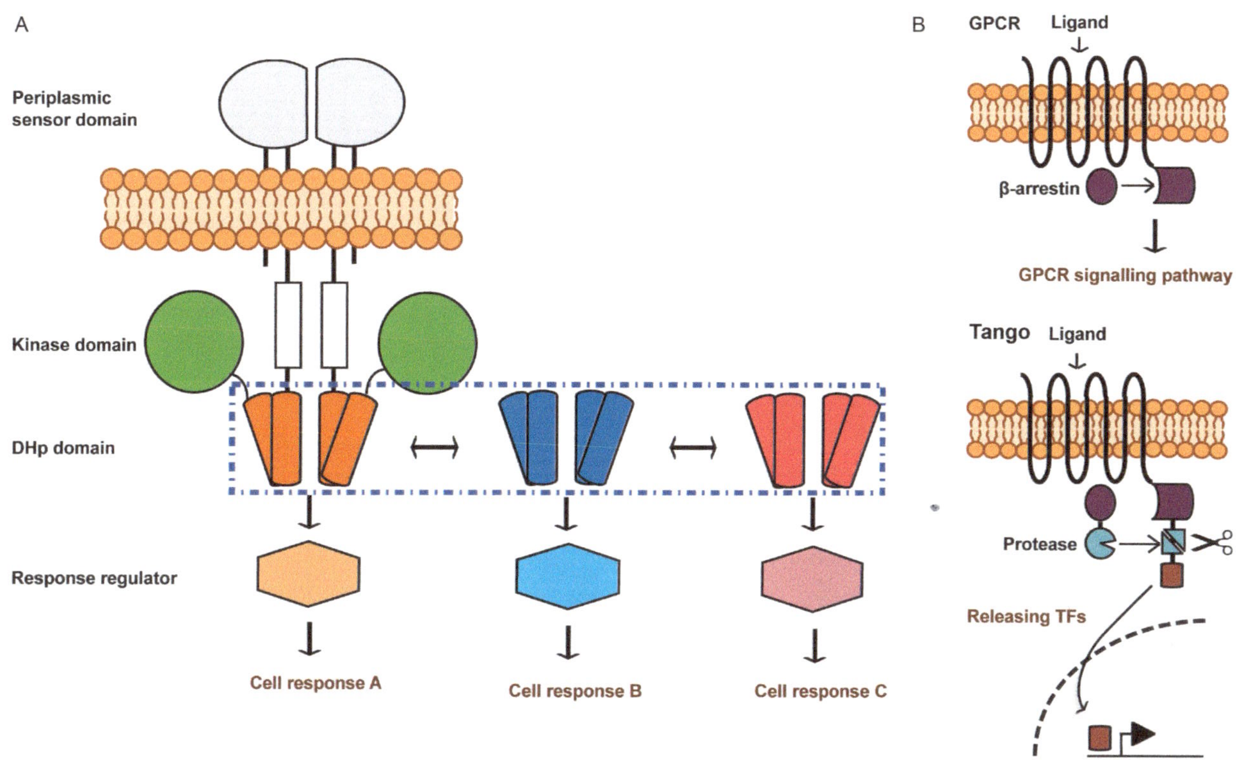

Fig. 2 Rewiring signal output. (A) Rewiring bacterial two-component systems. The signal output of TCS can be rewired by switching the dimerization and histidine phosphorylation domain (DHp) domains or fusing the C-terminal to the SH3 domain. (B) Tango reporter system. The tetracycline transactivator protein, tTA is linked to the C-terminal of a GPCR with a flexible linker containing a TEV protease cutting site. The TEV protease is linked to β-arrestin. When the ligand binds to the GPCR, the TEV protease and its cutting site are brought in proximity by the β-arrestin-GPCR interaction. Therefore tTA will be released from the membrane and activate downstream tTA-dependent reporter genes.

the nucleus and activate reporter genes (Barnea et al., 2008). This concept has also been applied for rewiring the signal output of receptor tyrosine kinases and steroid hormone receptors.

2.3 Scalable, modular synthetic receptors platforms with multiple inputs and outputs

Despite the recent advances in expanding receptors sensing repertoires and rewiring signaling output pathways, the development of fully modularized synthetic receptor scaffolds with the capability of arbitrarily assemble different sensor and actuator modules is still critical. Such platforms would support the development of programmable cells independently of the cell type and interference from endogenous signaling pathways, while enabling multiplex signal detection and processing. Several modular receptor platforms have been developed recently and are presented below.

2.3.1 SynNotch

Synthetic Notch (SynNotch) receptors are designed from the core regulatory domain of mechanosensitive Notch receptors (Chowdhury et al., 2016; Gordon et al., 2015; Kovall & Blacklow, 2010), using artificial sensor and transcriptional modules (Fig. 3A) (Morsut et al., 2016). These modular receptors can be activated when their synthetic sensor domain (e.g., a single-chain antibody) binds to antigens present on target cell surfaces. The mechanical force generated from cell–cell interaction triggers proteo-cleavage of the SynNotch receptor close to the transmembrane domain and releases the intracellular transcriptional module from the cell membrane. The transcription factor can then enter the nucleus and activate its target promoter. SynNotch receptors have been applied in various mammalian cell types (Morsut et al., 2016), especially to engineer T-cells with customized therapeutic behaviors (Roybal et al., 2016a). For example, T-cells cytotoxic activity can be tightly controlled by a two-receptor AND-gate using a SynNotch and a CAR (Roybal et al., 2016b). Interaction of the SynNotch receptor with one antigen triggers the expression of the CAR for targeting the second antigen, activating the T-cell response. In addition, programmed SynNotch-based intercellular signaling networks have been used to generate self-organizing multicellular structures resembling the ones observed during the early stages of organismal development (Johnson, March, & Morsut, 2017; Toda, Blauch, Tang, Morsut, & Lim, 2018). These studies suggest that the SynNotch platform could be useful for studying the first step of development as well as for tissue engineering applications.

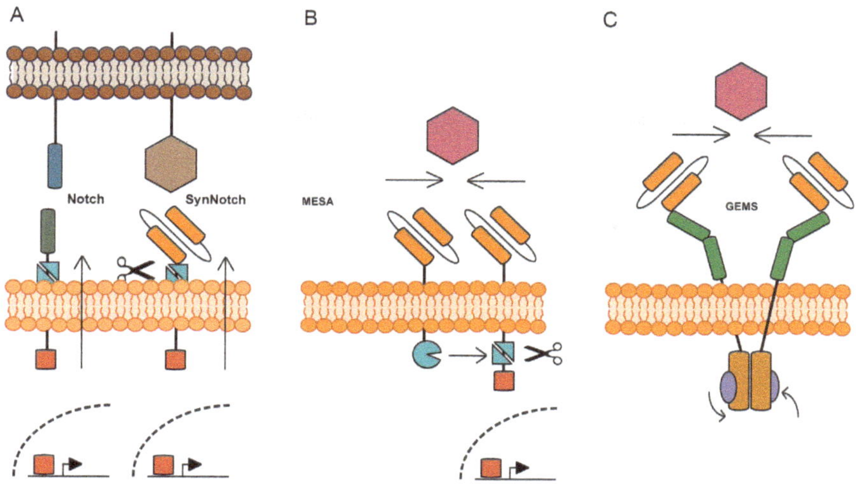

Fig. 3 Modular synthetic receptor platforms. (A) Synthetic Notch receptor (SynNotch). SynNotchs are ideal modular receptor platforms due to their highly switchable extracellular-sensing and intracellular-transcriptional modules. The interaction between SynNotch and cell-surface ligands triggers the intramembrane proteolysis and release of the actuator module from membrane. (B) Modular Extracellular Sensor Architecture (MESA). MESA contains two parts: a target chain (TC) with TEV protease cutting site-linked transcription factor (TF) and a protease chain (PC) fused to a TEV protease. The ligand binding will bring the TC and PC into close proximity and release the TF from membrane. (C) Generalized Extracellular Molecule Sensors (GEMS). The GEMS systems are based on the EpoR receptor, which forms an inactive dimer. Ligand binding triggers rotation of receptor subunits into an active state.

2.3.2 Modular extracellular sensor architecture (MESA)

Much like in the Tango system, the activation mechanism of receptors from the MESA family is based on ligand-induced receptor dimerization and proteolysis. MESA receptors are composed of two chains: the protease chain fused to a protease (e.g., TEV protease) and the target chain fused to the actuator module with a corresponding protease cleavage site. Both chains are connected to an extracellular scFv as sensing module. Dimerization of the sensing domain brings the protease in close proximity with its cognate target site and releases the actuator from the membrane, allowing it to transit to the nucleus to drive transcription (Fig. 3B) (Daringer, Dudek, Schwarz, & Leonard, 2014; Schwarz, Daringer, Dolberg, & Leonard, 2016). The actuator can be a transcription factor (Daringer et al., 2014) or a catalytically dead Cas9 protein (dCas9) fused to a transcriptional activation domain (Schwarz et al., 2016). Engineered human T cells with the MESA system can respond to the immunosuppressive signal vascular endothelial growth factor (VEGF) and produce the immunostimulatory factor IL-2 as a signal output (Schwarz et al., 2016).

2.3.3 Generalized extracellular molecule sensor (GEMS) platform

The design of the GEMS platform (Scheller, Strittmatter, Fuchs, Bojar, & Fussenegger, 2018) is based on erythropoietin (Epo) receptor scaffold. It can be adapted to various kinds of LBDs (e.g., scFv or VHH) and connected to at least four different endogenous signaling pathways (JAK/STAT, ERK/MAPK, phospholipase C (PLC) and PI3K/Akt) to drive user-defined signal outputs (Fig. 3C). Unlike MESA, transmembrane helices interactions keep the ligand-free form of GEMS in an inactive dimeric state. Upon ligand binding, each subunit of the dimeric synthetic receptor rotates around its own axis and triggers downstream signal activation. This stringent activation process reduces basal activity in the absence of ligand. Using natural kinase signaling cascades as signal outputs provides further signal amplification and leads to excellent signal-to-noise (S/N) ratios. The GEMS system can detect the presence of nicotine and prostate-specific antigen (PSA) at clinically relevant concentration.

3. Strategies to optimize synthetic receptors behaviors

Synthetic receptors should ideally have a low basal activity in the absence of ligand, and a high-fold change in reporter gene expression upon ligand binding. Different parameters can be tuned to achieve this goal, like the level of protein expression or receptors amino-acid sequence. A few examples are discussed below.

3.1 Next-generation Tango: The ChaCha system

In the original Tango design, the TEV protease and actuator module (tetracycline transactivator protein—tTa) with its TEV protease cleavage site were fused to an adaptor beta-arrestin and the C-terminus of a GPCR, respectively. However, the fusion of an actuator module to the C-terminus of a GPCR leads to "one ligand in, one actuator out" behavior, which limits the potential efficiency of one receptor molecule. Kipniss et al. (2017) designed a new system named ChaCha which differs from Tango in several manners (Fig. 4A): (i) Protease and actuator are inverted in Chacha (the protease is fused to the GPCR, and the actuator to arrestin). Consequently, the system output can be tuned in a more precise manner to avoid non-specific arrestin-GPCR interactions while conserving a sufficient enzymatic activity, leading better S/N ratio (from 1.2- to 4.4-fold); (ii) The protease cutting site used in Chacha is less efficiently cleaved by the protease, reducing non-specific cleavage and improving S/N ratio;

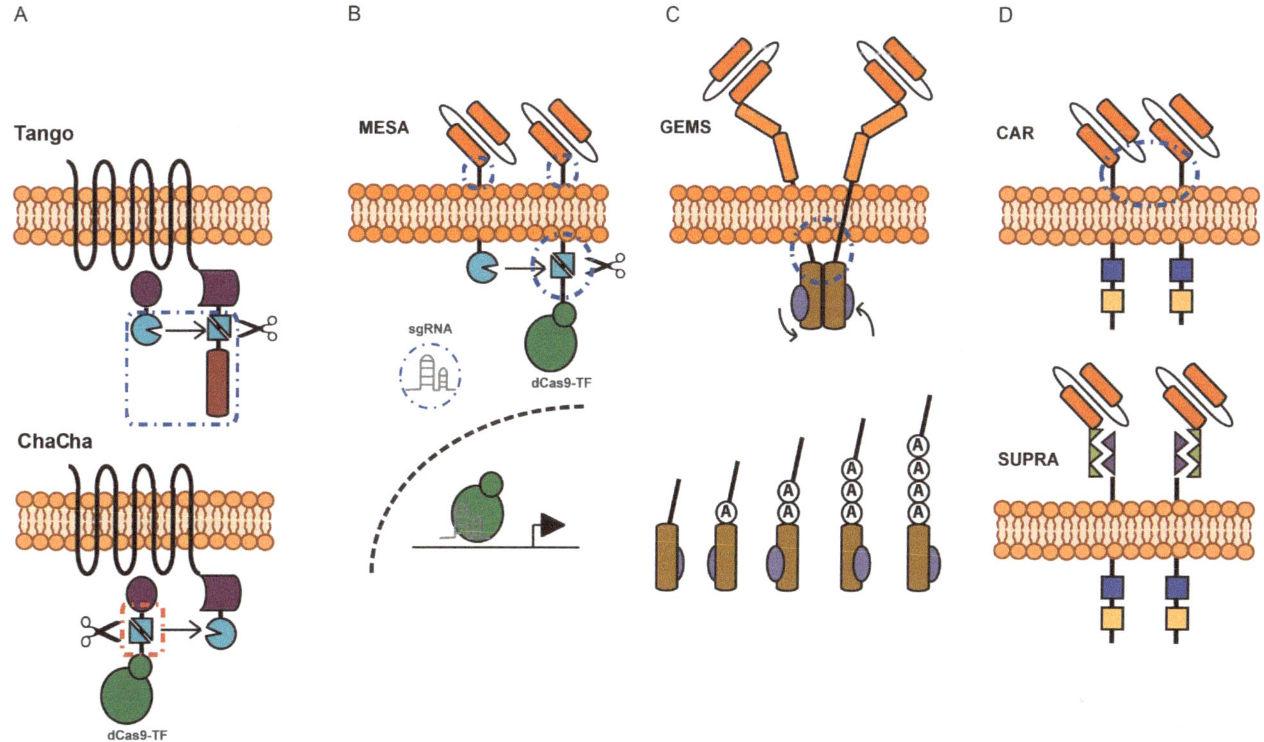

Fig. 4 Strategies for optimizing synthetic receptors behavior. (A) The CRISPR ChaCha system is an upgraded version of the Tango system. Switching the fusion partner of the protease to dCas9-TF provides to the ChaCha system a "one ligand in, multiple effectors out" feature. The use of the CRISPR-dCas9 system also provides flexibility in genome regulation. (B) Fine tuning of the MESA system is achieved by reducing the noise of the signal output by (i) using a non-optimal protease cutting site, reducing non-specific proteolysis and (ii) modifying the sgRNA sequences for optimized dCas9-TF gene regulation activity. (C) Fine tuning of the S/N ratio in the GEMS system is performed through the insertion of alanine residues in the juxtamembrane linker region to modify the rotation angles of each subunit. (D) The split, universal and programmable (SUPRA) CAR system is composed of a zipCAR (extracellular leucine zipper as sensing domain) and a zipFv (scFv fused to a cognate leucine zipper). The SUPRA system has various advantages, such as target switching, tuning signal strength, and logic processing of multiple input signals, without needing to re-engineer the whole T cell.

(iii) The actuator module was replaced in the ChaCha-GPCR format by a dCas9, improving output flexibility.

3.2 Optimizing expression levels

Even though the design concept of MESA—ligand-induced receptor dimerization is a simple and intuitive mechanism for synthetic receptor design, it still has a major limitation: controlling receptor expression levels is critical for system performance. In particular, overexpression of synthetic receptors leads to self-dimerization and therefore increases basal activity in the absence of ligand (Chang et al., 2018a). In addition, the type of ligand-binding domain could also greatly affect receptor behavior. For example, overexpression of an scFv molecule can lead to the formation of a dimer or higher-order of oligomers due to a phenomenon called "domain swapping" (Wörn & Plückthun, 2001). However, reducing the receptor expression level will also reduce the final signal output. MESA also adopted different strategies to optimize system performance (Daringer et al., 2014): similarly to the Tango and ChaCha system, MESA S/N ratio can be improved by (i) reducing proteo-cleavage efficiency (ii) changing the extra-cellular linker length; and (iii) tuning the signal output by downstream signal processing (Fig. 4B).

3.3 Optimizing behavior by varying linker regions

Unlike MESA, which uses ligand-induced receptor dimerization to bring two receptor subunits into close proximity, the dimerization of the GEMS ligand-sensing domain induces a rotation of each subunit from an inactive dimeric receptor to another orientation that activate downstream signaling. Insertion of alanine residues between the transmembrane domain and intracellular signaling domain influences the basal expression level and S/N ratio by modifying the rotation angle of the receptor subunit (100 degrees per alanine). In addition, different combinations of sensing domains and intracellular signaling domains will also influence the S/N ratio and ligand sensitivity (Fig. 4C).

3.4 Strategies for CARs optimization

The spacer regions of CARs connect the extracellular scFv to the transmembrane region and intracellular signal regulation domain. The length and sequence composition of the spacer regions have a great influence on CARs response to cell surface antigens (Castellarin, Watanabe, June,

Kloss, & Posey, 2018; Watanabe et al., 2016) or soluble ligands (Chang et al., 2018b). Mutational analysis of spacer sequences would be an effective, but challenging strategy to perform on a mammalian cell platform. Cho et al. designed a novel platform called the SUPRA (split, universal, and programmable) CAR system to bypass this challenge (Fig. 4D) (Chen, 2018; Cho, Collins, & Wong, 2018). SUPRA consists of two components: (i) a universal receptor, zipCAR, which is generated by fusing a leucine zipper to an extracellular domain and (ii) a tumor cell targeting scFv fused to another leucine zipper (zipFv). The zipFv binds to tumor surface antigen and activates zipCAR through leucine zipper interaction. SUPRA is not only able to fine-tune T-cell activation strength to reduce cell toxicity, but also able to use logic processing from multiple input signals. The fine tuning of the target specificity, signal strength, and multi-signal logic processing is achieved by changing the type, amount, and combination of isolated zipFv without further engineering the universal zipCAR.

4. Advantages and limitations of various receptor activation mechanisms

In summary, signal activation mechanisms applied to synthetic receptor design include mechanical forces, ligand-induced dimerization, and allosteric control.

The SynNotch design based mechanical force provides a fully modularized synthetic receptor design platform in which different kinds of sensor and actuator modules can be arbitrarily assembled. However, the SynNotch receptors can only be activated by cell surface targets and not by soluble ligands. The activation mechanism of CARs seems to be diverse: CAR has been shown to be activated through mechanical force (Hu & Butte, 2016; Ma & Finkel, 2010) as Notch receptors (Chowdhury et al., 2016) but also through soluble ligand-induced dimerization (Chang et al., 2018b). Further studies of this mechanism are required.

Ligand-induced receptor dimerization is the most common mechanism for synthetic receptor design. Chemically-inducible dimerization (CID) by well-known ligands such as rapamycin has been applied to illuminate numerous cellular signaling pathways by bringing proteins of interest into close proximity, membrane cross-linking, translocation and remote control pathway activity (DeRose, Miyamoto, & Inoue, 2013; Stanton, Chory, & Crabtree, 2018). Besides chemical compounds, receptors can be also triggered by soluble proteins (dimeric or monomeric with two epitopes) or cell surface multivalent antigens. In addition, different split-proteins can be used as an actuator module, such as split-T7 RNAP (Pu, Zinkus-Boltz, & Dickinson, 2017), split-Cas9 (Baeumler, Ahmed, & Fulga, 2017), split enzymes (Guo et al., 2016), or split fluorescent proteins (Cabantous et al., 2013). Recently, a human antibody-based chemically induced dimerizer system (AbCID) (Hill, Martinko, Nguyen, & Wells, 2018) has shown great potential to expand the sensing repertoire of dimerization-based receptor design against chemical compounds, such as synthetic antibodies targeting small molecule–protein complexes selected from phage display libraries. The versatile ligand-binding specificity of antibodies will drive the development of novel ligand-sensing modules.

However, self-activation due to receptor overexpression is still a big issue that needs to be solved in dimerization-based designs. Several strategies have been applied for improving the performance of dimerization-based systems, including fine tuning receptor expression levels, modifying linker length or sequence composition, tuning down the actuator activity (e.g., using a protease cutting site with lower cleavage propensity in Tango/ChaCha or MESA). In addition, optimizing sensor and actuator modules for high-solubility could limit the receptor propensity to aggregate in the absence of ligand.

The development of ligand-induced stabilization (Feng et al., 2015) and allosterically controlled synthetic receptor designs are based on protein engineering of functional variants by mutational scanning methods (Brandsen, Mattheisen, Noel, & Fields, 2018; Feng et al., 2015; Taylor et al., 2016). In addition to supporting the development of novel synthetic receptors, in depth exploration of protein fitness landscapes (Romero & Arnold, 2009) will help acquire further fundamental knowledge about protein folding and allosteric control mechanisms.

5. Perspectives

Here we have reviewed the current state of the art in synthetic receptor engineering. The different systems are collectively summarized in Table 1 along with their individual characteristics. To conclude, we will highlight future research directions that we believe to be critical for advancing the field of receptor engineering.

Table 1 Comparison of different synthetic receptor scaffold.

System	Species	Mechanism	Modularity	Output control	Sensitivity	Fold change	Stage	Reference
GEMS	Mammalian cell	Dimerization → Rotation	Input/output	Transcription	1 ng/mL to PSA	~35-fold to 40 ng/mL PSA	Lab	Scheller et al. (2018)
MESA	Mammalian cell	Dimerization	Input/output	Transcription	>25 ng/mL to VEGF	~5-fold to 250 ng/mL VEGF	Lab	Daringer et al. (2014), Schwarz et al. (2016)
SynNotch	Mammalian cell	Mechanical force	Input/output	Transcription	Cell surface antigen	5–20-fold	Lab	Morsut et al. (2016), Roybal et al. (2016a, 2016b), Toda et al. (2018)
Tango/ChaCha	Mammalian cell	Rewiring signal output by GPCR/arrestin interaction	Output	Transcription activity	~10 nM	5–10-fold	Lab	Barnea et al. (2008), Kipniss et al. (2017)
Synthetic bacteria two-component system	Bacteria	Rewiring signal output	Output	Transcription	Based on WT sensor	5–10-fold	Lab	Skerker et al. (2008), Whitaker et al. (2012)

Continued

Table 1 Comparison of different synthetic receptor scaffold.—cont'd

System	Species	Mechanism	Modularity	Output control	Sensitivity	Fold change	Stage	Reference
CAR	Mammalian cell	Mechanical force/Ligand induced dimerization	Input	T-cell signaling pathway	Cell surface target; ~5 ng/mL for soluble ligand		FDA approved for leukemia and lymphoma	Sadelain et al. (2013), June et al. (2018), Chang et al. (2018b)
Rewiring ligand sensing module (GFP/mCherry antibody)	Mammalian cell	Ligand induced dimerization	Input	Transcription	10–100 ng/mL	5–20-fold	Mice	Engelowski et al. (2018)
Rewiring ligand sensing module (VHH-Caffeine)	Bacteria, Mammalian cell	Ligand induced dimerization	Input	Transcription	0.1 μM	5–20-fold to 100 μM Caffeine	Lab	Chang et al. (2018a), Bojar et al. (2018)

RASSL	Mammalian cell	Engineering ligand binding pocket	Input	GPCR pathway	~ 5 nM	Alternated ligand specificity ~100-fold	Lab	Coward et al. (1998), Armbruster et al. (2007), Conklin et al. (2008), Smith et al. (2015), Campbell and Marchant (2018)
Allosteric engineering of transcription factor	Bacteria	Engineering ligand binding pocket and allosteric control mechanism	Input	Transcription activity regulated by TF	1–100 mM	>10-fold alternated ligand specificity at 10–100 mM ligand concentration	Lab	Taylor et al. (2016)

5.1 Studying signal transmission at the molecular level

Four different types of cross-membrane signal transmission mechanisms have been found in nature: dimerization, rotation, piston, and scissors (Matthews, Zoonens, & Engelman, 2006). To date, ligand-induced dimerization is still the most widely applied concept in synthetic receptor design. Further development of natural receptor scaffolds with different activation mechanisms will contribute to the expansion of the synthetic receptor toolbox. For example, the use of EpoR (which converts dimerization force into receptor rotation) in the GEMS system shows good S/N ratio and high modularity for different arbitrarily assembled sensors and actuator modules (Scheller et al., 2018). While detailed molecular-level studies of cytosolic transcription factors, such as the LacI repressor (Taylor et al., 2016), have had success, our knowledge about cross-membrane signal transmission is still limited. For instance, the length and sequence composition of the juxtamembrane linker of the *E. coli* pH stress sensor CadC are critical for its signal output performance and response to inputs (Buchner, Schlundt, Lassak, Sattler, & Jung, 2015; Chang et al., 2018a). However, the molecular determinants of this regulation are still ambiguous. In addition, linker regions in membrane receptors have a high structural flexibility and sequence diversity, complicating the study of signal transmission mechanisms by traditional structural biology methods. With current developments in synthetic biology, it is possible to design and construct novel synthetic receptor platforms to study their signal transmission mechanisms through directed evolution and high-throughput screening methods. These approaches can provide new tools to reveal the functional determinants and sequence-function relationship of cellular receptors.

In addition to signal transmission within individual receptor molecules, the interaction between receptors and other components of the signaling pathway is also critical to receptor performance. Most of current synthetic receptor designs are based on the assembly of natural components with limited optimization at the molecular level, mostly due to the limitation of library construction and screening in mammalian systems. By using yeast mating receptor Ste2 as a model system, Di Roberto et al. not only engineered a GPCR to improve its sensitivity through molecular evolution, but also got further insight into the underlying GPCR signaling networks (Di Roberto, Chang, Trusina, & Peisajovich, 2016). McClune et al. engineered orthogonal signaling pathways based on two-component system through mutational analysis of the interaction between HKs and RRs (McClune et al., 2019) Therefore, reconstruction of synthetic signaling pathways in higher transformation efficiency species such as yeast or *E. coli* can help decipher and optimize the interactions between its components.

5.2 Advancing the engineering of post-translational signaling circuits

Current synthetic receptor designs mostly use transcriptional regulation as an output. However, post-translational modification (PTM) is also an important control mechanism in natural signal transmission network (Salvesen, Hempel, & Coll, 2016). Synthetic signaling circuits based on PTM have been developed in recent years. Fernandez-Rodriguez et al. have built synthetic networks controlled by proteases and degrons (Fernandez-Rodriguez & Voigt, 2016). The group of Wendell Lim has pioneered the engineering of phospho-regulons to control cellular behavior (Gordley et al., 2016). Stein et al. developed a synthetic sensing and signal amplification platform based on proximity-induced activation of autoinhibited proteases (Stein & Alexandrov, 2014). These programmable protein circuits are majorly composed of various proteases or split-proteases, degrons and designed coiled coils (Fink et al., 2019; Gao, Chong, Kim, & Elowitz, 2018), and can implement Boolean logic functions in mammalian cells. Importantly, protease based genetic circuits are able to perform signal processing within minutes rather than hours, which further emphasize the advantage of designing genetic circuits based on PTM regulation (Fink et al., 2019). Further engineering of modular components with predictable composition properties will expand and strengthen PTM based genetic circuit design (Chassin et al., 2019).

To conclude, the field of synthetic receptors has greatly expanded in recent years, and has already provided powerful toolkits for interrogating and controlling cellular signaling with unprecedented precision and scale. Yet, the field is still in its infancy, and much work is needed to achieve a deeper molecular understanding of signaling and to define rational design processes for engineering synthetic receptors with predictable properties. While the task is daunting, sustained efforts of researchers building on continuous advances in synthetic biology will undoubtedly address these challenges in the near future.

Acknowledgments

We thank P. Voyvodic for providing feedback on this manuscript. We thank members of the synthetic biology and the CBS for constant, stimulating discussions. This work was supported by an ERC grant "COMPUCELL" to J.B. J.B also acknowledges continuous support from the INSERM and the Bettencourt-Schueller Foundation. The CBS acknowledges support from the French Infrastructure for Integrated Structural Biology (FRISBI) ANR-10-INSB-05-01.

References

Armbruster, B. N., Li, X., Pausch, M. H., Herlitze, S., & Roth, B. L. (2007). Evolving the lock to fit the key to create a family of G protein-coupled receptors potently activated by an inert ligand. *Proceedings of the National Academy of Sciences of the United States of America*, *104*, 5163–5168.

Baeumler, T. A., Ahmed, A. A., & Fulga, T. A. (2017). Engineering synthetic signaling pathways with programmable dCas9-based chimeric receptors. *Cell Reports*, *20*, 2639–2653.

Barnea, G., et al. (2008). The genetic design of signaling cascades to record receptor activation. *Proceedings of the National Academy of Sciences of the United States of America*, *105*, 64–69.

Bashor, C. J., & Collins, J. J. (2018). Understanding biological regulation through synthetic biology. *Annual Review of Biophysics*, *47*, 399–423.

Bermúdez-Humarán, L. G., & Langella, P. (2018). Live bacterial biotherapeutics in the clinic. *Nature Biotechnology*, *36*, 816–818.

Bober, J. R., Beisel, C. L., & Nair, N. U. (2018). Synthetic biology approaches to engineer probiotics and members of the human microbiota for biomedical applications. *Annual Review of Biomedical Engineering*, *20*, 277–300.

Bojar, D., Scheller, L., Hamri, G. C.-E., Xie, M., & Fussenegger, M. (2018). Caffeine-inducible gene switches controlling experimental diabetes. *Nature Communications*, *9*, 2318.

Bonnet, J., Subsoontorn, P., & Endy, D. (2012). Rewritable digital data storage in live cells via engineered control of recombination directionality. *Proceedings of the National Academy of Sciences of the United States of America*, *109*, 8884–8889.

Bonnet, J., Yin, P., Ortiz, M. E., Subsoontorn, P., & Endy, D. (2013). Amplifying genetic logic gates. *Science*, *340*, 599–603.

Brandsen, B. M., Mattheisen, J. M., Noel, T., & Fields, S. (2018). A biosensor strategy for E. coli based on ligand-dependent stabilization. *ACS Synthetic Biology*, 7, 1990–1999.

Brophy, J. A. N., & Voigt, C. A. (2014). Principles of genetic circuit design. *Nature Methods*, *11*, 508–520.

Buchner, S., Schlundt, A., Lassak, J., Sattler, M., & Jung, K. (2015). Structural and functional analysis of the signal-transducing linker in the pH-responsive one-component system CadC of Escherichia coli. *Journal of Molecular Biology*, *427*, 2548–2561.

Buskirk, A. R., & Liu, D. R. (2005). Creating small-molecule-dependent switches to modulate biological functions. *Chemistry & Biology*, *12*, 151–161.

Cabantous, S., et al. (2013). A new protein-protein interaction sensor based on tripartite split-GFP association. *Scientific Reports*, *3*, 2854.

Campbell, E. J., & Marchant, N. J. (2018). The use of chemogenetics in behavioural neuroscience: Receptor variants, targeting approaches and caveats. *British Journal of Pharmacology*, *175*, 994–1003.

Castellarin, M., Watanabe, K., June, C. H., Kloss, C. C., & Posey, A. D., Jr. (2018). Driving cars to the clinic for solid tumors. *Gene Therapy*, *25*, 165–175.

Chang, H.-J., Voyvodic, P. L., Zúñiga, A., & Bonnet, J. (2017). Microbially derived biosensors for diagnosis, monitoring and epidemiology. *Microbial Biotechnology, 10*, 1031–1035.

Chang, H.-J., et al. (2018a). A modular receptor platform to expand the sensing repertoire of bacteria. *ACS Synthetic Biology*, 7(1), 166–175. https://doi.org/10.1021/acssynbio.7b00266.

Chang, Z. L., et al. (2018b). Rewiring T-cell responses to soluble factors with chimeric antigen receptors. *Nature Chemical Biology, 14*, 317–324.

Chassin, H., et al. (2019). A modular degron library for synthetic circuits in mammalian cells. *Nature Communications, 10*, 2013.

Chen, Y. Y. (2018). Increasing T cell versatility with SUPRA CARs. *Cell, 173*, 1316–1317.

Chen, Y. Y., & Smolke, C. D. (2011). From DNA to targeted therapeutics: Bringing synthetic biology to the clinic. *Science Translational Medicine, 3*, 106ps42.

Chen, Z., et al. (2014). Incorporation of therapeutically modified bacteria into gut microbiota inhibits obesity. *The Journal of Clinical Investigation, 124*, 3391–3406.

Cheng, R. R., Morcos, F., Levine, H., & Onuchic, J. N. (2014). Toward rationally redesigning bacterial two-component signaling systems using coevolutionary information. *Proceedings of the National Academy of Sciences of the United States of America, 111*, E563–E571.

Cho, J. H., Collins, J. J., & Wong, W. W. (2018). Universal chimeric antigen receptors for multiplexed and logical control of T cell responses. *Cell, 173*, 1426–1438.e11.

Chowdhury, F., et al. (2016). Defining single molecular forces required for Notch activation using nano yoyo. *Nano Letters, 16*, 3892–3897.

Conklin, B. R., et al. (2008). Engineering GPCR signaling pathways with RASSLs. *Nature Methods, 5*, 673–678.

Courbet, A., Endy, D., Renard, E., Molina, F., & Bonnet, J. (2015). Detection of pathological biomarkers in human clinical samples via amplifying genetic switches and logic gates. *Science Translational Medicine*, 7, 289ra83.

Coward, P., et al. (1998). Controlling signaling with a specifically designed G_i-coupled receptor. *Proceedings of the National Academy of Sciences of the United States of America, 95*, 352–357.

Daringer, N. M., Dudek, R. M., Schwarz, K. A., & Leonard, J. N. (2014). Modular extracellular sensor architecture for engineering mammalian cell-based devices. *ACS Synthetic Biology, 3*, 892–902.

DeRose, R., Miyamoto, T., & Inoue, T. (2013). Manipulating signaling at will: Chemically-inducible dimerization (CID) techniques resolve problems in cell biology. *Pflügers Archiv, 465*, 409–417.

Di Roberto, R. B., Chang, B., Trusina, A., & Peisajovich, S. G. (2016). Evolution of a G protein-coupled receptor response by mutations in regulatory network interactions. *Nature Communications*, 7, 12344.

Endy, D. (2005). Foundations for engineering biology. *Nature, 438*, 449–453.

Engelowski, E., et al. (2018). Synthetic cytokine receptors transmit biological signals using artificial ligands. *Nature Communications, 9*, 2034.

Feng, J., et al. (2015). A general strategy to construct small molecule biosensors in eukaryotes. *Elife4*. e10606, https://doi.org/10.7554/eLife.10606.

Fernandez-Rodriguez, J., & Voigt, C. A. (2016). Post-translational control of genetic circuits using Potyvirus proteases. *Nucleic Acids Research, 44*, 6493–6502.

Fink, T., et al. (2019). Design of fast proteolysis-based signaling and logic circuits in mammalian cells. *Nature Chemical Biology, 15*, 115–122.

Gao, X. J., Chong, L. S., Kim, M. S., & Elowitz, M. B. (2018). Programmable protein circuits in living cells. *Science, 361*, 1252–1258.

Gordley, R. M., et al. (2016). Engineering dynamical control of cell fate switching using synthetic phospho-regulons. *Proceedings of the National Academy of Sciences of the United States of America, 113*, 13528–13533.
Gordon, W. R., et al. (2015). Mechanical allostery: Evidence for a force requirement in the proteolytic activation of Notch. *Developmental Cell, 33*, 729–736.
Guo, Z., et al. (2016). Engineered PQQ-Glucose dehydrogenase as a universal biosensor platform. *Journal of the American Chemical Society, 138*, 10108–10111.
Hammers, C. M., & Stanley, J. R. (2014). Antibody phage display: Technique and applications. *The Journal of Investigative Dermatology, 134*, 1–5.
Hauser, A. S., Attwood, M. M., Rask-Andersen, M., Schiöth, H. B., & Gloriam, D. E. (2017). Trends in GPCR drug discovery: New agents, targets and indications. *Nature Reviews Drug Discovery, 16*, 829–842.
Hill, S. M. (1998). Receptor crosstalk: Communication through cell signaling pathways. *The Anatomical Record, 253*, 42–48.
Hill, Z. B., Martinko, A. J., Nguyen, D. P., & Wells, J. A. (2018). Human antibody-based chemically induced dimerizers for cell therapeutic applications. *Nature Chemical Biology, 14*, 112–117.
Hu, K. H., & Butte, M. J. (2016). T cell activation requires force generation. *The Journal of Cell Biology, 213*, 535–542.
Hwang, I. Y., Koh, E., Kim, H. R., Yew, W. S., & Chang, M. W. (2016). Reprogrammable microbial cell-based therapeutics against antibiotic-resistant bacteria. *Drug Resistance Updates, 27*, 59–71.
Hwang, I. Y., et al. (2017). Engineered probiotic *Escherichia coli* can eliminate and prevent *Pseudomonas aeruginosa* gut infection in animal models. *Nature Communications, 8*, 15028.
Isabella, V. M., et al. (2018). Development of a synthetic live bacterial therapeutic for the human metabolic disease phenylketonuria. *Nature Biotechnology, 36*, 857–864.
Johnson, M. B., March, A. R., & Morsut, L. (2017). Engineering multicellular systems: Using synthetic biology to control tissue self-organization. *Current Opinion in Biomedical Engineering, 4*, 163–173.
June, C. H., O'Connor, R. S., Kawalekar, O. U., Ghassemi, S., & Milone, M. C. (2018). CAR T cell immunotherapy for human cancer. *Science, 359*, 1361–1365.
Keri, D., & Barth, P. (2018). Reprogramming G protein coupled receptor structure and function. *Current Opinion in Structural Biology, 51*, 187–194.
Kim, A. K., et al. (2016). Toward total synthesis of cell function: Reconstituting cell dynamics with synthetic biology. *Science Signaling, 9*, re1.
Kipniss, N. H., et al. (2017). Engineering cell sensing and responses using a GPCR-coupled CRISPR-Cas system. *Nature Communications, 8*, 2212.
Kovall, R. A., & Blacklow, S. C. (2010). Mechanistic insights into Notch receptor signaling from structural and biochemical studies. *Current Topics in Developmental Biology, 92*, 31–71.
Ladbury, J. E., & Arold, S. T. (2012). Noise in cellular signaling pathways: Causes and effects. *Trends in Biochemical Sciences, 37*, 173–178.
Lienert, F., Lohmueller, J. J., Garg, A., & Silver, P. A. (2014). Synthetic biology in mammalian cells: Next generation research tools and therapeutics. *Nature Reviews Molecular Cell Biology, 15*, 95–107.
Ma, Z., & Finkel, T. H. (2010). T cell receptor triggering by force. *Trends in Immunology, 31*, 1–6.
Mao, N., Cubillos-Ruiz, A., Cameron, D. E., & Collins, J. J. (2018). Probiotic strains detect and suppress cholera in mice. *Science Translational Medicine, 10*(445). eaao2586, https://doi.org/10.1126/scitranslmed.aao2586.
Mathur, M., Xiang, J. S., & Smolke, C. D. (2017). Mammalian synthetic biology for studying the cell. *The Journal of Cell Biology, 216*, 73–82.
Matthews, E. E., Zoonens, M., & Engelman, D. M. (2006). Dynamic helix interactions in transmembrane signaling. *Cell, 127*, 447–450.

McClune, C. J., Alvarez-Buylla, A., Voigt, C. A., & Laub, M. T. (2019). Engineering orthogonal signalling pathways reveals the sparse occupancy of sequence space. *Nature*, *574*, 702–706. https://doi.org/10.1038/s41586-019-1639-8.
Morsut, L., et al. (2016). Engineering customized cell sensing and response behaviors using synthetic Notch receptors. *Cell*, *164*, 780–791.
Pu, J., Zinkus-Boltz, J., & Dickinson, B. C. (2017). Evolution of a split RNA polymerase as a versatile biosensor platform. *Nature Chemical Biology*, *13*, 432–438.
Romero, P. A., & Arnold, F. H. (2009). Exploring protein fitness landscapes by directed evolution. *Nature Reviews Molecular Cell Biology*, *10*, 866–876.
Roybal, K. T., et al. (2016a). Engineering T cells with customized therapeutic response programs using synthetic notch receptors. *Cell*, *167*, 419–432. e16.
Roybal, K. T., et al. (2016b). Precision tumor recognition by T cells with combinatorial antigen-sensing circuits. *Cell*, *164*, 770–779.
Sadelain, M., Brentjens, R., & Rivière, I. (2013). The basic principles of chimeric antigen receptor design. *Cancer Discovery*, *3*, 388–398.
Salvesen, G. S., Hempel, A., & Coll, N. S. (2016). Protease signaling in animal and plant-regulated cell death. *The FEBS Journal*, *283*, 2577–2598.
Scheller, L., Strittmatter, T., Fuchs, D., Bojar, D., & Fussenegger, M. (2018). Generalized extracellular molecule sensor platform for programming cellular behavior. *Nature Chemical Biology*, *14*(7), 723–729. https://doi.org/10.1038/s41589-018-0046-z.
Schwarz, K. A., Daringer, N. M., Dolberg, T. B., & Leonard, J. N. (2016). Rewiring human cellular input–output using modular extracellular sensors. *Nature Chemical Biology*, *13*, 202.
Sedlmayer, F., Aubel, D., & Fussenegger, M. (2018). Synthetic gene circuits for the detection, elimination and prevention of disease. *Nature Biomedical Engineering*, *2*, 399–415.
Skerker, J. M., et al. (2008). Rewiring the specificity of two-component signal transduction systems. *Cell*, *133*, 1043–1054.
Smith, R. S., et al. (2015). Differential muscarinic modulation in the olfactory bulb. *The Journal of Neuroscience*, *35*, 10773–10785.
Sockolosky, J. T., et al. (2018). Selective targeting of engineered T cells using orthogonal IL-2 cytokine-receptor complexes. *Science*, *359*, 1037–1042.
Stanton, B. Z., Chory, E. J., & Crabtree, G. R. (2018). Chemically induced proximity in biology and medicine. *Science*, *359*(6380), eaao5902, https://doi.org/10.1126/science.aao5902.
Stein, V., & Alexandrov, K. (2014). Protease-based synthetic sensing and signal amplification. *Proceedings of the National Academy of Sciences of the United States of America*, *111*, 15934–15939.
Taylor, N. D., et al. (2016). Engineering an allosteric transcription factor to respond to new ligands. *Nature Methods*, *13*, 177–183.
Toda, S., Blauch, L. R., Tang, S. K. Y., Morsut, L., & Lim, W. A. (2018). Programming self-organizing multicellular structures with synthetic cell-cell signaling. *Science*, *361*, 156–162.
Watanabe, N., et al. (2016). Fine-tuning the CAR spacer improves T-cell potency. *Oncoimmunology*, *5*, e1253656.
Whitaker, W. R., Davis, S. A., Arkin, A. P., & Dueber, J. E. (2012). Engineering robust control of two-component system phosphotransfer using modular scaffolds. *Proceedings of the National Academy of Sciences of the United States of America*, *109*, 18090–18095.
Wörn, A., & Plückthun, A. (2001). Stability engineering of antibody single-chain Fv fragments. *Journal of Molecular Biology*, *305*, 989–1010.
Wu, C.-Y., Roybal, K. T., Puchner, E. M., Onuffer, J., & Lim, W. A. (2015). Remote control of therapeutic T cells through a small molecule-gated chimeric receptor. *Science*, *350*, aab4077.

CHAPTER NINE

A suite of bioassays to evaluate CREB inhibitors

Bingbing X. Li*, Xiangshu Xiao*
Program in Chemical Biology, Department of Chemical Physiology and Biochemistry, Knight Cancer Institute, Oregon Health and Science University, Portland, OR, United States
*Corresponding authors: e-mail address: lib@ohsu.edu; xiaoxi@ohsu.edu

Contents

Abstract

The cyclic-AMP response element binding protein (CREB) is an important nuclear transcription factor and has been shown to be overexpressed and/or over-activated in many different cancer types, suggesting that targeting CREB is a novel approach for developing cancer therapies. Our lab discovered the first cell-permeable small molecule inhibitor of CREB, from which we further developed a potent CREB inhibitor with *in vivo* anti-cancer activity. In this article, we detailed our biochemical and cell-based bioassays to assess different small molecule CREB inhibitors.

1. Introduction

The cyclic-AMP response element (CRE) binding protein (CREB) is a critical nuclear transcriptional factor participating in a wide range of cellular processes including cellular proliferation and differentiation (Shaywitz & Greenberg, 1999). In unstimulated cells, CREB resides in the nucleus to bind CRE sequences in the genome. Upon receiving extracellular stimuli (e.g., growth factors, hormones), CREB becomes phosphorylated at Ser133 by several kinases such as protein kinase A (PKA), protein kinase B (PKB/Akt), mitogen-activated protein kinases (MAPKs) and p90

Methods in Enzymology, Volume 633
ISSN 0076-6879
https://doi.org/10.1016/bs.mie.2019.11.002

ribosomal S6 kinase (p90RSK) (Xiao, Li, Mitton, Ikeda, & Sakamoto, 2010). Phosphorylated CREB (pCREB) at Ser133 (Mayr & Montminy, 2001) can bind CREB-binding protein (CBP) and its paralog p300 through the interaction between kinase-inducible domain (KID) in CREB and the KID-interacting domain (KIX) in CBP/p300 (Radhakrishnan et al., 1997). The resulting CREB/CBP complex will further recruit other transcriptional machinery to turn on CREB-dependent gene transcription (Cardinaux et al., 2000). On the other hand, protein phosphatases including protein phosphatase 1 (PP1) (Hagiwara et al., 1992), protein phosphatase 2A (PP2A) (Wadzinski et al., 1993) and phosphatase and tensin homolog (PTEN) (Gu et al., 2011) can dephosphorylate CREB to inactivate CREB-mediated gene transcription. In cancer cells, the kinases phosphorylating CREB are often over-activated and/or overexpressed while the phosphatases dephosphorylating CREB are frequently inactivated or deleted (McConnell & Wadzinski, 2009; Song, Salmena, & Pandolfi, 2012). So it comes to no surprise that CREB over-expression and/or over-activation was found in many cancer types. Thus, targeting CREB may represent a unique strategy for developing novel cancer therapies (Xiao et al., 2010).

For the reasons mentioned above, we have been interested in developing small molecule CREB inhibitors since 2009 (Jiang, Li, Xie, Delaney, & Xiao, 2012; Li & Xiao, 2009; Li et al., 2014; Li, Yamanaka, & Xiao, 2012; Xiao et al., 2010; Xie, Li, Broussard, & Xiao, 2013; Xie et al., 2015; Xie, Li, & Xiao, 2013, 2017). We discovered the first cell-permeable small molecule inhibitors of CREB (Li & Xiao, 2009) and the most potent CREB inhibitor with *in vivo* anti-cancer efficacy (Xie et al., 2015, 2017). The chemical structures of two representative CREB inhibitors, naphthol AS-E (or alternatively called **XXK06)** and **666-15**, are shown in Fig. 1. Critical to the discovery and development of these inhibitors is the

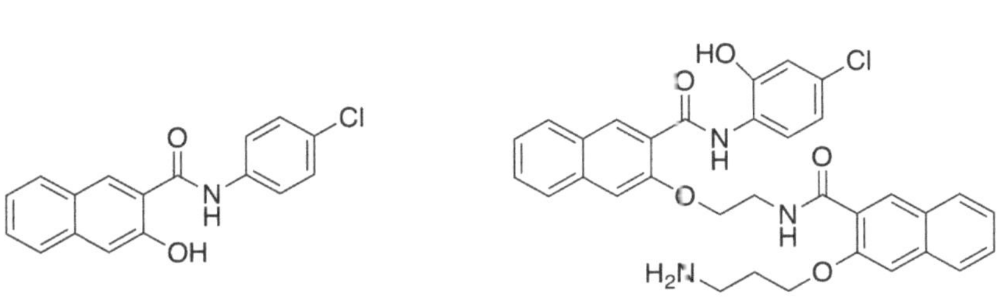

naphthol AS-E (XXK06) **666-15**

Fig. 1 Chemical structures of **XXK06** and **666-15**. **XXK06** was used to evaluate inhibition of KIX-KID interaction while **666-15** was used to inhibit CREB-mediated gene transcription in living cells.

availability of a suite of bioassays to guide the optimization of small molecule inhibitors. For this purpose, we developed two orthogonal assays to assess small molecule inhibitors of CREB. The first one is to assess the biochemical CREB-CBP interaction using a split *Renilla* luciferase assay and the second one is cell-based transcription reporter assay using *Renilla* luciferase as the reporter (Li & Xiao, 2009).

The native *Renilla luciferase* (*rluc*) gene was cloned from marine organism *Renilla reniformis* (Sea pansy) (Lorenz, McCann, Longiaru, & Cormier, 1991; Matthews, Hori, & Cormier, 1977). Subsequent optimization efforts by Promega scientists developed a synthetic *rluc* gene for enhanced expression in mammalian cells to be used as a genetic reporter enzyme. The bioluminescence signal produced by *Renilla* luciferase enzyme is derived from decarboxylative oxidation reaction of coelenterazine substrate to coelenteramide (Matthews et al., 1977). Unlike firefly luciferase, this bioluminescence reaction is ATP- and Mg^{2+}-independent (Matthews et al., 1977). It has been shown that firefly luciferase is more prone to be directly inhibited by compounds from the chemical libraries than *Renilla* luciferase (Auld et al., 2008), which is at least partially due to the requirements of ATP and Mg^{2+} for the former. In this case, chemical molecules that can bind to the ATP-binding pocket or chelate divalent metal ions will inhibit firefly luciferase activity directly. This vulnerability creates potential artifact in assessing chemical inhibitors of certain biological pathways when firefly luciferase is used as a reporter (Bakhtiarova et al., 2006). Alternatively, distinct chemical entities can be deliberately developed to inhibit firefly luciferase (Liu et al., 2012).

Bimolecular complementation is a unique method to assess protein-protein interactions (Kerppola, 2008). In this approach, a protein (e.g., a fluorescent protein or luciferase) is split into two halves. Then the N-terminal half is fused with one protein of interest (bait) and the C-terminal half is fused with the binding partner of the protein of interest (prey). On their own, these fusions lose the activity of the protein to be split. When the bait and prey bind together, this will reconstitute the activity of the original protein to be split (e.g., fluorescent activity or luciferase activity) (Kerppola, 2008; Ozawa, Kaihara, Sato, Tachihara, & Umezawa, 2001). In this article, we describe our split *Renilla* luciferase assay to monitor CREB-CBP interaction both in living cells and in biochemical assays. To assess the functional effect of small molecules in inhibiting CREB-mediated gene transcription, we also describe our cell-based transcription reporter assay using *Renilla* luciferase as a transcription reporter. Expression of *Renilla* luciferase is under the control of three tandem copies of CRE in the promoter region.

2. Protocols

2.1 Split *Renilla* luciferase assay

2.1.1 *Optimization of split* Renilla *luciferase assay to monitor KIX-KID interaction*

Renilla luciferase can be split into N-terminal portion (RLucN, aa2-229) and C-terminal portion (RLucC, aa230-314) to monitor protein-protein interactions (Paulmurugan & Gambhir, 2003). RLucN and RLucC themselves do not have any luciferase activity. One portion of *Renilla* luciferase was fused with the KID from CREB (100–161) while the other portion was fused with the KIX domain from CBP (553–679). Upon stimulation, phosphorylated KID binds KIX to bring the *Renilla* luciferase fragments together to restore an active *Renilla* luciferase. If a compound inhibits KID-KIX binding interaction, the amount of active *Renilla* luciferase activity should be diminished. It is expected that the relative orientation between *Renilla* luciferase fragments and KIX/KID is critical to have a functional *Renilla* luciferase complementation assay. Based on the three-dimensional NMR protein structure of the KID bound to KIX (Radhakrishnan et al., 1997), it would be desirable to have a pair of constructs where the two halves of RLuc are fused to the N-terminus of KIX and the C-terminus of KID or N-terminus of KID and C-terminus of KIX. Therefore, a total of four possible pairs (A-D, B-C, E-F, G-H) were designed (Fig. 2). In each of the constructs, a nuclear localization sequence (NLS) was added at the C-terminal end to facilitate nuclear localization because endogenous CREB and CBP are both localized inside the nucleus.

2.1.1.1 Procedures

1. The mammalian expression constructs A-H were prepared according to the standard molecular cloning approach. The sequence of each construct was verified by Sanger sequencing. Note, whereas we prepared these constructs by molecular cloning, these and other similar constructs can also be prepared by gene synthesis with a reasonable cost.
2. 1 million HEK 293T cells were plated into one well of a 6-well plate with Dulbecco's Modified Eagle's Medium (DMEM) media containing 10% fetal bovine serum (FBS) and non-essential amino acids. The cells were allowed to attach to the bottom of the plate for overnight in a humidified incubator at 37 °C with 5% CO_2.

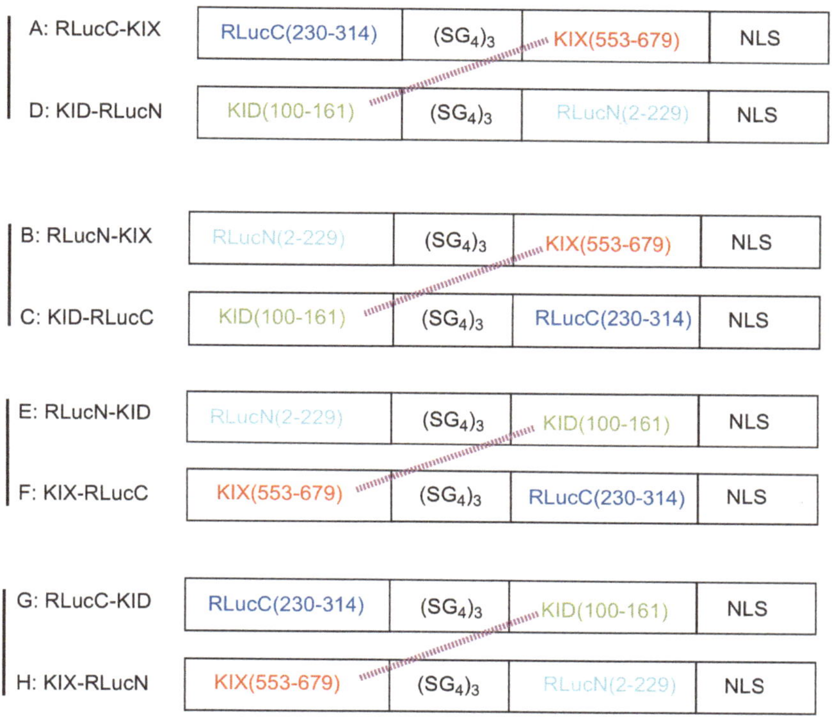

Fig. 2 Schematic diagrams of four possible pairs of the split *Renilla* luciferase assay constructs to monitor KIX-KID interaction. In each of the construct, a nuclear localization sequence (NLS) was added at the C-terminus to facilitate nuclear localization of the expressed proteins in mammalian cells. $(SG)_4$ was inserted as a flexible linker in each of the construct.

3. The cells were transfected with each combination of the plasmids (A-D, B-C, E-F, G-H) by Lipofectamine™ 2000 (Life Technologies) according to manufacturer's protocol. We used 1 μg of each plasmid for each well of the cells. One well of untransfected cells were also prepared.
4. Three hours later, the transfected cells were collected and resuspended in phenol red-free DMEM media with 10% FBS. The cells were then replated into a 96-well plate at a density of 3×10^4 cells/well in 100 μL of media. The cells were then allowed to attach to the bottom of the wells for overnight. The untransfected cells were also plated into the 96-well plate for background measurements in step 6.
5. The cells were then treated with different concentrations of forskolin (Fsk) for 1–2 h in a humidified incubator at 37 °C with 5% CO_2.

6. Freshly prepared coelenterazine (Nanolight technology) solution in PBS (pH 7.4, 20 μg/mL, 100 μL) was then added to each well. The luciferase activity was then measured in a plate reader (Packard Fusion, Perkin Elmer). Note, the aqueous coelenterazine solution is unstable and should be prepared immediately before use. However, an ethanol solution of coelenterazine at 1 mg/mL can be prepared and stored at −80 °C in the dark for at least 6 months.
7. The raw luciferase measurement data were converted into relative fold of induction, which is derived from (Fsk-treated sample reading-background reading)/(DMSO-treated sample reading-background reading). The background reading was from untransfected cells treated with DMSO only. Representative results are shown in Fig. 3.
8. Based on the results in Fig. 3, we concluded that the pair A-D was the best among the four pairs because this pair gave the maximal fold of induction by Fsk under the identical conditions. This A-D pair was used for the rest of the split *Renilla* luciferase assay to monitor KIX-KID interaction.

2.1.2 Inhibition of KIX-KID interaction by XXK06 to be monitored by split Renilla *luciferase assay in living cells*

With the optimized A-D pair to monitor KIX-KID interaction in living HEK 293T cells, this assay can be adapted to assess and screen small molecule

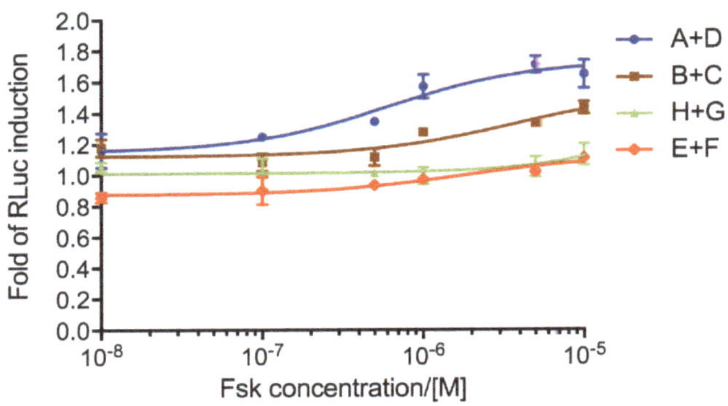

Fig. 3 Optimization of split *Renilla* luciferase assay in HEK 293T cells to monitor KIX-KID interaction. HEK 293T cells were transfected with different pairs of plasmids. The cells were then treated with different concentrations of Fsk (10, 5, 1, 0.5, 0.1, and 0.01 μM) for 1–2 h. The resulting *Renilla* luciferase activity was measured *in situ* after the addition of coelenterazine to a final concentration of 10 μg/mL with a plate reader. The fold of RLuc induction = (Fsk-treated sample reading-background reading)/(DMSO-treated sample reading-background reading), where the background reading was from untransfected cells treated with DMSO only.

inhibitors of KIX-KID interaction (Li & Xiao, 2009; Xie, Li, Broussard, & Xiao, 2013). To illustrate this capability, we used cell-permeable small molecule **XXK06** (Fig. 1) as an example (Li & Xiao, 2009). For this assay, the steps 1–4 in Section 2.1.1.1 were followed. Then **XXK06** was added at different concentrations and incubated with the cells for 1 h at 37 °C followed by addition of a fixed concentration of Fsk at 2.5 μM. The cells were further incubated for 2 h before the measurement of *Renilla* luciferase activity as in step 6 in Section 2.1.1.1. Representative results are shown in Fig. 4.

2.1.3 *Inhibition of KIX-KID interaction by XXK06 to be monitored by split* Renilla *luciferase assay* in vitro

The cell-based split *Renilla* luciferase assay described above is very useful in assessing if the inhibitors to be investigated are cell-permeable and can inhibit KIX-KID interaction in living cells. However, a biochemical *in vitro* assay would be desirable to investigate if the inhibitors are indeed directly inhibiting the KIX-KID interaction instead of other cellular events that might also lead to inhibition of KIX-KID interaction in the cells (e.g., phosphorylation of KID). In this respect, we also developed a biochemical version of the split *Renilla* luciferase assay to monitor KIX-KID interaction using the A-D pair (Li & Xiao, 2009).

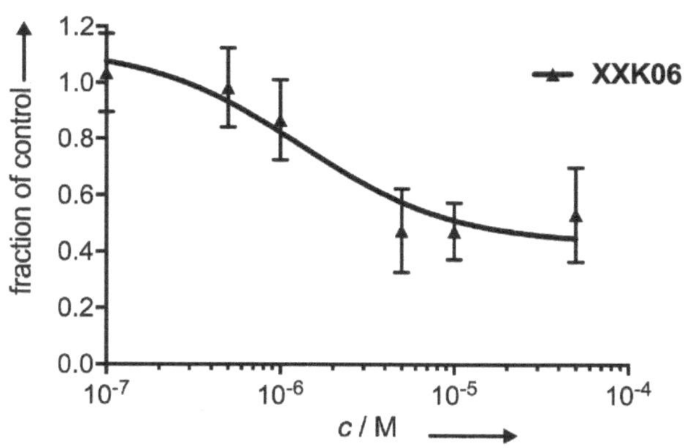

Fig. 4 **XXK06** inhibits KIX-KID interaction in living HEK 293T cells. The cells were transfected with the A-D pair of plasmids. Then the cells were treated with increasing concentrations of **XXK06** for 1 h followed by addition of Fsk (2.5 μM) for 2 h. The measured *Renilla* luciferase activity was processed as in Fig. 3 and further normalized to 1.0 for the cells treated with Fsk only. *Modified from reference Li, B. X., & Xiao, X. (2009). Discovery of a small-molecule inhibitor of the KIX-KID interaction.* Chembiochem: A European Journal of Chemical Biology, 10, *2721–2724. Copyright (2009) of Wiley-VCH.*

2.1.3.1 Procedures

1. Preparation of recombinant His_6-tagged A protein from *E. coli*. An *E. coli* expression plasmid for His_6-tagged A was prepared by sub-cloning A into pET-15b vector using standard molecular cloning approach. The sequence of the resulting plasmid was verified by Sanger sequencing. The plasmid was then transformed into BL21(DE3)pLysS competent cells (Promega) for protein expression. In this strain of *E. coli*, the expression of protein of interest can be induced by isopropyl β-D-1-thiogalactopyranoside (IPTG). The purification of His_6-A was carried out using standard affinity fast protein liquid chromatography with a Ni^{2+} column. After purification, the protein was dialyzed into the storage buffer [50 mM Tris (pH 8.0), 100 mM KCl, 12.5 mM $MgCl_2$, 1 mM EDTA, 0.05% Tween-20] (Xiao, Yu, Lim, Sikder, & Kodadek, 2007) and stored at −80 °C.
2. Preparation of D from HEK 293T cells. Five million HEK 293T cells were plated into a 10-cm plate with DMEM media containing 10% FBS and non-essential amino acids. The cells were allowed to attach to the bottom of the plate for overnight in a humidified incubator at 37 °C with 5% CO_2. The cells were then transfected with 6 μg plasmid D by Lipofectamine™ 2000 and OptiMEM (Life Technologies) for 3 h according to the manufacturer's protocol. Then the cells were incubated in DMEM media containing 10% FBS and non-essential amino acids for 20 h, when Fsk (5 μM) was added for 2 h. The cells were then collected by scraping and washed with cold PBS (5 mL) twice. The cell pellets were lysed in 600 μL 1 × *Renilla* luciferase lysis buffer (Promega) supplemented with protease inhibitor cocktail (Roche Applied Science), phosphatase inhibitor cocktail (Pierce) and 1 mM phenylmethylsulfonyl fluoride (PMSF) on ice for 20 min. Then the lysates were cleared by centrifuging at 14,000 rpm for 15 min at 4 °C. The concentration of the cleared lysates was determined by Protein assay Dye Reagent Concentrate (Bio-Rad). The concentration was usually about 1 mg/mL.
3. To assess the formation of KIX-KID complex using the split *Renilla* luciferase assay, His_6-A (15 ng) from step 1 and D-containing HEK 293T cell lysates from step 2 (0.75 μg) were mixed together in 1 × *Renilla* luciferase lysis buffer (30 μL, Promega) and incubated on ice for overnight. Then 5 μL of the incubation mixture was mixed with 30 μL of coelenterazine solution in PBS (pH 7.4, 10 μg/mL) and the luciferase activity was measured immediately using FB12 single tube luminometer (Berthold). Representative results are shown in Fig. 5A.

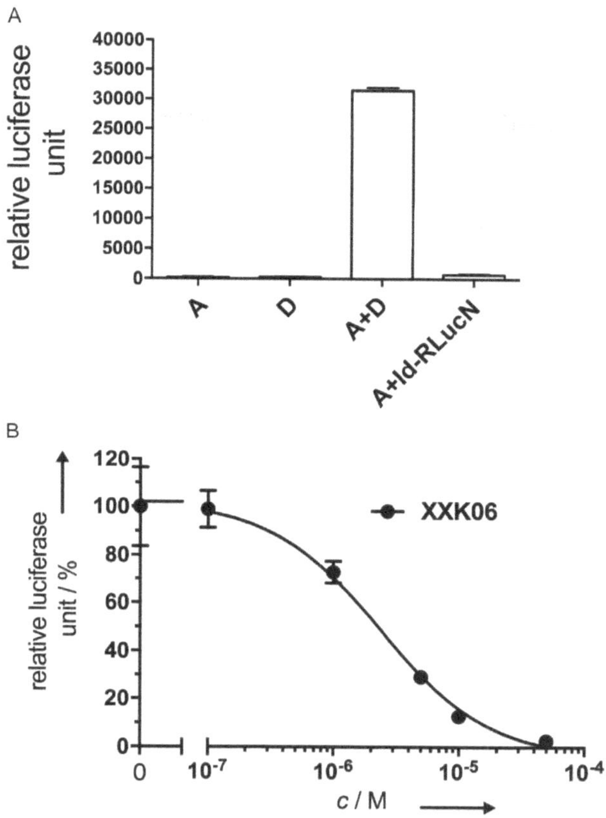

Fig. 5 **XXK06** inhibits KIX-KID interaction *in vitro*. (A) KIX-KID interaction effectively reconstitutes active *Renilla* luciferase in the split *Renilla* luciferase assay. The recombinant fusion protein A was mixed with different RLucN-containing HEK 293T cell lysates on ice for overnight. Then the luciferase activity was measured in a tube luminometer upon addition of coelenterazine. (B) **XXK06** inhibits KIX-KID interaction *in vitro*. The recombinant fusion protein A was mixed with D-containing HEK 293T cell lysates in the presence of different concentrations of **XXK06** on ice for overnight. Then the remaining luciferase activity was measured in a tube luminometer upon addition of coelenterazine. *Modified from reference Li, B. X., & Xiao, X. (2009). Discovery of a small-molecule inhibitor of the KIX-KID interaction.* Chembiochem: A European Journal of Chemical Biology, 10, *2721–2724. Copyright (2009) of Wiley-VCH.*

As expected, neither A nor D alone showed appreciable luciferase activity. However, when A and D were mixed together, robust reconstitution of luciferase activity was achieved. As a control, when A was mixed with a non-binding pair inhibitor of DNA binding Id-RLucN (Paulmurugan & Gambhir, 2003), only minimal luciferase activity was observed, demonstrating the specificity of the assay (Fig. 5A). We have observed that the 1 × *Renilla* luciferase lysis buffer is a preferred buffer for this assay because this buffer contains proprietary components that can suppress the auto-bioluminescent activity of coelenterazine. In addition, the presence of mild detergent in this buffer can decrease the chance of artifact from pan-assay interference compounds (PAINS) when assessing small molecule inhibitors (Baell & Nissink 2018).

4. To evaluate inhibition of KIX-KID interaction by **XXK06** using the *in vitro* split *Renilla* luciferase assay, the assay mixture in step 3 was prepared in the presence of different concentrations of **XXK06**. And the mixture was incubated on ice for overnight. Then the remaining luciferase activity was measured by adding 5 μL of the incubation mixture to 30 μL of coelenterazine solution in PBS (pH 7.4, 10 μg/mL). The measurement was performed with FB12 single tube luminometer (Berthold). Representative results are shown in Fig. 5B. When setting up these assays, we typically set up the assays in 96-well PCR plates.

2.2 CREB transcription reporter assay

The oncogenic and other function of CREB is mainly through its transcription activity to transcribe its target genes in the cells. To further assess if a small molecule can inhibit CREB's transcription activity inside the cells, we developed a cell-based transcription reporter assay (Li & Xiao, 2009). For this assay, the key reagent needed is a CREB reporter construct. While firefly luciferase is a commonly used transcription reporter, we adopted *Renilla* luciferase as the reporter because it is less prone to be inhibited by general small molecules for the reasons mentioned above (Auld et al., 2008). To make this construct responsive to CREB activation, the promoter for this construct contains three tandem copies of consensus CRE sequences (GCACCAGACAGTGACGTCAGCTGCCAGATCCCATGGCCGT CATACTGTGACGTCTTTCAGATGGGAGAAC) (Alluri, Liu, Yu, Xiao, & Kodadek, 2006). In this section, we describe protocols to ensure that the reporter construct is responsive to CREB activation and illustrate the inhibition of CREB's transcription activity using **666-15** (Li et al., 2016; Xie et al., 2015) as a small molecule inhibitor.

2.2.1 Procedures

1. The reporter construct named as CRE-RLuc was created by swapping the firefly luciferase gene in CRE-FLuc (gift of Tom Kodadek, Scripps Florida) with a synthetic *Renilla* luciferase gene from pRL-SV40

(Promega) using standard molecular cloning approach. The sequence of the construct was verified by Sanger sequencing. Alternatively, this construct can also be prepared using standard gene synthesis approach.

2. HEK 293T cells were plated into a 6-well plate at 1 million cells/well in DMEM media containing 10% FBS and non-essential amino acids. The cells were allowed to attach to the bottom of the plate for overnight in a humidified incubator at 37 °C with 5% CO_2.
3. The cells were transfected with 1 μg CRE-RLuc plasmid by Lipofectamine™ 2000 following manufacturer's protocol.
4. Three–five hours later, the transfected cells were collected and replated into a whole 96-well plate at ~10,000 cells/well. The cells were then allowed to attach to the bottom of the wells for overnight.
5. The cells were treated with different concentrations of **666-15** (5, 1, 0.5, 0.1, 0.05, 0.01, 0.001, and 0 μM) for 30 min, when Fsk (10 μM) was added to stimulate CREB's transcription activity.
6. The cells were then further incubated for 5–6 h. Then the media were removed and 30 μL 1 × *Renilla* luciferase lysis buffer (Promega) was added. The plate was shaken at room temperature for 20 min to lyse the cells. Alternatively, the plate can be frozen at −80 °C for later measurements.
7. The *Renilla* luciferase activity in each well was measured using the Renllia Luciferase Assay System (Promega) with 5 μL of the cell lysates from each well in the FB12 single tube luminometer (Berthold). While a plate reader capable of measuring bioluminescence can also be technically used to measure the *Renilla* luciferase activity, we found that the single tube luminometer typically generate more consistent results.
8. To normalize the *Renilla* luciferase activity from each well, the protein concentration of the lysates was determined with Protein Assay Dye Reagent Concentrate (Bio-Rad) following manufacturer's protocol. The *Renilla* luciferase activity from each well was normalized by dividing the raw *Renilla* luciferase activity by the amount of proteins used to give relative light units per microgram protein (RLU/μg).
9. Representative results are shown in Fig. 6. In Fig. 6A, we showed that the construct was very responsive to CREB stimulation by Fsk. Upon Fsk stimulation, more than 20-fold stimulation was observed. We have observed that the fold of stimulation was variable across different experiments, which is at least partially due to the variable transfection efficiency for each experiment. However, Fsk should be able to robustly stimulate CREB's transcription activity. If this stimulation is not observed, additional trouble-shooting efforts are needed. When

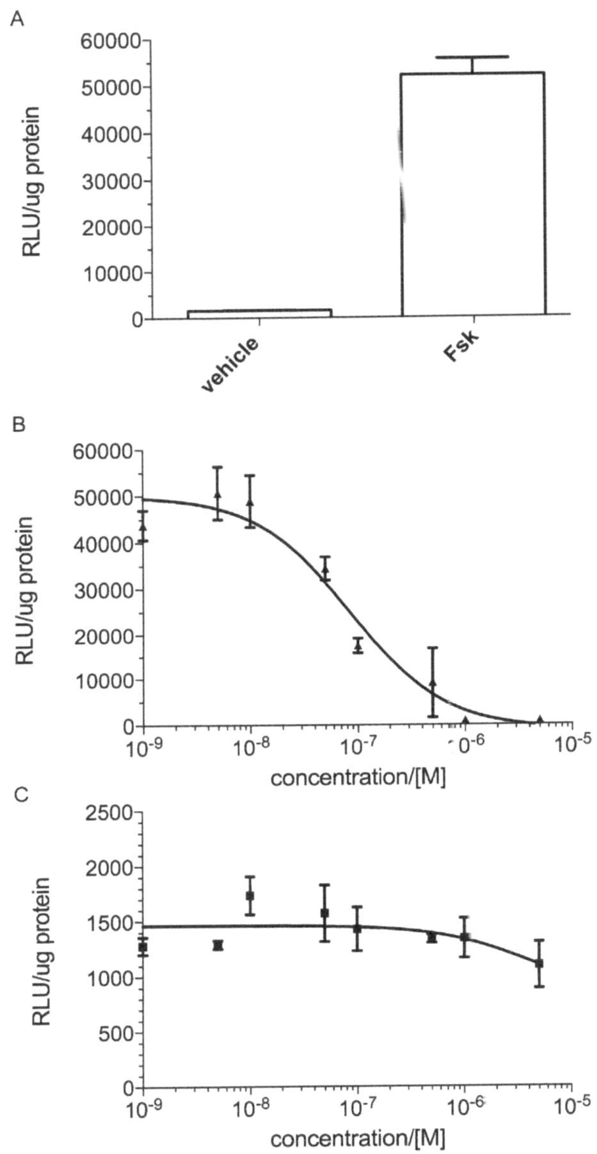

Fig. 6 **666-15** potently inhibited CREB-mediated gene transcription in HEK 293T cells. (A) Fsk stimulated CREB-dependent gene transcription. HEK 293T cells were transfected with CRE-RLuc reporter construct. Then the cells were treated with or without Fsk (10 μM) for 6 h. Then resulting luciferase activity was measured and analyzed as described in Section 2.2.1. (B) **666-15** inhibited CREB-dependent gene transcription. The experiments were the same as in (A) except that the cells were pretreated with increasing concentrations of **666-15** for 30 min followed by treatment with Fsk (10 μM) for 6 h before luciferase measurement. (C) **666-15** did not appreciably inhibit basal CREB activity. The experiments were performed in the same way as in (B) except that Fsk treatment was omitted. The luciferase activity was all normalized to the protein content and expressed as relative luciferase unit (RLU)/μg protein. *Modified from reference Li, B. X., Gardner, R., Xue, C., Qian, D. Z., Xie, F., Thomas, G., et al. (2016). Systemic inhibition of CREB is well-tolerated in vivo.* Scientific Reports, *6, 34513.*

666-15 was preincubated with the cells prior to the addition of Fsk, dose-dependent inhibition of CREB-mediated gene transcription was observed as shown in Fig. 6B. Importantly, if Fsk was not added to stimulate CREB's transcription activity, no obvious inhibition of the basal activity could be observed by **666-15** (Fig. 6C). Perhaps, the residual activity in the absence of Fsk was due to the preexisting *Renilla* luciferase synthesized prior to the addition of **666-15**. In fact, during the ~5.5 h drug treatment incubation, no significant increase of *Renilla* luciferase was typically observed if Fsk was not added. This is again to demonstrate the importance of Fsk stimulation effect before assessing different chemical inhibitors.

3. Conclusions

CREB is perhaps one of the best studied transcription factors. It is involved in multiple cellular functions. A large number of studies have been focusing on the understanding of how CREB regulates these cellular functions (Shaywitz & Greenberg, 1999). Numerous studies have shown that CREB is critically important in the pathogenesis of multiple types of cancers (Conkright & Montminy, 2005; Xiao et al., 2010). Interestingly, a group of different rare cancer types harbors a fusion between Ewing sarcoma gene (EWS) and CREB or its closely related transcription factor activating transcription factor 1 (ATF1) (Tsukamoto et al., 2013; Wang et al., 2009; Zucman et al., 1993). The resulting fusion EWS-CREB or EWS-ATF1 encodes a constitutively active transcription factor independent of phosphorylation (Fujimura et al., 1996) to drive the development of these rare cancers (Straessler et al., 2013; Yamada et al., 2013). Therefore, targeting CREB represents an appealing direction for developing novel anti-cancer therapies (Xiao et al., 2010). Without properly designed robust assays, it will be very difficult to evaluate the putative small molecule inhibitors. The biochemical and cell-based *Renilla* luciferase assays described here will provide effective and reliable methods to assess different small molecule inhibitors of CREB to derive potential candidates for further preclinical and clinical development. The inhibitors to be developed will also help further our understanding of CREB's functions in different biological pathways (Kang et al., 2015).

Acknowledgments

This work was supported by NIH R01GM122820, R21CA220061, Oregon Health & Science University School of Medicine, Office of the Technology Transfer and Business

Development, Oregon Clinical and Translational Research Institute, Medical Research Foundation of Oregon and Susan G. Komen for the Cure (KG100458). We thank Prof. Tom Kodadek (Scripps Florida) for sharing CRE-FLuc plasmid and Prof. Sanjiv Gambhir (Stanford University) for sharing Id-RLucN plasmid.

References

Alluri, P., Liu, B., Yu, P., Xiao, X. S., & Kodadek, T. (2006). Isolation and characterization of coactivator-binding peptoids from a combinatorial library. *Molecular BioSystems*, *2*(11), 568–579.

Auld, D. S., Southall, N. T., Jadhav, A., Johnson, R. L., Diller, D. J., Simeonov, A., et al. (2008). Characterization of chemical libraries for luciferase inhibitory activity. *Journal of Medicinal Chemistry*, *51*(8), 2372–2386.

Baell, J. B., & Nissink, J. W. M. (2018). Seven year itch: Pan-assay interference compounds (PAINS) in 2017-utility and limitations. *ACS Chemical Biology*, *13*(1), 36–44.

Bakhtiarova, A., Taslimi, P., Elliman, S. J., Kosinski, P. A., Hubbard, B., Kavana, M., et al. (2006). Resveratrol inhibits firefly luciferase. *Biochemical and Biophysical Research Communications*, *351*(2), 481–484.

Cardinaux, J. R., Notis, J. C., Zhang, Q., Vo, N., Craig, J. C., Fass, D. M., et al. (2000). Recruitment of CREB binding protein is sufficient for CREB-mediated gene activation. *Molecular and Cellular Biology*, *20*(5), 1546–1552.

Conkright, M. D., & Montminy, M. (2005). CREB: The unindicted cancer co-conspirator. *Trends in Cell Biology*, *15*(9), 457–459.

Fujimura, Y., Ohno, T., Siddique, H., Lee, L., Rao, V. N., & Reddy, E. S. P. (1996). The EWS-ATF-1 gene involved in malignant melanoma of soft parts with t(12;22) chromosome translocation, encodes a constitutive transcriptional activator. *Oncogene*, *12*(1), 159–167.

Gu, T., Zhang, Z., Wang, J., Guo, J., Shen, W. H., & Yin, Y. (2011). CREB is a novel nuclear target of PTEN phosphatase. *Cancer Research*, *71*(8), 2821–2825.

Hagiwara, M., Alberts, A., Brindle, P., Meinkoth, J., Feramisco, J., Deng, T., et al. (1992). Transcriptional attenuation following cAMP induction requires PP-1-mediated dephosphorylation of CREB. *Cell*, *70*(1), 105–113.

Jiang, M., Li, B. X., Xie, F., Delaney, F., & Xiao, X. (2012). Design, synthesis, and biological evaluation of conformationally constrained analogues of naphthol AS-E as inhibitors of CREB-mediated gene transcription. *Journal of Medicinal Chemistry*, *55*(8), 4020–4024.

Kang, X., Lu, Z., Cui, C., Deng, M., Fan, Y., Dong, B., et al. (2015). The ITIM-containing receptor LAIR1 is essential for acute myeloid leukaemia development. *Nature Cell Biology*, *17*(5), 665–677.

Kerppola, T. K. (2008). Bimolecular fluorescence complementation (BiFC) analysis as a probe of protein interactions in living cells. *Annual Review of Biophysics*, *37*, 465–487.

Li, B. X., Gardner, R., Xue, C., Qian, D. Z., Xie, F., Thomas, G., et al. (2016). Systemic inhibition of CREB is well-tolerated in vivo. *Scientific Reports*, *6*, 34513.

Li, B. X., & Xiao, X. (2009). Discovery of a small-molecule inhibitor of the KIX-KID interaction. *Chembiochem: A European Journal of Chemical Biology* *10*, 2721–2724.

Li, B. X., Xie, F., Fan, Q., Barnhart, K. M., Moore, C. E., Rheingold, A. L., et al. (2014). Novel type of prodrug activation through a long-range O,N-acyl transfer: A case of water-soluble CREB inhibitor. *ACS Medicinal Chemistry Letters*, *5*(10), 1104–1109.

Li, B. X., Yamanaka, K., & Xiao, X. (2012). Structure-activity relationship studies of naphthol AS-E and its derivatives as anticancer agents by inhibiting CREB-mediated gene transcription. *Bioorganic and Medicinal Chemistry*, *20*(23), 6811–6820.

Liu, Y., Fang, J., Cai, H., Xiao, F., Ding, K., & Hu, Y. (2012). Identification and synthesis of substituted pyrrolo[2,3-d]pyrimidines as novel firefly luciferase inhibitors. *Bioorganic and Medicinal Chemistry*, *20*(18), 5473–5482.

Lorenz, W. W., McCann, R. O., Longiaru, M., & Cormier, M. J. (1991). Isolation and expression of a cDNA encoding Renilla reniformis luciferase. *Proceedings of the National Academy of Sciences of the United States of America*, *88*(10), 4438–4442.

Matthews, J. C., Hori, K., & Cormier, M. J. (1977). Purification and properties of Renilla reniformis luciferase. *Biochemistry*, *16*(1), 85–91.

Mayr, B., & Montminy, M. (2001). Transcriptional regulation by the phosphorylation-dependent factor CREB. *Nature Reviews: Molecular Cell Biology*, *2*(8), 599–609.

McConnell, J. L., & Wadzinski, B. E. (2009). Targeting protein serine/threonine phosphatases for drug development. *Molecular Pharmacology*, *75*(6), 1249–1261.

Ozawa, T., Kaihara, A., Sato, M., Tachihara, K., & Umezawa, Y. (2001). Split luciferase as an optical probe for detecting protein-protein interactions in mammalian cells based on protein splicing. *Analytical Chemistry*, *73*(11), 2516–2521.

Paulmurugan, R., & Gambhir, S. S. (2003). Monitoring protein – protein interactions using split synthetic Renilla luciferase protein-fragment-assisted complementation. *Analytical Chemistry*, *75*, 1584–1589.

Radhakrishnan, I., Perez-Alvarado, G. C., Parker, D., Dyson, H. J., Montminy, M. R., & Wright, P. E. (1997). Solution structure of the KIX domain of CBP bound to the transactivation domain of CREB: A model for activator: Coactivator interactions. *Cell*, *91*(6), 741–752.

Shaywitz, A. J., & Greenberg, M. E. (1999). CREB: A stimulus-induced transcription factor activated by a diverse array of extracellular signals. *Annual Review of Biochemistry*, *68*, 821–861.

Song, M. S., Salmena, L., & Pandolfi, P. P. (2012). The functions and regulation of the PTEN tumour suppressor. *Nature Reviews: Molecular Cell Biology*, *13*(5), 283–296.

Straessler, K. M., Jones, K. B., Hu, H., Jin, H., van de Rijn, M., & Capecchi, M. R. (2013). Modeling clear cell sarcomagenesis in the mouse: Cell of origin differentiation state impacts tumor characteristics. *Cancer Cell*, *23*(2), 215–227.

Tsukamoto, Y., Nakata, Y., Futani, H., Fukunaga, S., Kajimoto, K., & Hirota, S. (2013). A rare case of clear cell sarcoma with 4 types of EWSR1-ATF1 fusions detected not in primary site but in metastatic site. *Pathology, Research and Practice*, *209*(12), 803–807.

Wadzinski, B. E., Wheat, W. H., Jaspers, S., Peruski, L. F., Lickteig, R. L., Johnson, G. L., et al. (1993). Nuclear-protein phosphatase-2A dephosphorylates protein kinase a phosphorylated CREB and regulates CREB transcriptional stimulation. *Molecular and Cellular Biology*, *13*(5), 2822–2834.

Wang, W. L., Mayordomo, E., Zhang, W., Hernandez, V. S., Tuvin, D., Garcia, L., et al. (2009). Detection and characterization of EWSR1/ATF1 and EWSR1/CREB1 chimeric transcripts in clear cell sarcoma (melanoma of soft parts). *Modern Pathology*, *22*(9), 1201–1209.

Xiao, X., Li, B. X., Mitton, B., Ikeda, A., & Sakamoto, K. M. (2010). Targeting CREB for cancer therapy: Friend or foe. *Current Cancer Drug Targets*, *10*, 384–391.

Xiao, X., Yu, P., Lim, H. S., Sikder, D., & Kodadek, T. (2007). Design and synthesis of a cell-permeable synthetic transcription factor mimic. *Journal of Combinatorial Chemistry*, *9*(4), 592–600.

Xie, F., Li, B. X., Broussard, C., & Xiao, X. (2013). Identification, synthesis and evaluation of substituted benzofurazans as inhibitors of CREB-mediated gene transcription. *Bioorganic and Medicinal Chemistry Letters*, *23*(19), 5371–5375.

Xie, F., Li, B. X., Kassenbrock, A., Xue, C., Wang, X., Qian, D. Z., et al. (2015). Identification of a potent inhibitor of CREB-mediated gene transcription with efficacious in vivo anticancer activity. *Journal of Medicinal Chemistry*, *58*(12), 5075–5087.

Xie, F., Li, B. X., & Xiao, X. (2013). Synthesis and evaluation of an O-aminated naphthol AS-E as a prodrug of CREB-mediated gene transcription inhibition. *Letters in Organic Chemistry*, *10*, 380–384.

Xie, F., Li, B. X., & Xiao, X. (2017). Design, synthesis and biological evaluation of regioisomers of 666-15 as inhibitors of CREB-mediated gene transcription. *Bioorganic and Medicinal Chemistry Letters*, *27*, 994–998.

Yamada, K., Ohno, T., Aoki, H., Semi, K., Watanabe, A., Moritake, H., et al. (2013). EWS/ATF1 expression induces sarcomas from neural crest-derived cells in mice. *Journal of Clinical Investigation*, *123*(2), 600–610.

Zucman, J., Delattre, O., Desmaze, C., Epstein, A. L., Stenman, G., Speleman, F., et al. (1993). EWS and ATF-1 gene fusion induced by t(12,22) translocation in malignant melanoma of soft parts. *Nature Genetics*, *4*(4), 341–345.

CHAPTER TEN

Identification of lamins as the molecular targets of LBL1 using a clickable photoaffinity probe

Xiangshu Xiao, Bingbing X. Li*

Program in Chemical Biology, Department of Chemical Physiology and Biochemistry, Knight Cancer Institute, Oregon Health and Science University, Portland, OR, United States

*Corresponding author: e-mail address: lib@ohsu.edu

Contents

Abstract

Phenotypic screening is a powerful approach to discover small molecules targeting pathways or disease biology with complex genetic causes. Following the initial discovery of these small molecules is their target identification, which is at the cornerstone in addressing their biological and clinical utility. Yet, finding the needle in the haystack remains a challenge. Nuclear lamins are type V intermediate filament proteins that form a filamentous structure underneath the inner nuclear envelope to support the mechanical stability of the mammalian cell nucleus. They also participate a myriad of other cellular signaling processes with incompletely understood molecular mechanisms. Small molecules that can directly bind to nuclear lamins will be incredible tools to address lamins' roles in different aspects of biology. However, these small molecules did not exist until recently. We previously discovered an acylpyrroloquinazoline called **LBL1** that selectively killed breast cancer cells without harming normal human cells. To help understand the mechanism of action of **LBL1**, we recently took an unbiased chemical proteomics approach to identify its direct binding targets from the entire human cellular proteome. In this chapter, we describe our detailed methods to identify and validate lamins as the direct targets of **LBL1**. In this approach, we developed a clickable photoaffinity probe called **LBL1-P** that contains acylpyrroloquinazoline,

Methods in Enzymology, Volume 633
ISSN 0076-6879
https://doi.org/10.1016/bs.mie.2019.02.038

trifluoromethyldiazirine and alkyne groups. Furthermore, we described a fluorescence microscopic method to validate that **LBL1** directly targets lamin A in living cells. When properly designed, this approach should be broadly applicable to other bioactive small molecules.

1. Introduction

Phenotypic screening remains a powerful method to identify bioactive molecules to modulate various biological processes despite our increased understanding of their underlying molecular mechanisms (Swinney & Anthony, 2011; Wagner & Schreiber, 2016). The chemical tools or probes derived from the phenotypic screens represent useful additions to the arsenal at our disposal to further study the mechanisms of biological processes in question or provide starting points for developing therapeutic agents with clinical benefits. Albeit useful, these chemical tools often suffer from poor mechanistic understanding because the molecular target(s) that these chemical tools directly bind to are not evident from the initial phenotypic screening assays (Moffat, Vincent, Lee, Eder, & Prunotto, 2017). Therefore, identification of the molecular targets that the small molecules can directly bind to is of paramount importance to further understand their mechanism of action (Wagner & Schreiber, 2016). These mechanistic investigations could further add our understanding of the disease biology and/or cellular signaling pathways, complementing the knowledge garnered through genetic strategies by overexpression and/or suppression.

The cytoskeleton of mammalian cells is composed of macrofilaments from tubulins, intermediate filaments (IF), and microfilaments from actins (Fletcher & Mullins, 2010; Fuchs & Cleveland, 1998; Robert, Hookway, & Gelfand, 2016). While only a few proteins make up the macrofilaments and microfilaments, a large number of proteins (>70) form IFs with diameters of ~10 nm (Hesse, Magin, & Weber, 2001). Among these proteins, nuclear lamins are type V IF proteins (Aebi, Cohn, Buhle, & Gerace, 1986; Burke & Stewart, 2013). In mammals, there are three lamin genes (*LA*, *LB1* and *LB2*) encoding four major homologous proteins (LA, LC, LB1 and LB2). LA and LC are alternative splicing products of a single gene *LA* (Dechat, Adam, Taimen, Shimi, & Goldman, 2010; Lin & Worman, 1993). *LB1* and *LB2* are two different genes (Dechat et al., 2010). All the lamins are highly homologous in sequence and predicted structure. They contain a central, α-helical coiled-coil rod domain (Strelkov, Schumacher, Burkhard, Aebi, & Herrmann, 2004), which is flanked by a non α-helical N-terminal head and a C-terminal tail (Dechat et al., 2010). Unlike other cytosolic IF proteins, lamins contain a nuclear localization sequence (NLS) and immunoglobulin (Ig) domain with a two-layered sandwich composed of anti-parallel β-strands (Dhe-Paganon, Werner, Chi, & Shoelson, 2002; Ruan et al., 2012). This unique structural organization allows lamins to function not only as nuclear scaffold proteins, but also as signaling proteins by interacting with other proteins (Dechat et al., 2000; Gerace & Tapia, 2018; Shumaker et al., 2008; Zastrow, Flaherty, Benian, & Wilson, 2006; Zastrow, Vlcek, & Wilson, 2004).

While numerous small molecules (e.g., latrunculin, phalloidin, paclitaxel, vinblastine) have been identified to target actins and tubulins, which provides potent and efficacious cancer therapies as well as molecular tools to study these filaments (Jordan & Wilson, 1998, 2004), few or no small molecules have been known to interact with IF proteins (Bargagna-Mohan et al., 2007; Lee et al., 2016; Li, Chen, Chao, David, & Xiao, 2018). This dearth of small molecules to target IF proteins is largely due to their unique structures and a lack of biochemical assays appropriate for high throughput screening (Burke & Stewart, 2013). However, we believe that small molecules targeting IF proteins can provide equally powerful tools to further study their biology as well as providing novel starting points to develop potential therapies for a variety of diseases including cancer. In this regard, forward chemical genetic strategy (Bredel & Jacoby, 2004) might be an excellent alternative to identify such small molecules to target IF proteins.

We previously described the discovery of a novel acylated pyrroloquinazoline compound as a non-toxic agent to selectively kill breast cancer cells (Chen, Kassenbrock, Li, & Xiao, 2013). **LBL1** is a prototype in this series of compounds (Fig. 1A) (Chen et al., 2013). However, its direct molecular targets were unknown at the time of discovery. Initial targeted approaches failed to identify its mechanism of action. These include investigating transcription factors [cAMP-response element binding protein (CREB) (Chen et al., 2013), nuclear factor NF-κB, Hippo pathway], enzymes [dihydrofolate reductase (DHFR) (Chen et al., 2013), protein kinase A (PKA), DNA methyltransferase 1 (DNMT1), hexokinase 2 (HK2)] and membrane proteins [G-protein coupled receptor 119 (GPR119), voltage-gated sodium channel Nav1.7, calcitonin gene-related peptide (CGRP) receptor, metabotropic glutamate receptor 2 (mGluR2)]. To identify the direct protein target(s) that **LBL1** could bind to in the entire cellular proteome in an unbiased way, we developed a chemical proteomics

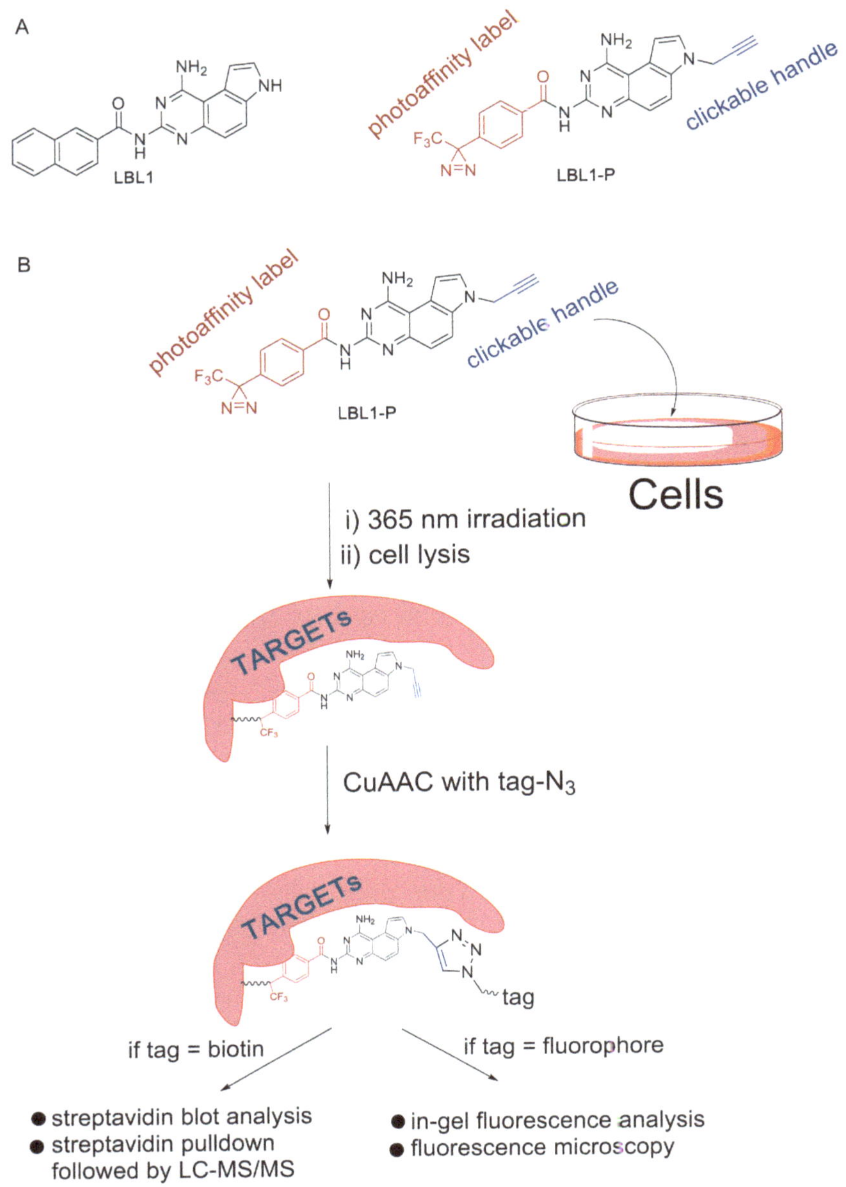

Fig. 1 Overview of target identification using a clickable photoaffinity probe **LBL1-P**. (A) Chemical structures of **LBL1** and **LBL1-P**. The clickable photoaffinity probe **LBL1-P** contains bioactive acylpyrroloquinazoline core, trifluoromethyldiazirine as the photo-crosslinking moiety and a clickable alkyne group. (B) A schematic diagram of target identification using the clickable photoaffinity probe **LBL1-P** in intact cells. In this approach, the living cells could be treated with **LBL1-P** in the presence or absence

approach using a clickable photoaffinity probe **LBL1-P** (Fig. 1A) (Li, Chen, Chao, David, et al., 2018; Li, Chen, Chao, Zheng, & Xiao, 2018). In this approach, **LBL1-P** was designed to contain three functional units: (1) the acylpyrroloquinazoline core to bind to its target(s); (2) a trifluoromethyldiazirine group as a photocrosslinking moiety to covalently capture any targets that the acylpyrroloquinazoline core could bind to upon UV irradiation; and (3) a clickable alkyne handle allowing copper-catalyzed azide alkyne cycloaddition (CuAAC) click reaction with a tagged azide for various downstream applications (Fig. 1B). These applications may include enriching proteins for target identification by mass spectroscopy, SDS-PAGE analysis and fluorescence microscopic analysis (Li, Chen, Chao, Zheng, et al., 2018).

In this chapter, we describe the details of our efforts in identifying nuclear lamins as the molecular targets of **LBL1** using the clickable photoaffinity probe **LBL1-P**. These results established **LBL1** (lamin-binding ligand 1) as the first small molecule known to directly bind nuclear lamins.

2. Protocols

2.1 Identification of lamins as the targets of LBL1 by photocrosslinking

2.1.1 Materials and equipment needed

Materials and Reagents

MDA-MB-468 breast cancer cell line (ATCC)

LBL1 (see Chen et al., 2013) for its synthesis)

LBL1-P (see Li, Chen, Chao, David, et al., 2018 for its synthesis)

---

of competitor **LBL1**. Then the cells would be irradiated at 365 nm to initiate covalent crosslinking between **LBL1-P** and the protein targets that it could directly interact with in the cells. Once crosslinked, the cells would be lysed and the lysates would be subjected to Cu-catalyzed azide alkyne cycloaddition (CuAAC) reaction with biotin-N_3 or fluorophore-N_3. When biotin-N_3 was used, the clicked lysates could be analyzed directly by SDS-PAGE with streptavidin blotting. Or alternatively, the clicked lysates could be purified by streptavidin pulldown and the enriched proteins could be analyzed by LC-MS/MS to identify the proteins. When a fluorescent azide was used, the clicked lysates could be directly visualized by SDS-PAGE with a fluorescent gel scanner. Or alternatively, the CuAAC could be performed directly in the cells and then the cells could be visualized under a fluorescence microscope to investigate the localization of the probe inside the cells.

Phosphate buffered saline [PBS, Na_2HPO_4 (4.3 mM), KH_2PO_4 (1.4 mM), KCl (2.7 mM), NaCl (137 mM), pH 7.4]
Dulbecco's Modified Eagle Medium (DMEM, Gibco)
Fetal bovine serum (FBS, Hyclone)
Non-essential amino acids (Life Technologies)
Pre-cast SDS-PAGE gel (Bio-rad)
Nonfat dry milk
Streptavidin-horseradish peroxidase (HRP) conjugate (GE Healthcare)
Tris-buffer saline with Tween-20 [TBST, Tris (50 mM), NaCl (150 mM), Tween-20 (0.1%), pH 7.5]
Sodium dodecyl sulfate (SDS)
BCA protein assay kit (Pierce)
Biotin-azide (catalog # AZ104, Click Chemistry Tools)
Rhodamine-azide (AZ109, Click Chemistry Tools)
Copper (II) sulfate
Tris(2-carboxyethyl)phosphine (TCEP)
Tris[(1-benzyl-1H-1,2,3-triazol-4-yl)methyl]amine (TBTA)
6 × SDS-PAGE sample buffer [Tris (300 mM, pH 6.8), SDS (12%), glycerol (60%), bromophenol blue (0.6%), 2-mercaptoethanol (5%)]
Polyvinyl difluoride (PVDF) membrane
Chemiluminescent substrates for HRP (e.g., Clarity™, Bio-rad)
Tissue culture-treated plates (10-cm, 15-cm and 6-well)
Dithiothreitol (DTT)
Iodoacetamide
Streptavidin agarose beads (Thermo-Fisher)
Spin-X Centrifuge Tube filter (Corning #8163)
Biotin
Ammonium bicarbonate

Equipment

CO_2 incubator
UV crosslinker (FB-UVXL-1000, Fisher Scientific)
Table-top centrifuge
Probe sonicator
Gel electrophoresis and transfer system (e.g., mini-PROTEAN Tetracell, Bio-rad)
Plate reader capable of reading UV–Vis absorbance
ChemiDoc™ MP imaging system
Vortex mixer
Heat block

2.1.2 Procedures

2.1.2.1 Analytical scale for streptavidin blot analysis

1. Plate 1 million MDA-MB-468 cells in DMEM media containing 10% FBS and non-essential amino acids into 6-well plates and allow the cells to attach for overnight in a humidified incubator at 37 °C with 5% CO_2.
2. The cells in each well were washed once with PBS (1 mL). Then PBS (1.5 mL) was added to each well.
3. The cells were treated with different concentrations of **LBL1** for 10 min at 37 °C. Then different concentrations of **LBL1-P** were added to the cells and incubated at 37 °C for 30 min. We typically designed the experiments to include **LBL1** at 5 × the concentration of **LBL1-P** as a competitor. Often a series of titrations should be included. Inclusion of **LBL1** as the competitor was critical to discern potential targets because only those that could be competed were relevant targets for **LBL1**.
4. The cells were then cooled to 4 °C in the dark for 5 min.
5. The cells were then irradiated at 365 nm for 5 min. Make sure the plate lid was removed from the plate. We typically fixed the distance between the lamp and cells at 5–10 cm. The UV reaction time needs to be empirically optimized for each specific case.
6. The cells were collected by scraping and centrifuged in a table-top centrifuge at 8000 × RPM for 5 min at 4 °C. The supernatant was removed and the cell pellets were washed twice with 1 mL ice-cold PBS each. The cell pellets were then lysed in 100 μL PBS with 1%SDS. We typically sonicated the cell lysates with a probe sonicator to breakdown the jelly genomic DNA for easy pipetting. The cell lysates were centrifuged in a table-top centrifuge at 14,000 × RPM for 10 min and the supernatant was saved as the cell lysates. The protein concentration of the cell lysates was determined by a BCA protein assay kit (Pierce).
7. Equal amount (e.g., 10 μg) of the protein lysates from each condition was taken for CuAAC with a biotin-azide. Add appropriate amount of PBS so that each reaction's volume was the same across different conditions. For the CuAAC click reaction, we generally premixed the components in the order listed in Table 1, right before use. A bulk mixture can be prepared for all the reactions. Add appropriate amount of the click reaction cocktail into the protein lysates and mix well. The reaction mixture was incubated at room temperature in the dark for 1.5 h.
8. Add 6 × SDS-PAGE sample buffer into each click reaction mixture. The samples were then loaded into a pre-cast SDS-PAGE gel for

Table 1 Preparation of CuAAC reaction cocktail.

Reagent	Stock concentration (mM) (solvent)	Dilution factor	Working concentration (μM)	Amount for a 20 μL reaction (μL)
Azide	10 (DMSO)[a]	1:200	50	0.1
TCEP	50 (H_2O)[b]	1:50	1000	0.4
TBTA	10 (DMSO)[a]	1:100	100	0.2
$CuSO_4$	50 (H_2O)[b]	1:50	1000	0.4

[a]The DMSO stock solution should be stored at −20 °C.
[b]This solution should be prepared freshly for each use.

electrophoresis. A gradient gel appropriate for a large range of molecular weights is recommended.

9. Once the bromophenol blue reached the front end of the gel, transfer the proteins from the gel to a PVDF membrane following manufacturer's protocol.
10. The PVDF membrane was then blocked in 5% nonfat dry milk in TBST for 1 h at room temperature.
11. The membrane was washed two times with TBST for 5 min each at room temperature. Then it was incubated with streptavidin-HRP in TBST (1:5,000) at room temperature for 1 h.
12. The membrane was washed three times with TBST for 5 min each. Then the biotinylated proteins on the PVDF membrane were detected using appropriate chemiluminescent substrates following manufacturer's protocol. A typical result is shown in Fig. 2. It is expected that **LBL1-P** should label its targets in a dose-dependent manner. Furthermore, the relevant targets for **LBL1** should be competed by excess **LBL1**.
13. Alternatively, if the lysates were clicked with a rhodamine-N_3, the clicked signal could also be directly visualized in the SDS-PAGE gel using a fluorescent gel scanner (e.g., ChemiDoc™ MP, Bio-rad). If desired, additional cell fractionation method could also be employed to investigate which cell compartment the probe-labeled proteins are resided in. Please see reference (Li, Chen, Chao, David, et al., 2018) for additional details.

2.1.2.2 Preparative scale for proteomic analysis

1. Plate 3 × 15-cm plates of MDA-MB-468 cells and allow them to grow to ~80% confluence. For each additional condition, prepare another set of 3 × 15-cm plates of cells.

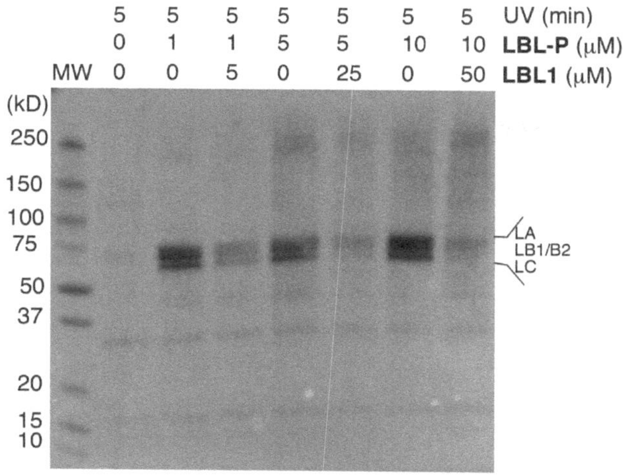

Fig. 2 Typical streptavidin blot results of MDA-MB-468 cells treated with different concentration of **LBL1** and/or **LBL1-P** after UV irradiation at 365 nm. The cells were treated with indicated concentrations of **LBL1** and/or **LBL1-P**. Then the cells were processed for CuAAC click reaction with biotin-N_3. The whole cell lysates were then separated on a SDS-PAGE gel followed by electro-transfer to a PVDF membrane. The biotinylated proteins were detected with streptavidin-HRP. *Reprinted with permission from reference Li, B. X., Chen, J., Chao, B., David, L. L., & Xiao, X. (2018). Anticancer pyrroloquinazoline LBL1 targets nuclear lamins.* ACS Chemical Biology, 13, *1380–1387. Copyright (2018) American Chemical Society.*

2. Treat and process the cells as described in steps 2–7 in Section 2.1.2.1. It is expected that ~6 mg of cell lysates will be obtained from each condition with 3 × 15-cm plates of MDA-MB-468 cells. The lysates were then clicked with biotin-N_3 at the final lysate concentration of 1 mg/mL. We typically ran the reactions at a scale of 500 μL per reaction. The total reaction mixture was anticipated to be 6 mL and will be used for the next steps. If the final volume is different, the amount of each reagent needs proportional adjustment.
3. MeOH (6 mL) and $CHCl_3$ (1.5 mL) were added to each of the click reaction mixture. If the volume of the reaction mixture was different, the ratio of PBS:MeOH:$CHCl_3$ should be maintained at 4:4:1. The mixture was shaken vigorously for 30 s and centrifuged at 8000 × *g* for 10 min.
4. The liquid was removed and the protein solids were gently washed with 3 × 3 mL MeOH/$CHCl_3$ (1:1) without breaking the protein solid pellets.

5. MeOH (1.6 mL) was added to the protein solids and the mixture was sonicated with a probe sonicator to fully resuspend the solids. Then $CHCl_3$ (0.4 mL) was added and the mixture was shaken vigorously for 30 s. The mixture was centrifuged at 14,000 × RPM for 10 min to pellet the solids.
6. The protein solids were resuspended in 1.5 mL PBS with 8 *M* urea and 2% SDS and sonicated with a probe sonicator to dissolve the solids.
7. The solution was centrifuged at 14,000 × RPM to remove any undissolved solids. It is critical that the undissolved solids were removed at this step. Otherwise, they would significantly contaminate the bound proteins derived from step 14 (see below).
8. The protein solution was reduced by adding DTT (the stock was 1 M in H_2O and should be made freshly) to a final concentration at 10 mM. The solution was incubated at 55 °C for 30 min.
9. Then iodoacetamide (the stock was 625 mM in H_2O and should be made freshly) was added to a final concentration at 25 mM. The solution was tumbled in the dark at room temperature for 30 min.
10. The solution was diluted 10 × with PBS to a final [SDS] = 0.2%.
11. 50 μL of PBS pre-washed streptavidin beads were added to the solution and the mixture was tumbled at room temperature for 2 h.
12. The beads were pelleted by centrifuging at 700 × *g* for 2 min. The supernatant was removed.
13. The beads were washed with 3 × 1 mL 1% SDS/PBS followed by 6 × 1 mL PBS. All the centrifuging steps were done at 3000 × RPM for 2 min in a table-top centrifuge at room temperature. To prevent potential contaminations, the washes and all the following steps should be performed on a clean bench with minimal contaminations. We typically performed these steps in a laminar flow hood.
14. The bound proteins were eluted with 75 μL PBS containing 1% SDS and 1 mM biotin by heating at 95 °C for 5 min.
15. The eluted proteins were collected by a Spin-X Centrifuge Tube filter.
16. The beads were further washed with 3 × 100 μL NH_4HCO_3 (100 mM).
17. The elution and washings were combined and lyophilized. The residue was subjected to trypsin digestion and liquid chromatography coupled to tandem mass spectroscopy analysis (LC-MS/MS) to identify the bound proteins. Both label-free spectra counting and tandem mass tag (TMT) labeling can be used to quantify the relative amount of each protein under different experimental groups. The relevant targets for

LBL1 should have lower abundance in the group with competitor than the group without competitor. Appropriate statistics should be applied to detect significance.

2.2 Validation of lamin A as the target of LBL1 by fluorescence microscopy imaging

2.2.1 Materials and equipment needed (in addition to those listed in Section 2.1.1)

MDA-MB-231 cells (ATCC)
Tissue culture-treated plates (10-cm and 24-well)
Cover slips (e.g., catalog # 12-545-81, Fisher Scientific)
Poly-L-lysine (catalog # P4707, Sigma)
Formaldehyde (37%)
Triton X-100
Bovine serum albumin (BSA)
Click-iT cell reaction cocktail (Life Technologies)
anti-LA (catalog # SAB4200236, Sigma)
Donkey anti-mouse IgG conjugated with Alexa Fluor 488 (Jackson ImmunoResearch Laboratory)
ProLong Gold Anti-Fade reagent with 4′,6-diamidino-2-phenylindole (DAPI) (Life Technologies)
Fluorescence microscope (e.g., ApoTome, Zeiss)
Imaris software package

2.2.2 Procedures

1. Put a sterile cover slip into each well of a 24-well plate. The cover slip was treated with poly-L-lysine (500 μL) for 4 h at 37 °C. Then the poly-L-lysine solution was removed and the cover slip was further washed with 3 × 500 μL sterile ddI H_2O.
2. Plate MDA-MB-231 cells (10,000/well) in DMEM media containing 10% FBS and non-essential amino acids into the 24-well plate containing the cover slips prepared in step 1 and allow the cells to attach for overnight.
3. The cells were washed once with PBS (500 μL). Then PBS (500 μL) was added to each well.
4. The cells were treated with 0 or 1 μM **LBL1** for 20 min followed by treatment with 0 or 0.1 μM **LBL1-P** for another 20 min at 37 °C.
5. Remove 200 μL of the PBS solution from each well and cool the plate to 4 °C in the dark. Then the plate was irradiated at 365 nm for 5 min.

6. The PBS solution was removed and the cells were washed twice with ice-cold PBS (1 mL). For each wash, allow incubating with PBS on ice for 5 min.
7. The cells were then fixed with 3.7% formaldehyde in PBS (0.5 mL) for 15 min at room temperature. Then the cells were washed twice with 0.5 mL PBS containing 3% BSA. For each wash, allow incubating with the PBS solution at room temperature for 5 min.
8. The cells were then permeabilized with a solution (0.5 mL) of PBS containing 0.5% triton X-100 for 20 min at room temperature. Then the cells were washed twice with 0.5 mL PBS containing 3% BSA. For each wash, allow incubating with the PBS solution at room temperature for 5 min.
9. The cells were incubated with Click-iT cell reaction cocktail supplemented with rhodamine-N_3 (1 μM) for 30 min at room temperature. At the end of the click reaction, the cells were washed twice with 0.5 mL PBS containing 3% BSA. For each wash, allow incubating with the PBS solution at room temperature for 5 min. For the fluorescence imaging analysis, we found that the Click-iT cell reaction cocktail gave lower background signal than the reagents we prepared in Table 1. In addition, the rhodamine-N_3 concentration to be used needs to be optimized in each case due to its hydrophobic nature and potential non-specific binding.
10. The cells were then incubated with anti-LA (1:400) in PBS containing 3% BSA and 0.5% triton X-100 for overnight at 4 °C. The cells were washed twice with 0.5 mL PBS containing 3% BSA and 0.5% triton X-100. For each wash, allow incubating with the PBS solution at room temperature for 5 min.
11. The cover slips were incubated with donkey anti-mouse IgG conjugated with Alexa Fluor 488 (1:1000) in PBS containing 3% BSA and 0.5% triton X-100 for 1 h at room temperature. The cells were washed twice with 0.5 mL PBS containing 3% BSA and 0.5% triton X-100. For each wash, allow incubating with the PBS solution at room temperature for 5 min.
12. The cover slips were then mounted to glass slides with ProLong Gold Anti-Fade reagent with DAPI. The coverslips were then imaged with a fluorescence microscope (ApoTome, Zeiss). Typical results are shown in Fig. 3. It is expected that the **LBL1-P** signal should co-localize with anti-LA signal if LA were the target of **LBL1-P**. Furthermore, inclusion of **LBL1** as a competitor should compete **LBL1-P**'s labeling without affecting anti-LA signal.

13. An additional useful strategy to validate that **LBL1** binds to lamins in living cells is to use specialized cells whose lamin protein expression was knocked out or silenced. The essentially same steps described above can be followed. Typical results from these studies can be found in reference (Li, Chen, Chao, Zheng, et al., 2018), where mouse embryonic fibroblasts without lamins were used.

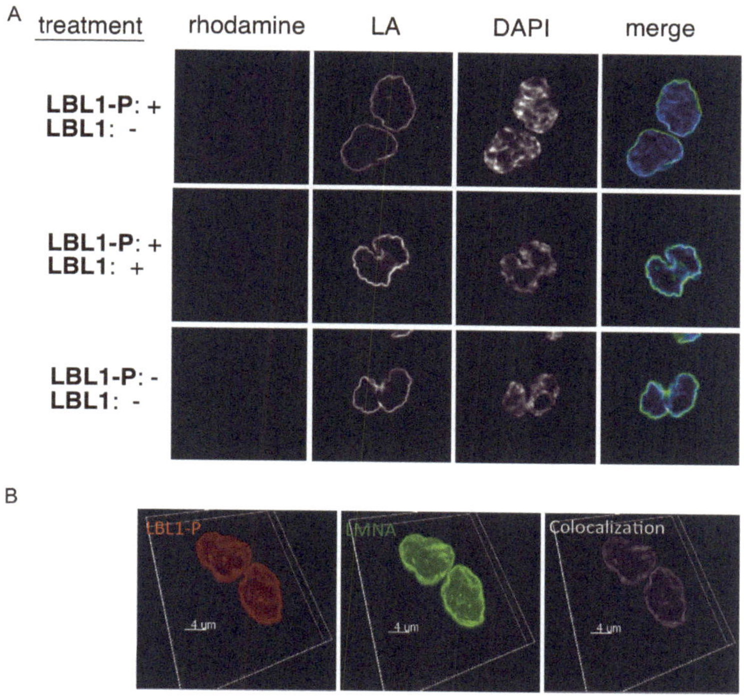

Fig. 3 Typical fluorescence micrographs of MDA-MB-231 cells treated with **LBL1** (0 or 1 μM) and/or **LBL1-P** (0 or 0.1 μM). The cells on coverslips were treated as indicated. Then the cells were fixed, permeabilized and clicked with a rhodamine-N_3. After the click reaction, the cells were further stained with anti-LA by immunofluorescence. Then the cells were examined under a fluorescence microscope. The cell nucleus was marked with DAPI. (A) Representative fluorescence micrographs of MDA-MB-231 cells treated with different compound combinations. (B) The cells from the top panel in (A) were deconvoluted using *Z*-stacks of images. The colocalization signal was calculated from a new channel in the Coloc™ module in the Imaris imaging analysis software package. *Reprinted with permission from reference Li, B. X., Chen, J., Chao, B., David, L. L., & Xiao, X. (2018). Anticancer pyrroloquinazoline LBL1 targets nuclear lamins.* ACS Chemical Biology, 13, *1380–1387. Copyright (2018) American Chemical Society.*

3. Conclusions

Target identification of bioactive small molecules remains a critical challenge in chemical biology research (Burdine & Kodadek, 2004). Over the past two decades, many different chemical biology approaches have been developed to facilitate target identification in an unbiased way from the complex biological milieu (Ziegler, Pries, Hedberg, & Waldmann, 2013). For example, the bioactive small molecules can be directly immobilized to solid-phase matrixes to precipitate and enrich their binding proteins from crude cell lysates, which can then be identified by either Edman degradation or protein mass spectroscopy. The binding receptor for immunosuppressive macrolide FK506 was identified through this method and found to be FK506-binding protein (FKBP) (Harding, Galat, Uehling, & Schreiber, 1989). Another chemical proteomics approach is called drug affinity responsive target stability (DARTS) (Lomenick et al., 2009; Brett Lomenick, Olsen, & Huang, 2010). DARTS is based on the principle that small molecule's binding to its target proteins will enhance their proteolytical stability against proteases, which can be analyzed by quantitative protein mass spectroscopy. Protein translation initiation factor eIF4A was found to be a target for resveratrol using DARTS method (Lomenick et al., 2009). Yet another chemical proteomics approach related to DARTS is based on the melting temperature shift of target proteins upon small molecule binding. This approach, called cellular thermal shift assay (CETSA) (Martinez Molina et al., 2013), was used to identify nucleophosmin (NPM) as a potential target for a simple natural product hordenine (Park, Ha, Koo, Park, & Park, 2017), which has bioactivities including stimulation of gastrin release and inhibition of norepinephrine uptake (Barwell, Basma, Lafi, & Leake, 1989; Yokoo et al., 1999). A more recently evolved approach for targeted protein degradation called proteolysis-targeting chimeras (PROTACs) can also be potentially engineered for target deconvolution (Lai & Crews, 2017). In addition to the chemical biology approaches, other molecular biology and bioinformatics strategies have also been successfully utilized to identify receptors for bioactive small molecules. For example, the receptor for capsaicin was found to be TRPV1 (transient receptor potential cation channel subfamily V member 1) through expression cloning (Caterina et al., 1997) and the target for the anticancer indenoisoquinoline was found to be human topoisomerase I through bioinformatic COMPARE analysis (Kohlhagen, Paull, Cushman, Nagafuji, & Pommier, 1998; Marzi et al., 2018).

For all the chemical proteomics strategies mentioned above, it is important to verify that the chemical probes to be designed and synthesized for target identification retain the bioactivity of the original compound although decorating the original compound with different chemical handles will most likely result in decreased bioactivity. For the approach with clickable photoaffinity probes, the substitution position of the two required chemical groups, i.e., photocrosslinking unit and clickable handle, can have a huge effect on the targets to be identified because these changes can sometimes alter the potential binding target profiles and/or the potential reactivity of the generated reactive carbene species (Niphakis et al., 2015). In this regard, we found that **LBL1-P** retained the biological activity expected for **LBL1**, but with reduced potency (Li, Chen, Chao, David, et al., 2018). As the first small molecule known to directly target nuclear lamins, **LBL1** shall be a useful tool to help dissect the roles of lamins in many aspects of lamin biology including mechanotransduction, stem cell maintenance and DNA repair (Butin-Israeli, Adam, Goldman, & Goldman, 2012; Gerace & Tapia, 2018). In the case of DNA repair, we recently discovered that LA has a unique role in modulating the protein stability of DNA recombinase Rad51 to regulate homologous recombination repair of DNA double-strand breaks (Li, Chen, Chao, Zheng, et al., 2018). The covalent nature of the photocrosslinking unit may make the approach using clickable photoaffinity probes particularly applicable to hydrophobic compounds such as bioactive lipid molecules, whose receptors are likely to be membrane-bound proteins (Alkayed et al., 2018; Niphakis et al., 2015). We envision that this approach should be a powerful one to identify the targets of many different bioactive small molecules.

Acknowledgments

This work was supported by NIH R01CA211866 and partially by R01GM122820. Oregon Health & Science University School of Medicine, Office of the Technology Transfer and Business Development, Oregon Clinical and Translational Research Institute, Murdock Charitable Trust, Medical Research Foundation of Oregon and Lloyd Fund partially supported this research. We thank the Eli Lilly's Open Innovation Drug Discovery (OIDD) team for sharing initial profiling results of **LBL1** against different biological targets.

References

Aebi, U., Cohn, J., Buhle, L., & Gerace, L. (1986). The nuclear lamina is a meshwork of intermediate-type filaments. *Nature*, *323*(6088), 560–564.

Alkayed, N. J., Cao, Z., Qian, Z. Y., Nagarajan, S., Liu, X., Nelson, J., et al. (2018). *Bidirectional control of coronary vascular resistance by eicosanoids via a novel GPCR*. bioRxiv.

Bargagna-Mohan, P., Hamza, A., Kim, Y. E., Khuan Abby Ho, Y., Mor-Vaknin, N., Wendschlag, N., et al. (2007). The tumor inhibitor and antiangiogenic agent withaferin A targets the intermediate filament protein vimentin. *Chemistry and Biology*, *14*(6), 623–634.

Barwell, C. J., Basma, A. N., Lafi, M. A., & Leake, L. D. (1989). Deamination of hordenine by monoamine oxidase and its action on vasa deferentia of the rat. *Journal of Pharmacy and Pharmacology*, *41*(6), 421–423.

Bredel, M., & Jacoby, E. (2004). Chemogenomics: An emerging strategy for rapid target and drug discovery. *Nature Reviews Genetics*, *5*(4), 262–275.

Burdine, L., & Kodadek, T. (2004). Target identification in chemical genetics: The (often) missing link. *Chemistry and Biology*, *11*(5), 593–597.

Burke, B., & Stewart, C. L. (2013). The nuclear lamins: Flexibility in function. *Nature Reviews Molecular Cell Biology*, *14*(1), 13–24.

Butin-Israeli, V., Adam, S. A., Goldman, A. E., & Goldman, R. D. (2012). Nuclear lamin functions and disease. *Trends in Genetics*, *28*(9), 464–471.

Caterina, M. J., Schumacher, M. A., Tominaga, M., Rosen, T. A., Levine, J. D., & Julius, D. (1997). The capsaicin receptor: A heat-activated ion channel in the pain pathway. *Nature*, *389*(6653), 816–824.

Chen, J., Kassenbrock, A., Li, B. X., & Xiao, X. (2013). Discovery of a potent anti-tumor agent through regioselective mono—Acylation of 7-pyrrolo[3,2-]quinazoline-1,3-diamine. *Medicinal Chemistry Communications*, *4*(9), 1275–1282.

Dechat, T., Adam, S. A., Taimen, P., Shimi, T., & Goldman, R. D. (2010). Nuclear lamins. *Cold Spring Harbor Perspectives in Biology*, *2*(11), a000547.

Dechat, T., Korbei, B., Vaughan, O. A., Vlcek, S., Hutchison, C. J., & Foisner, R. (2000). Lamina-associated polypeptide 2alpha binds intranuclear A-type lamins. *Journal of Cell Science*, *113*, 3473–3484.

Dhe-Paganon, S., Werner, E. D., Chi, Y. I., & Shoelson, S. E. (2002). Structure of the globular tail of nuclear lamin. *Journal of Biological Chemistry*, *277*(20), 17381–17384.

Fletcher, D. A., & Mullins, R. D. (2010). Cell mechanics and the cytoskeleton. *Nature*, *463*(7280), 485–492.

Fuchs, E., & Cleveland, D. W. (1998). A structural scaffolding of intermediate filaments in health and disease. *Science*, *279*(5350), 514–519.

Gerace, L., & Tapia, O. (2018). Messages from the voices within: Regulation of signaling by proteins of the nuclear lamina. *Current Opinion in Cell Biology*, *52*, 14–21.

Harding, M. W., Galat, A., Uehling, D. E., & Schreiber, S. L. (1989). A receptor for the immunosuppressant FK506 is a cis-trans peptidyl-prolyl isomerase. *Nature*, *341*(6244), 758–760.

Hesse, M., Magin, T. M., & Weber, K. (2001). Genes for intermediate filament proteins and the draft sequence of the human genome: Novel keratin genes and a surprisingly high number of pseudogenes related to keratin genes 8 and 18. *Journal of Cell Science*, *114*(Pt. 14), 2569–2575.

Jordan, M. A., & Wilson, L. (1998). Microtubules and actin filaments: Dynamic targets for cancer chemotherapy. *Current Opinion in Cell Biology*, *10*(1), 123–130.

Jordan, M. A., & Wilson, L. (2004). Microtubules as a target for anticancer drugs. *Nature Reviews Cancer*, *4*(4), 253–265.

Kohlhagen, G., Paull, K., Cushman, M., Nagafuji, P., & Pommier, Y. (1998). Protein-linked DNA strand breaks induced by NSC 314622, a novel noncamptothecin topoisomerase I poison. *Molecular Pharmacology*, *54*, 50–58.

Lai, A. C., & Crews, C. M. (2017). Induced protein degradation: An emerging drug discovery paradigm. *Nature Reviews. Drug Discovery*, *16*(2), 101–114.

Lee, S. J., Jung, Y. S., Yoon, M. H., Kang, S. M., Oh, A. Y., Lee, J. H., et al. (2016). Interruption of progerin-lamin A/C binding ameliorates hutchinson-gilford progeria syndrome phenotype. *Journal of Clinical Investigation*, *126*(10), 3879–3893.

Li, B. X., Chen, J., Chao, B., David, L. L., & Xiao, X. (2018). Anticancer pyrroloquinazoline LBL1 targets nuclear lamins. *ACS Chemical Biology*, *13*, 1380–1387.

Li, B. X., Chen, J., Chao, B., Zheng, Y., & Xiao, X. (2018). A lamin-binding ligand inhibits homologous recombination repair of DNA double-strand breaks. *ACS Central Science*, *4*(9), 1201–1210.

Lin, F., & Worman, H. J. (1993). Structural organization of the human gene encoding nuclear lamin A and nuclear lamin C. *Journal of Biological Chemistry*, *268*(22), 16321–16326.

Lomenick, B., Hao, R., Jonai, N., Chin, R. M., Aghajan, M., Warburton, S., et al. (2009). Target identification using drug affinity responsive target stability (DARTS). *Proceedings of the National Academy of Sciences of the United States of America*, *106*(51), 21984–21989.

Lomenick, B., Olsen, R. W., & Huang, J. (2010). Identification of direct protein targets of small molecules. *ACS Chemical Biology*, *6*(1), 34–46.

Martinez Molina, D., Jafari, R., Ignatushchenko, M., Seki, T., Larsson, E. A., Dan, C., et al. (2013). Monitoring drug target engagement in cells and tissues using the cellular thermal shift assay. *Science*, *341*(6141), 84–87.

Marzi, L., Agama, K., Murai, J., Difilippantonio, S., James, A., Peer, C. J., et al. (2018). Novel fluoroindenoisoquinoline non-camptothecin topoisomerase I inhibitors. *Molecular Cancer Therapeutics*, *17*(8), 1694–1704.

Moffat, J. G., Vincent, F., Lee, J. A., Eder, J., & Prunotto, M. (2017). Opportunities and challenges in phenotypic drug discovery: An industry perspective. *Nature Reviews. Drug Discovery*, *16*(8), 531–543.

Niphakis, M. J., Lum, K. M., Cognetta, A. B., 3rd, Correia, B. E., Ichu, T. A., Olucha, J., et al. (2015). A global map of lipid-binding proteins and their ligandability in cells. *Cell*, *161*(7), 1668–1680.

Park, H., Ha, J., Koo, J. Y., Park, J., & Park, S. B. (2017). Label-free target identification using in-gel fluorescence difference via thermal stability shift. *Chemical Science*, *8*(2), 1127–1133.

Robert, A., Hookway, C., & Gelfand, V. I. (2016). Intermediate filament dynamics: What we can see now and why it matters. *BioEssays*, *38*(3), 232–243.

Ruan, J., Xu, C., Bian, C., Lam, R., Wang, J. P., Kania, J., et al. (2012). Crystal structures of the coil 2B fragment and the globular tail domain of human lamin B1. *FEBS Letters*, *586*(4), 314–318.

Shumaker, D. K., Solimando, L., Sengupta, K., Shimi, T., Adam, S. A., Grunwald, A., et al. (2008). The highly conserved nuclear lamin Ig-fold binds to PCNA: Its role in DNA replication. *Journal of Cell Biology*, *181*(2), 269–280.

Strelkov, S. V., Schumacher, J., Burkhard, P., Aebi, U., & Herrmann, H. (2004). Crystal structure of the human lamin A coil 2B dimer: Implications for the head-to-tail association of nuclear lamins. *Journal of Molecular Biology*, *343*(4), 1067–1080.

Swinney, D. C., & Anthony, J. (2011). How were new medicines discovered? *Nature Reviews. Drug Discovery*, *10*(7), 507–519.

Wagner, B. K., & Schreiber, S. L. (2016). The power of sophisticated phenotypic screening and modern mechanism-of-action methods. *Cell Chemical Biology*, *23*(1), 3–9.

Yokoo, Y., Kohda, H., Kusumoto, A., Naoki, H., Matsumoto, N., Amachi, T., et al. (1999). Isolation from beer and structural determination of a potent stimulant of gastrin release. *Alcohol and Alcoholism*, *34*(2), 161–168.

Zastrow, M. S., Flaherty, D. B., Benian, G. M., & Wilson, K. L. (2006). Nuclear titin interacts with A- and B-type lamins in vitro and in vivo. *Journal of Cell Science*, *119*(Pt. 2), 239–249.

Zastrow, M. S., Vlcek, S., & Wilson, K. L. (2004). Proteins that bind A-type lamins: Integrating isolated clues. *Journal of Cell Science*, *117*(Pt. 7), 979–987.

Ziegler, S., Pries, V., Hedberg, C., & Waldmann, H. (2013). Target identification for small bioactive molecules: Finding the needle in the haystack. *Angewandte Chemie (International Ed. in English)*, *52*(10), 2744–2792.

CHAPTER ELEVEN

REX technologies for profiling and decoding the electrophile signaling axes mediated by Rosetta Stone proteins

Marcus J.C. Long[a], Daniel A. Urul[b,c], Yimon Aye[b,*]
[a]Beverley East Riding of Yorkshire, United Kingdom
[b]Institute of Chemical Sciences and Engineering, École Polytechnique Fédérale de Lausanne, Lausanne, Switzerland
[c]Department of Chemistry and Chemical Biology, Cornell University, Ithaca, NY, United States
*Corresponding author: e-mail address: yimon.aye@epfl.ch

Contents

Methods in Enzymology, Volume 633
ISSN 0076-6879
https://doi.org/10.1016/bs.mie.2019.02.039

Abstract

It is now clear that some cysteines on some proteins are highly tuned to react with electrophiles. Based on numerous studies, it is also established that electrophile sensing underpins rewiring of several critical signaling processes. These electrophile-sensing proteins, or privileged first responders (PFRs), are likely critically relevant for drug design. However, identifying PFRs remains a challenging and unsolved problem, despite the development of several high-throughput methods to ID proteins that react with electrophiles. More importantly, we remain unable to rank how different PFRs identified under different conditions relate to one another, in terms of sensing or signaling capacity. Here we evaluate different methods to assay sensing functions of proteins and discuss these methods in the context of developing a "ranking scheme." Based on theoretical and experimental evidence, we propose that T-REX—the only targeted-electrophile delivery tool presently available—is a reliable method to rank PFRs. Finally, we address to what extent electrophile sensing and downstream signaling are correlated. Based on our current data, we observe that such behaviors are indeed correlated. It is our hope that through this manuscript researchers from various arms of the stress signaling fields will focus on developing a quantitative understanding of precision electrophile labeling.

1. Introduction

The cell is a powerhouse of oxidative chemistry. Most classic cellular processes taught in schools and universities involve oxidation of biological substrates, such as glucose (Gatenby & Gillies, 2004) or pyruvate (Owen, Kalhan, & Hanson, 2002), to generate energy equivalents. These processes are uniquely chemically selective and challenging. Generations have rightly marveled at their intricacy. Interestingly, many of these processes generate reactive chemical intermediates as part of their cycles. Several of these electrophiles, such as fumarate (Blatnik, Frizzell, Thorpe, & Baynes, 2008) and methyl glyoxal (Bollong et al., 2018), modify proteins. There is also evidence that mechanisms exist to regulate these processes (Menard, Maughan, & Vigoreaux, 2014). Additionally, metabolic processes are not actually 100% efficient; seepage of electrons can generate reactive oxygen species (ROS) (Jastroch, Divakaruni, Mookerjee, Treberg, & Brand, 2010), upregulating oxidative stress that can in turn generate reactive electrophilic species (RES) (Pizzimenti, Toaldo, Pettazzoni, Dianzani, & Barrera, 2010). Finally, the cell contains regulated enzymes intended to upregulate both ROS and RES as part of coordinated signaling pathways that are essential for fitness (Parvez, Long, Poganik, & Aye, 2018). Thus, it is unsurprising that there is a wealth of evidence that ROS and RES can function as cellular signals. For instance, it is clear that ROS can stimulate growth of both non-transformed (Carol & Dolan, 2006) and transformed cells (Kumari, Badana, Gavara, Gugalavath, & Malla, 2018; Zhou, Shao, & Spitz, 2014). Indeed, it is believed that ROS is a critical part of transformation processes, contributing to the mutator phenotype (Sallmyr, Fan, & Rassool, 2008) and stimulating growth through modulation of specific proteins (Kang, Lee, & Lee, 2015). Mature cancer cells also maintain a high ROS environment, indicating that they are "addicted to ROS." RES—the focus of this manuscript—have also been linked to modulation of cell signaling processes. However, there is much less of consensus in terms of magnitude and signs of responses observed (Parvez et al., 2018). Furthermore, known modified proteins are much less readily linked to observed phenotypes.

Emerging data indicate that some RES sensors can change their function upon RES modification, in the process altering signal flux (Castro, Jung, Grune, & Siems, 2017). These proteins must engage the specific RES and also couple that modification to rewiring flux through (a) specific pathway(s). In some cases, this information exchange can occur through precise interchange between the RES chemical code and conventional PTMs, such as phosphate and ubiquitin (Zhao, Long, Wang, Zhang, & Aye, 2018). We have called such proteins "Rosetta Stones," as they serve as a mediator or "dragoman" to translate biological information. The discovery of such proteins implies that electrophile sensing and the modification of signaling outputs in some cases may be linked. However, as we describe below, RES signaling can be conceptualized into a two-step process. The first step involves engaging the reactive chemical (sensing). The second step leads to changes in pathway flux (signaling). We thus break the discussion of the relevant methods to interrogate RES-regulated biological processes into these two parts, re-uniting them at the end. Although the manuscript centers on native lipid-derived bioactive RES, 4-hydroxynonenal (HNE), the methods and/or underpinning ideas are transposable to studying other RES beyond HNE including synthetic covalent small-molecules.

2. High throughput methods to ID RES-sensitive proteins

2.1 Bolus RES-exposure methods: Leveraging activity-based protein profiling (ABPP)

The Cravatt laboratory has thoughtfully disclosed a version of isoTOP-ABPP (Weerapana et al., 2010) that provides a measure of HNE occupancy on a subset of proteins (Fig. 1A) (Wang et al., 2014). The protocol involves

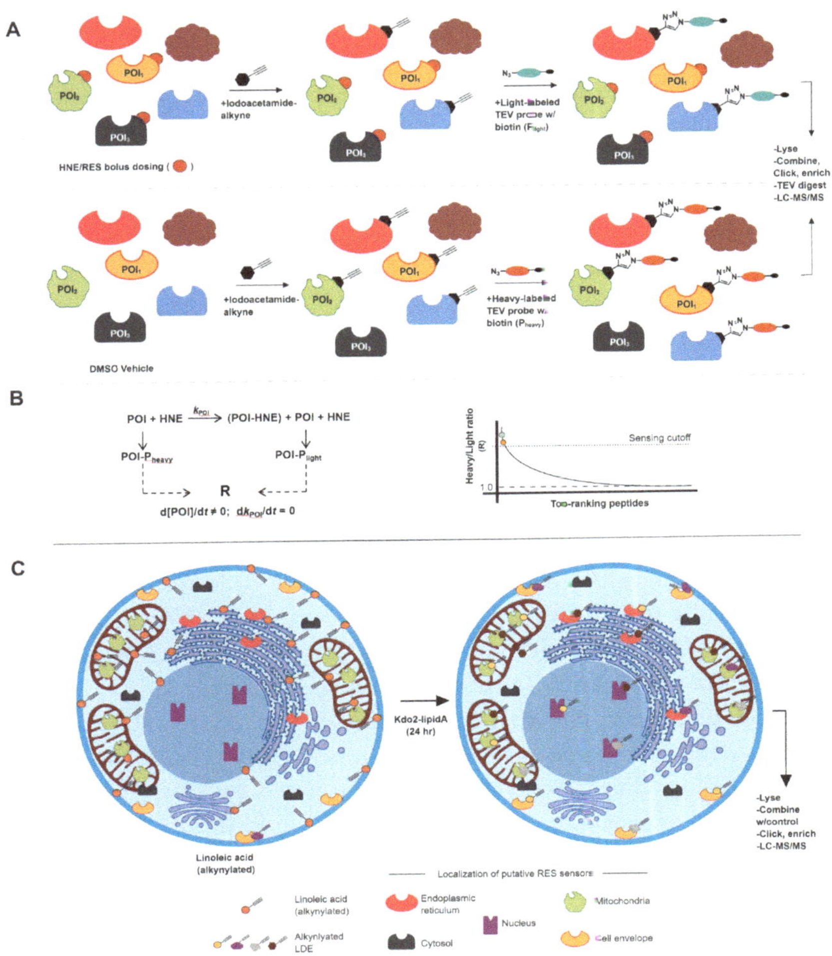

Fig. 1 Representative methods of high-throughput (HT) identification (ID) of RES-sensing proteins. (A) Competitive activity-based protein profiling (ABPP) adapted to target ID of HNE-sensing proteins. Treatment of lysates [originating from cells treated with either HNE (100 μM, 1 h) or DMSO] with alkynylated activity probe (alkynylated iodoacetamide; 100 μM, 1 h), followed by Click coupling with light or heavy TEV-cleavable biotin tags (P_{light} and P_{heavy}), respectively, streptavidin enrichment, and downstream MS/MS analysis, altogether enable *indirect* report of HNE sensitivity (Wang, Weerapana, Blewett, & Cravatt, 2014). (B) ABPP ranks proteins by "R value," a ratio of heavy/light probe modification ($POI\text{-}P_{light}/POI\text{-}P_{heavy}$) that serves as a readout of the intrinsic kinetics of HNE labeling by a specific protein (k_{POI}), assuming constant POI-concentration, [POI] (Wang et al., 2014). (C) ID of HNE sensors built on utilization of pseudo-endogenous generation of HNE from alkynylated precursors. Incorporation of linoleic acid (LA) into phospholipids, followed by Kdo2-lipidA-activated autoxidation (see text for details), reveals potential key players in stress-response (Beavers et al., 2017).

an "activity" assay in which cysteines are labeled with an affinity probe, such as alkyne-functionalized iodoacetamide (IA), that can be subsequently conjugated with biotin-azide. Subsequently, the biotinylated proteins are affinity purified and specific cysteine-containing *probe-modified* peptides are analyzed by digest MS. However, because of technical limitations (either in the mass spectrometry technique, the probe treatment or elsewhere), typically only 700–2000 cysteines (Lawson et al., 2015) are detected (from a total of >200,000 possible unique cysteines in humans). The assay is then repeated on lysates of cells (or simply lysates) treated with a specific electrophile, such as HNE. By addition of an isotopomer of the alkyne-functionalized affinity probe above to the HNE-treated lysate, then performing streptavidin enrichment/digest/elution/MS on a 1:1 mixture of HNE (or any other electrophile)-treated and -untreated samples, the amount of HNEylation on specific cysteines detected by the ABPP probe can be measured.

Assessing HNEylation is achieved simply by examining the ratio of isotopomeric cysteine-containing peptide to the canonical congener. This readout is possible because the probe is retained in the modified fragment. The readout for the experiment is typically referred to as "R," the ratio of light to heavy isotopomeric probe-modified fragments observed in the MS (Fig. 1B). We will discuss the wider utility of the R value later. However, it is worthwhile mentioning that R is not only a function of HNE occupancy on the cysteine in question. As ABPP is essentially an activity assay, any modification to the enzyme in question, including off-target changes in association or locale or oxidation, could reduce labeling by the probe, and hence affect R. These concerns are intrinsic to the method because ABPP uses bolus treatment of electrophiles that is known to, for instance, modulate redox homeostasis, and activate apoptosis pathways. Testifying to the utility of this venerable method, five specific hyperactive cysteine sensors of HNE (100 μM), such as ZAK(C22) and RTN4(C1001), have been identified (Wang et al., 2014). In the case of ZAK(C22), this residue was validated as being sensitive to HNE labeling in vitro.

2.2 Quasi-endogenous, prolonged exposure: Simulating endogenous RES-biosynthesis

The recent innovation from the Marnett laboratory is built on feeding of cells with linoleic acid (LA), a precursor to lipid-derived electrophiles (Beavers et al., 2017). LA was modified with a terminal alkyne; then, cells were stimulated with LA(alkyne) at micromolar concentrations for 24h, followed by the saccharolipid glycan, Kdo2-lipid A, for another 24h

(Fig. 1C). A parallel experiment using non-alkyne-functionalized-LA in combination with SILAC-based proteomics analysis enables quantitative target ID. 3300 proteins were identified as HNE sensitive from 3816 total proteins using this approach. The data outputs are similar to those obtained by direct-capture methods (which run on the premise that HNEylated proteins typically contain a reactive aldehyde post Michael adduction) reported by the Wang laboratory (Chen et al., 2018). Both methods saw significant variations across the data sets obtained over three runs. One conclusion from these data was that mitochondrial proteins were the most reactive under these conditions. However, how much this organelle specificity is due to localized stress generation, or implicit sensing capacity, is unknown.

2.3 Transient, low-occupancy approach: Genome-wide target-ID of kinetically-privileged native RES sensors (G-REX)

One of the key components likely leading to variability among the aforementioned RES-target-ID methods is that the electrophile or a precursor is administered from outside the cells giving no control over entry or stability of the electrophile or its metabolites. Thus, it is difficult to calculate the intracellular concentration of HNE (or its various metabolite adducts) at any time. Furthermore, the concentration of HNE is likely cell-dependent (a function of local density, health of cells at the specific time, for example), and locale-dependent (differences in concentration between nucleus and cytosol, for instance). To allow calibration of the dosage of RES released, and control timing of release and duration of exposure, our laboratory recently introduced G-REX (Zhao et al., 2018). G-REX uses a photocaged version of a lipid-derived electrophile such as HNE (alkyne-functionalized) that is capable of binding stoichiometrically to an artificial protein, HaloTag. Halo expression level thus controls precisely the amount of RES available inside the cell (Fig. 2). The half-life for photouncaging of the RES is <1 min (Lin et al., 2015; Parvez et al., 2016). Subsequent biotin-azide coupling and enrichment proteomics protocols enable target ID of kinetically-privileged substoichiometric HNE-sensors. SILAC- or TMT-based variants of proteomics approaches render quantitative ranking of the targets from G-REX. Most notably, individual G-REX hits can be directly validated for target- and RES-specific pathway modulation using T-REX approach (which we discuss in Section 4).

We have modeled release rate, known Halo protein expression levels (5–10 μM in HEK293T at least) (Zhao et al., 2018) and known detoxification rates of HNE: based on these data, we expect that the maximum HNE

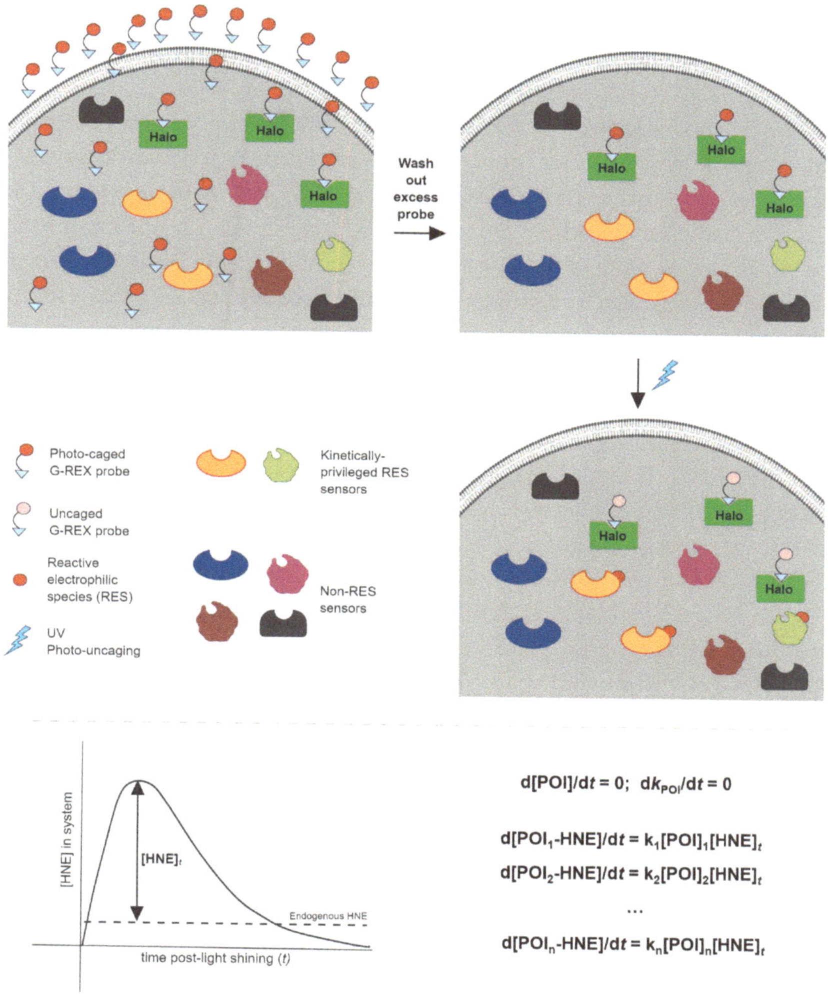

Fig. 2 G-REX mines native RES-sensors with high spatiotemporal resolution and controlled RES chemotype and dosage. G-REX is built on direct capture of native sensors in vivo. Following treatment of live specimens with the specific photocaged REX-probe and washing away the unbound excess probe, the concentration of RES released in the cell (or a specific organelle/tissue), upon subsequent photouncaging at a preordained time, is at maximum equal to the concentration of HaloTag in vivo (Liu, Long, & Aye, 2019; Zhao et al., 2018). Locale of the RES release is determined by where Halo is expressed (also see Fig. 3). G-REX thus more closely mimics native electrophilic signaling and offers a user-controlled RES-sensor ID. By utilizing a specific RES/covalent-ligand such as HNE as a chemical-signal input, a quantitative G-REX output can be obtained ([POI-HNE], where POI is the putative protein sensor). Further downstream work is required to more accurately assess the relative rates of RES-reactivity across the top hits, since the output is affected by both intrinsic kinetic properties k_n and endogenous concentration of the individual sensor protein-of-interest $[POI]_n$. as well as [HNE] at a given time t.

concentration obtainable under G-REX is approximately 1 μM (Liu et al., 2019). Critically, this concentration is close to that reported to be relevant under signaling conditions (Parvez et al., 2018).

Furthermore, classical methods above involving bolus RES-dosing and/or prolonged RES-exposure times likely lead to identification of poorly-reactive proteins as HNE-sensors or loss-of-function sensors. G-REX complements these aforementioned methods by capturing native sensors undergoing low-RES-occupancy modifications, thus minimizing toxicity. G-REX also enables *direct* capture of the sensors, thereby circumventing the reliance on proxies, such as IA and associated shortcomings, such as false negatives. We consider these aspects further in the later sections.

Halo can be sent to specific locales in cells by fusing a localization tag, such as nuclear localization sequence (Tomoshige, Naito, Hashimoto, & Ishikawa, 2015). Thus, G-REX could open doors to probing locale-specific sensing of electrophiles as well. The Halo protein is expressed in the cell of interest, currently through transient transfection, although likely other methods of transgene delivery are possible. Critically, the combination of the photocaged Halo-targetable ligand and Halo, ensure that a calibrated amount of HNE can be released rapidly to the cells, and executed so at the preordained time and space (Fig. 3).

2.4 Comparison of sensor ID methods

2.4.1 *ABPP offers high-resolution information on small numbers of potentially-sensing cysteines*

The methods above are quite different in their outputs and set up. ABPP and related incarnations (Qian & Weerapana, 2017) give accurate insight into a fraction of cellular cysteines. The accuracy of the information provided for these specific profiled residues is very useful, accurate and highly discriminatory. However, it is worth noting that the method is blind to how other proteins are modified/changed. Since the method involves indirect target-ID (using affinity probes such as IA), the change in reactivity observed could also be ascribed to changes in protein integrity/stability and not a direct modification. Critically, known targets of HNE, such as Keap1, are not detected in the HNE-sensor-profiling by ABPP, nor are residues other than cysteines (such as Lys and His), which are susceptible to HNE. Alternative methods to probe other nucleophilic residues are appearing (Hacker et al., 2017; Poganik & Aye, 2018). We note, however, that the ABPP approach is a hugely useful technique, as it is widely applicable, highly accurate and can be conducted on many different systems (including endogenous signaling

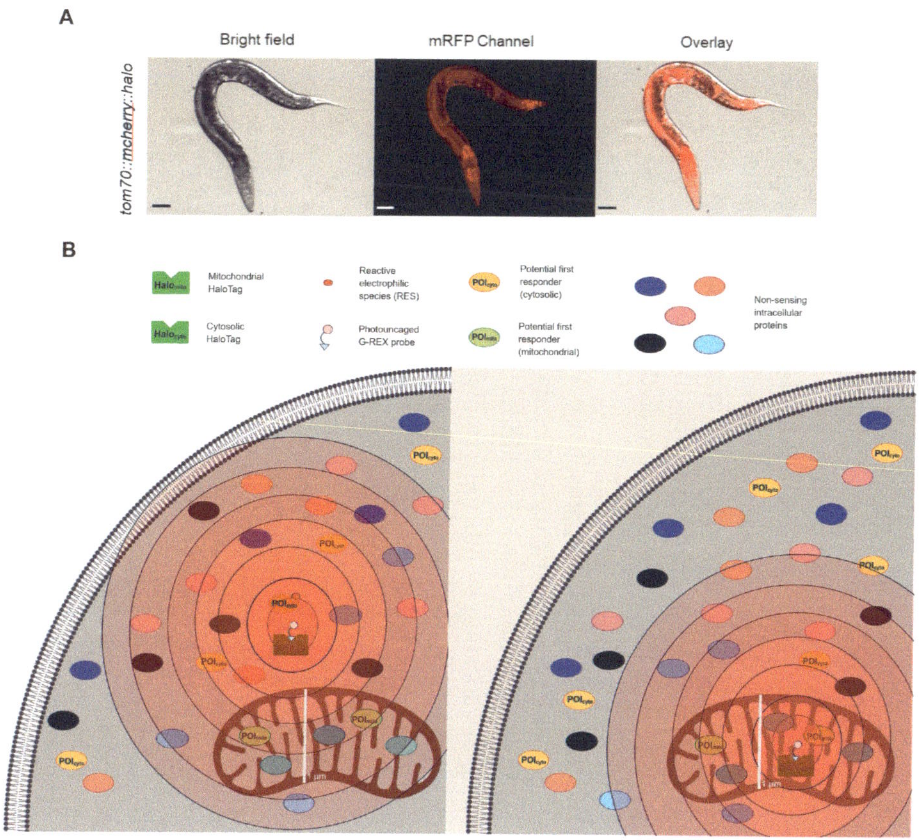

Fig. 3 See legend on next page.

electrophiles, electrophilic drugs, metabolites, etc.). These capabilities are available because the "chemical biology" steps are all performed post processing and small-molecule covalent ligands are directly exposed to the specimen of interest.

2.4.2 Directness of the readout

isoTOP-ABPP method does not directly readout RES-modifications, such as HNEylation (Fig. 1B), and thus is prone to some artifacts that need to be carefully considered. On the other hand, precursor-feeding methods using LA/Kdo2-lipid A likely generate exclusively alkyne-tagged lipid electrophiles; however, there is a significant number of these species that can be derived from LA (Fig. 1C), which may or may not overlap in their protein targets. The increased chemodiversity of signals generated could help explain why over 80% of proteins detected by in this method were scored as labeled.

The recently reported approach by Wang et al., and related long-established direct-capture strategies directly measure protein carbonylation. But many modifications other than HNE can lead to protein carbonylation, including modification by other electrophiles (Wong, Bansal, Marcocci, & Suzuki, 2012) and oxidative modifications (Fan et al., 2009) that are upregulated under bolus dosing. Furthermore, some HNEylation events also likely do not retain the aldehyde function, either due to redox

Fig. 3 Expanding G-REX toward locale-specific sensing and target ID in complex live models. (A) REX-probes have been shown to be compatible with complex living systems such as *Danio rerio* (Long et al., 2017) and *C. elegans* (Long et al., 2018), rendering G-REX platform amenable to mining novel sensors in these animals One of the on-going applications of G-REX involves locale-specific mining of RES-sensors at the live organismal level; e.g., organelle-specific HaloTag expression [mitochondrial outer membrane (OM) localization via OM-anchoring domain of Tom70, OM-translocase from *Saccharomyces cerevisiae*] can be achieved in complex model organisms such as *C. elegans*. Scale bars are 50 μm. (B) G-REX integrating localized expression of Halo (such as to mitochondria, right) explores region-specific RES-sensors that may have otherwise been missed through ubiquitous/whole-cell HaloTag expression (left). Concentric shaded rings in the diagram depict different diffusion distances for a RES such as HNE. *Note*: for locale-specific G-REX, these distances should be as low as possible: e.g., for mitochondria (at cross-section diameters reportedly up to 1 μm), longer linear diffusion distances would reduce organelle specificity. *Panel (A): Reprinted (adapted) with permission from Long, M. J. C., Urul, D. A., Chawla, S., Lin, H. Y., Zhao, Y., Haegele, J. A., et al. (2018). Precision electrophile tagging in* Caenorhabditis elegans. Biochemistry, *57(2), 216–220. https://doi.org/10.1021/acs.biochem.7b00642. Copyright (2017) American Chemical Society.*

modification (McMahon, Lamont, Beattie, & Hayes, 2010), or because cross-links can subsequently occur (Xu, Liu, Mayank, & Sayre, 1999), especially under HNE-excess or bolus-exposure conditions. G-REX is the only method thus far that *directly* mines HNE sensors and also one that allows for native signal to freely bind to and react with (in any mode: i.e., Michael addition, Schiff-base formation; and via any residue: i.e., Cys, Lys, His, or any other) native privileged sensors. The only modification within G-REX pertains to alkyne functionalization (for the ease of target ID through affinity proteomics). However, other, less intrusive modifications such as isotope labeling of the native RES chemical signal, combined with autoradiography-based analysis, may also be used. G-REX enables all of these features because a RES is photouncaged in situ and the RES-modification experiment lasts for only a few minutes. Notably, any concerns with alkyne-unit modification can be easily validated using T-REX system (vide infra) for pathway studies, which uses a genuine native RES with no modification.

Indeed, LA/Kdo2-lipid-A-based feeding method and G-REX both make use of the terminal alkyne-tagged HNE (HNE-alkyne). Other methods use native RES (with no modification). Many years of characterization of HNE-alkyne has found little to no difference between HNE and HNE-alkyne in terms of target spectra. However, as our understanding of HNE changes, it is possible that differences between HNE and HNE-alkyne may appear. Furthermore, the reliance on alkynes can limit generality as many molecules may not lend themselves to alkyne tagging without loss/depletion of chemical analogy/function integrity, and thus target ID, to an extent that is largely unpredictable (e.g., fumarate).

2.4.3 The occupancy threshold

Less than 1% of the 1000 cysteines profiled were assigned as highly HNE sensitive in the original work using ABPP. A similar conclusion was reached by Wang's ABPP approach, although around 10-fold more proteins were identified as sensors (around 5% of the data set). The notion that few proteins are HNE sensors stands in contrast to bolus HNE-exposure-based *direct enrichment* methods that indicate that huge swathes (around 25%) of the proteome are "HNE sensitive." The most likely reason for these discrepancies is that enrichment strategies have a low threshold for scoring a hit, whereas ABPP has a high threshold. Of course, it is possible that the cysteines covered within the ABPP data set do not represent "the best" HNE sensors, but it is probably more likely that the ratio of HNE sensors detected by ABPP are reflective of the actual numbers of sensors in the genome. The principal reason for this conclusion is that the direct enrichment approaches (e.g., those reported by the Wang and Marnett laboratories) do not measure total labeling, but state simply that a protein is enriched relative to a defined background (i.e., occupancy could be very low). Thus, they have relatively low thresholds for success. On the other hand, the R value in the original isoTOP-ABPP approach (Fig. 1B) is a measure of "total" occupancy.

Intriguingly, both modes of scoring sensors could be relevant. Since we do not know what occupancy is required for signaling, nor do we know if high-extent of labeling is even possible biologically (for instance, if locale-specific sensing occurs, or if there is dominant-negative regulation of the pathway following HNEylation of a specific protein). Thus it is also possible that the threshold set by the original ABPP is too stringent.

2.4.4 A note on stimulation methods, such as G-REX and the precursor-feeding approach

The feeding method is built on "endogenous HNE/electrophile" production. However, we stress that a stimulator is required to force generation of electrophiles. It is unknown to what extent these conditions are relevant to typical endogenous signaling methods, and also what effects the precursor and the stimulator treatments have on cells independent of RES generation. Although G-REX is referred to as a non-bolus dosing strategy, G-REX does use excess of a xenobiotic molecule whose bioactivity it is difficult to assess completely. We have, however, conducted quite rigorous examination into the biological activity of the G-REX probe: it is not toxic to cells, and does not affect the development of embryonic fish or *Caenorhabditis* elegans. To this end, it is a significant improvement on bolus dosing.

G-REX uses the alkyne handle on HNE to allow biotinylation of proteins tagged with HNE during the G-REX method. As an enrichment strategy, G-REX does not require that HNE occupancy be high to score as a hit. However, what separates G-REX from other *direct capture* methods is that a small amount of HNE is released inside cells within a very brief period of time. Thus, G-REX likely labels the best sensors (i.e., those that can intercept HNE even though it is present for a "brief spell"), and should not pick up as much collateral damage as in overload methods. It remains to be rigorously assessed how many sensors G-REX is able to capture. However, it is likely that G-REX may require numerous replicates to identify the majority of sensors under specific conditions. Hits ID'ed by G-REX have certainly been shown to react rapidly with HNE in vitro, and via T-REX validation platform (vide infra), hinting at the validity of this assumption. Unfortunately, G-REX does not provide a simple parameter to rank proteins currently, so it is not as directly quantitative as ABPP methods. As alluded to in Section 2.3, coupling G-REX with SILAC/TMT makes quantitative profiling possible. However, as there is no internal standard, it is unlikely that a parameter as simple as the R factor will ever be derived from G-REX.

2.4.5 Proteome modification

G-REX is the only method that requires some genetic modification/modification of the proteome of the tested cells. It remains unknown how expression of the Halo transgene affects the cell's innate regulation. However, we acknowledge that manipulation is a significant limitation, since, for some cell lines, transfection is difficult, and viral expression systems can be variable and often give lower expression than transient transfection. Furthermore, although there is no evidence Halo can perturb cells, worms, or fish, it is essentially impossible to prove cells are not affected in some minute way.

2.5 No perfect HT-sensor screen: Downstream validation is the essence of target ID

Preceding sections have discussed varied approaches to ID RES-sensors using HNE as a representative RES, with differing strengths and weaknesses. As for most high-throughput (HT) screens, the data should not be taken purely at face value. We strongly suggest performing downstream validation of labeling and investigation into activity/function alterations, both in vitro and in cells/animals. Although, probing target-specific HNEylation was impossible for many years, we recently disclosed techniques to study ramifications of specific on-target labeling that are applicable to live cells, fish, and worms. We will discuss how to go about validating and understanding HT data below. With keen interests to establish a holistic understanding of native electrophile sensing/signaling, we also consider how to develop ranking systems for HNE/RES-sensitivity in this manuscript.

3. Ranking individual protein's RES-sensitivity

It has become a leitmotif of modern research that HT-profiling [PAR-clip (Spitzer et al., 2014), ChIP-seq (Park, 2009), and so on] is used to generate huge lists of potential sensors, interactors, or regulators, etc.

Much of the time, these lists are too large to follow up upon in the entirety, and the assay outputs do not necessarily allow unambiguous ranking (Friedersdorf & Keene, 2014) and correlation to biochemical/functional properties such as a rate or binding affinity (K_d) (She et al., 2017). This is certainly true for all the electrophile sensor ID methods above; RES-sensor ID methods that score ~80% of observed proteins as RES-sensors indicate that RES-sensing is such a common thread between different proteins that this property is part of the intrinsic fabric of the proteome. However, we believe that such conclusions are a consequence of the low bar set by enrichment experiments coupled with the intrinsic electrophile reactivity of cysteine. Thus, focusing more on kinetics—which describe occupancy over a given period of time—is of major relevance for successful functional validations in RES-sensing.

To enable this focus-shift to occur, a parameter to rank HNE-sensitivities of proteins *in their native environment* is needed. Ideally, such a parameter would be protein and small molecule concentration-independent. One way to rank reactivity is to calculate second order HNE adduction rates for each protein. However, RES-sensitivity assessment in vitro—although widely performed—should be interpreted carefully, due to vastly-different contexts/parameters that exist between in vitro and in vivo scenarios underlying non-enzymatic RES-modifications. True concentrations of RES and/or POI, and involvement of associating/potentially-competing binding partners are difficult to model in isolated systems, and any differences in protein conformational states/dynamics and microenvironments (e.g., pH, redox state) could alter the sensing properties. Accordingly, second-order rate constants derived from in vitro experimentation may constitute an ideal ranking scheme.

The R value from the original ABPP method, or the equivalent L/H ratio from the related methods, is indeed linked to kinetic sensitivity of proteins to HNE. However, it is unknown how these parameters correspond to second-order rate constant for reaction of HNE with the protein of interest. Experiments have shown that R is not a function of protein expression, at least for a few proteins. However, R is a complex function of multiple rates/concentrations/metabolic processes. Furthermore, when R is derived from bolus dosing of whole cells, further confounding factors can cause issues. For instance, HNE occupancy on many proteins is likely high under these conditions. On the other hand, none of the direct capture/enrichment approaches including G-REX, and feeding methods, provide a similar easily extractable parameter to rank proteins.

4. Pathway validations at individual protein level

4.1 Conceptualizing electrophile signaling

Once a protein is modified by HNE, its properties *may* be changed. One can consider four possible outcomes: no change to the protein, gain of (new) function (Li & Babu, 2018), loss/decrease of an existing function (Ségalat, 2007), amplification of an existing function (Ségalat, 2007; Whitaker-Menezes et al., 2011). Any process that elicits a change in the protein function can change/modulate cell decision-making, and hence can be considered "electrophile signaling." In reality, the likelihood of each change in function bringing about meaningful effects changes as a function of occupancy *and* mechanism. Gain of function can have extreme effects on cellular processes even with low occupancy (i.e., any protein ID'ed by any method above could affect cell signaling if the protein *were* to function through gain of function). Conversely, stoichiometric loss of or twofold amplification of function typically requires significant (>50%) occupancy to elicit similar effects (Huang, Lee, Marcotte, & Hurles, 2010). It remains unclear if such high occupancies are obtainable under "normal" signaling conditions, although based on our assessment of the methods above, we feel that it is unlikely that this happens for a huge amount of proteins under redox signaling conditions. Multi-fold-amplification of existing function and dominant negative loss of function can exert significant phenotypes at modest occupancy and thus could contribute to "endogenous" signaling (many of the proteins ID'ed above by any method could affect signaling if they *were* to function through either process). Thus, it is critical that sensors be individually evaluated for occupancy and mechanism using a (relatively) unbiased approach.

4.2 Tracking downstream signal propagation

Having established a HNE(or other electrophile)-sensing protein using an HT screen, the next typical experiment is to evaluate how HNE affects protein function. In the early days of the field, these experiments would have been carried out exclusively in vitro. However, as we discussed above, it is important to try to understand how electrophile sensing affects the specific protein in cells, or ideally organisms.

Unfortunately, proteins that have been identified as being HNE labeled by electrophiles under bolus HNE-exposure, and validated as HNE sensitive in vitro, typically do not appear to contribute significantly to the overall response observed under bolus dosing. For instance, Pin1, Zak, and PKCθ, have been identified as sensors of different electrophiles under bolus dosing. However, in cells, knockdown of these proteins has a meager effect on the overall phenotypic outputs. One way to interpret these outcomes is that redox signaling is a complex phenomenon and it cannot be ascribed to a single protein. However, another point of view is that we do not really understand redox signaling in cells, and the bolus dosing regimens applied to understand redox signaling are (part of) the challenge in that they do not model these processes in terms of locale of generation, duration, or target spectrum effectively (Fig. 4).

4.3 T-REX: Validation and monitoring downstream signaling

T-REX is a chemical-biology method used to model protein-specific HNEylation/RES-adduction. The method proceeds via "breaking of the tether," wherein a native diffusible reactive species is released in the vicinity of a specific POI. The POI then has the first refusal for reaction with the RES prior to the electrophile diffusing away from the POI and being metabolized or distributed across the whole (reactive) proteome, as occurs under G-REX. The setup is achieved by expressing Halo-protein as a genetic fusion with the POI. When such cells are treated with Halo-targetable photocaged precursor to HNE, namely, Ht-PreHNE (the same probe used in G-REX), or a similar photocaged probe, followed by washout of the probe, on-demand delivery of the RES of interest within the coordination shell of the POI is achieved. Our data have shown that for proteins that are kinetically-privileged RES sensors, HNE (or another RES chemotype under study) can be intercepted by the POI prior to diffusion of the delivered RES from the solvent cage.

The principal evidence for a pseudo-intramolecular delivery mechanism is the fact that when Halo and Keap1 (the best HNE sensor POI we have so far characterized under T-REX conditions) are separated, there is no HNEylation of Keap1 following T-REX-assisted release of HNE from the Halo protein *that is not fused to Keap1* (Fig. 5). Thus we can conclude that when Halo and Keap1 are fused together, the mechanism of HNEylation of Keap1 differs from the case when Halo and Keap1 are separately expressed (Lin et al., 2015; Parvez et al., 2016).

We interpret this difference to mean that HNE and Halo-Keap1 exist in the same solvent cage. Evidence that this is a solvent cage and not some other

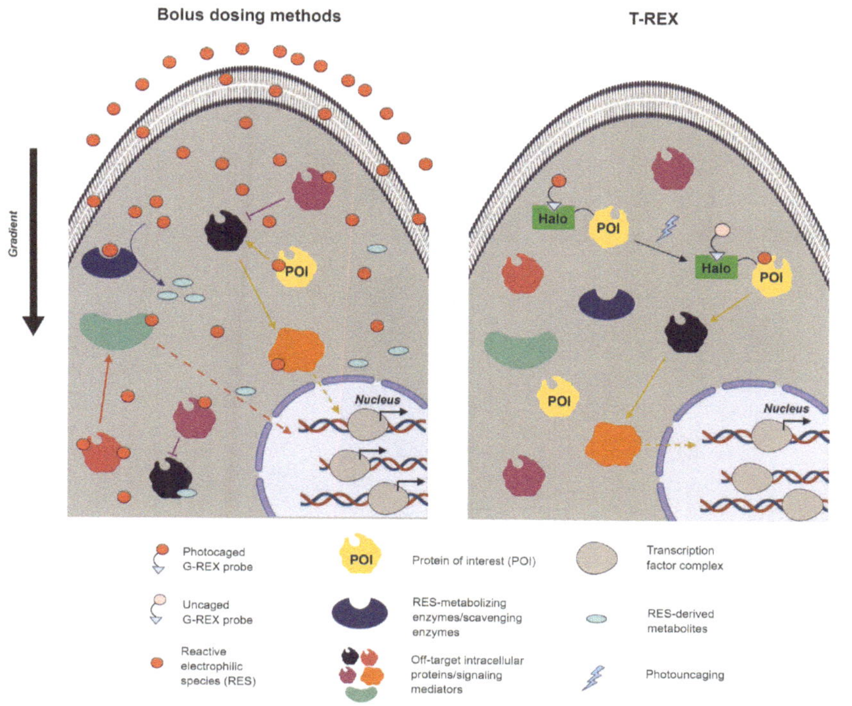

Fig. 4 T-REX is a unique and necessary tool for directly interrogating and validating signaling outputs exclusively arising from POI-specific RES-modifications. (Left) Bulk HNE exposure to cells/animals—while representing a rapid and facile means to ID candidate RES-sensors—render challenges in studying downstream consequences of specific sensor RES-modification. These difficulties are in part due to: off-target RES-binding and associated toxicity and secondary responses, etc., following RES-saturation of first-responding sensor cysteines; uncontrolled conversion of RES in cells into other reactive metabolites that can adduct additional proteins; and perturbation of cellular redox balance. (Right) T-REX enables selective labeling of the target protein-of-interest (POI) in the cell or animals, through controlled release of RES from a photocaged precursor bound to Halo fused to POI. T-REX provides a clearer insight to a given POI—RES-pair-specific modulation of signaling output as well as the kinetics of RES-sensor modification. *Note*: The REX-probe used in G-REX (see Fig. 2) and T-REX is identical for a given RES of interest.

intermediate can be found from the fact that Halo-Keap1 and Keap1-Halo have the same HNEylation extent as each other (and modification site also remains the same in both cases, i.e., N- or C-terminal Halo fusion) (Parvez et al., 2016). From the experiment using unfused Keap1 and Halo, we can also infer that there is minimal labeling of other non-halo fused proteins under T-REX conditions.

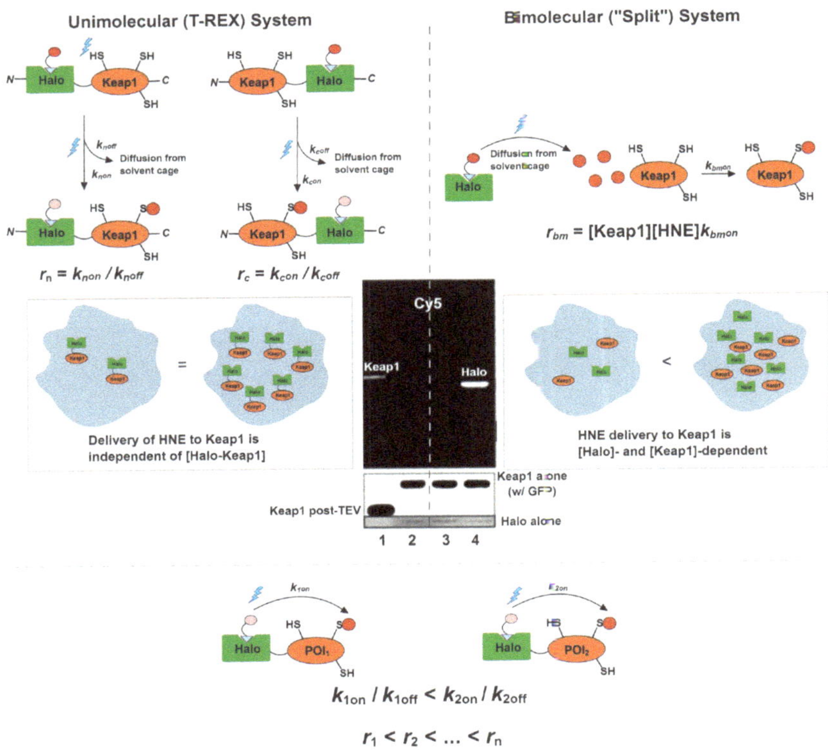

Fig. 5 T-REX analyses HNE-sensing kinetics of individual POIs (Left). In the T-REX model, the release of HNE and subsequent covalent modification happens in a pseudo-intramolecular fashion within the solvent cage surrounding Halo-POI. As such, the observed rate of this reaction (r_{nterm} or r_{cterm}; "nterm" and "cterm" designating, respectively, N- or C-terminal Halo-fusion) are independent of POI (here, Keap1) or released-RES concentration, and are instead governed by the off-rate of the released RES, here HNE (diffusing into the environment; k_{noff}) and the on-rate of covalent modification of the POI (k_{non}) (Parvez et al., 2016). (Right) In a bimolecular system (so-called split system, where Halo and POI are separately expressed), this is not the case (Lin, Haegele, Disare, Lin, & Aye, 2015). Here, both the sensor-POI (Keap1) and the released RES can vary in concentration, and the rate itself (r_{bm}; "bm" designating bimolecular) is concentration-dependent. Further, release of the RES into the cell from Halo enables the measured rate to be influenced by HNE-binding to off-targets. The "Cy5 gel" illustrates a representative in-gel fluorescence data set from T-REX targeting efficiency analysis, whereby labeling is evaluated by Cy5-azide click coupling that reports on alkyne signal (either within the photocaged probe or the released RES that, respectively, bind Halo and RES-sensor-protein). The blots below the Cy5 gel correspond to western blots confirming the presence of indicated protein in the lysate. Lane 1 corresponds to data originating from cells expressing the Halo-Keap1 that were subjected to full T-REX

(Continued)

4.4 T-REX as a means to rank RES-sensing efficiencies

Delivery efficiency in T-REX is thus a function of a pseudo-intramolecular reaction between HNE and Halo-POI in the solvent cage, and there is negligible contribution from the second-order reaction between HNE and Halo-POI. The second term that gives rise to delivery efficiency is therefore diffusion of HNE from the solvent cage, which is also a pseudo first-order process. From this reasoning, we can define that the delivery efficiency is actually a ratio of two pseudo-first-order rate constants multiplied by the Halo-Keap1·HNE encounter complex (Dinkova-Kostova, Holtzclaw, & Kensler, 2005), which is constant for each term and hence cancels. It thus stands to reason that delivery efficiency is an unbiased measure of HNE labeling rate that could potentially be used as a scale to rank RES-sensing propensity (Fig. 5).

Indeed, T-REX is transposable in terms of both RES-targeting efficiency and downstream pathway-modulation in both cells (high expression of Halo-POI) and fish/worms (lower Halo-POI expression) (Long et al., 2017, 2018; Zhao et al., 2018), which offers some credence to this postulate. Of course, ignoring how Halo affects a specific POI's RES-reactivity, there could be other aspects of this simple system that do not commute to a "real" second-order rate constant. However, by carefully scaling against the relative second-order rates of reaction between HNE and POI derived using traditional kinetic experiments, the hypothesis is testable. However, it would be most informative to test this postulate against numerous POIs with various delivery efficiencies (where the expression of these proteins also

Fig. 5—Cont'd conditions, indicating HNE(alkyne)-modification of Keap1 (note: "post TEV cleavage" designates samples that were subjected to TEV-protease-assisted separation of Halo and Keap1 post cell lysis prior to Click coupling with Cy5-azide). Lanes 2 and 3 correspond to a bimolecular system (i.e., the "split" system) where Halo and Keap1 are expressed separately (each fused to GFP) in cells. Data from these samples show no HNE(alkyne)-modification of Keap1. Lane 4 corresponds to the same bimolecular system harvested without any photouncaging, demonstrating the "photocaged-probe-loaded" HaloTag. (Bottom) The measured output of T-REX kinetics is largely intrinsic to the POI itself (specifically k_{on}), enabling the "targeting efficiency" parameter derived from T-REX analysis to be used for "ranking" of differential sensitivity to a given RES (if $k_{1on} < k_{2on}$, then corresponding $r_1 < r_2$; thus, rates can be used to rank sensors) (see text for details). *Reprinted (adapted) with permission from Lin, H. Y., Haegele, J. A., Disare, M. T., Lin, Q., & Aye, Y. (2015). A generalizable platform for interrogating target- and signal-specific consequences of electrophilic modifications in redox-dependent cell signaling.* Journal of the American Chemical Society, 137*(19), 6232–6244. doi:10.1021/ja5132648. Copyright (2015) American Chemical Society.*

differs in cells). Currently, we do not have access to such a range of proteins identified under T-REX. We hope that as we perform more G-REX experiments, we will be able to achieve such a range of HNE-sensitive proteins of varied biological properties/functions.

However, based on the series of proteins that we have characterized by T-REX thus far, we start to see some trends. Keap1 is our best HNE sensor (around 35% delivery efficiency) we have so far characterized (Lin et al., 2015; Parvez et al., 2016). Keap1 is a master regulator of the antioxidant response (AR). Upon HNEylation of Keap1, AR upregulation is expected. T-REX-assisted HNE delivery to Keap1 alone in cells gives at least as high levels of AR than prolonged bolus HNE-dosing of cells. On the other hand, Akt3, another HNE sensor identified under T-REX conditions, is labeled moderately well by HNE (20%) and gives ~30% of the maximal pathway-output possible (Long et al., 2017). Similarly, Ube2V2 gives ~15% delivery efficiency, and HNEylation on this protein upregulates DNA damage response (DDR), achieving ~30% of the maximal response of the system (Zhao et al., 2018).

Based on these limited data thus far, the larger magnitude of signal output, relative to a control, is certainly found in systems with better T-REX-delivery efficiency, which is at least reassuring (Fig. 6). Nevertheless, it is important to recognize that maximum signal outputs are dependent on the tolerance of the assay, the nature of the positive control used, toxicities of the positive control, mechanism of signaling (both intrinsic to the target protein and in off-target effects), all of which constitute challenging parameters to precisely evaluate. It is for this reason that focusing on labeling extent is likely to be retained as a yardstick for whether a POI is sensitive and to rank the magnitude of signal outputs, although, we stress that precise mechanistic analysis of signaling aspects offered by T-REX is important to be carried out.

One criticism that can be levied at T-REX is that it is hard to know whether the occupancy levels assessed under T-REX are obtainable under normal signaling conditions. To answer this question, we need to understand more about what endogenous redox signaling is, and there is little in the way of answers to that question yet. T-REX ultimately can only show sufficiency of signaling at a given ligand occupancy and mode of signaling mechanism. Although through the knowledge of these two aspects, it could be possible to propose which proteins are most likely to be involved in modulating a specific signal transduction pathway.

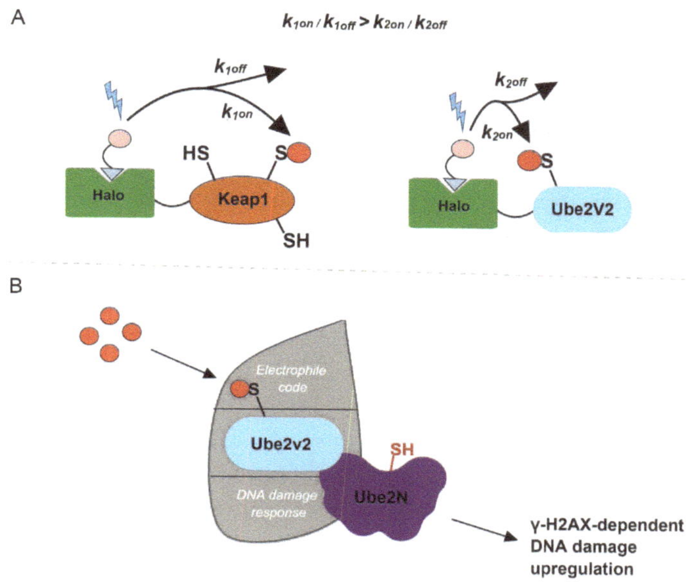

Fig. 6 A case study of the recent discovery pipeline enabled by REX technologies. (A) Ube2V2 was uncovered as a sensing candidate using inaugural G-REX approach (Zhao et al., 2018). Through T-REX-assisted evaluation of relative RES-sensitivity (i.e., targeting efficiency/site-occupancy, see text for discussion), Ube2v2 was less sensitive to HNE than Keap1. (B) Ube2V2 translates RES modification of a sensing cysteine (black; distinguished from the red catalytic cysteine on Ube2N) into a ubiquitin code through allosteric stimulation of Ube2N activity, consequently modulating DNA damage response signaling. This type of signal translation and propagation enables Ube2V2 to act as a "Rosetta Stone" proteins, converting the chemical language of electrophiles to that of canonical signal, ubiquitin (Zhao et al., 2018). This mode-of-action by HNE on a specific protein functions differently to the direct modulation of Keap1 function.

5. Case study: Ube2V2/Ube2V1

We discuss one case study that shows the progression from G-REX HT-screen to T-REX downstream validation, ultimately uncovering a new regulatory mechanism of DNA damage repair (Zhao et al., 2018; Zhou et al., 2014). We stress that any of the above target-ID platforms, in place of G-REX, could in principle be coupled with the T-REX-pathway investigation platform. However, because G-REX and T-REX deploy the same small-molecule photocaged probe, and because HaloTagged-human and mouse genes are readily available commercially, T-REX validation is immediately accessible following G-REX target ID.

We began by screening for low molecular-weight proteins that are potentially HNE-sensors using G-REX. We thus identified two related HNE sensors, the catalytically-inactive E2 ubiquitin-conjugating proteins, Ube2V1 and Ube2V2. Neither protein had been identified as HNE sensitive previously. T-REX validation of these two hit proteins showed that both were HNE sensitive, giving relatively low occupancy (5–15%) under T-REX. Ube2V2 was highly sensitive to HNE in vitro (Fig. 7).

Importantly, both these proteins associate with catalytically-active E2 ubiquitin-conjugating enzyme, UbE2N. Association of either Ube2V2 or Ube2V1 to UbE2N stimulates UbE2N's enzymatic activity. In cells, association of Ube2V2 or Ube2V1 with UbE2N controls different signaling pathways, DDR and NFKB signaling, respectively. Intriguingly, UbE2N is an enzyme known to be sensitive to synthetic RES (at its catalytic cysteine). Critically, even though UbE2N is a known nucleophilic protein, UbE2N, in vitro and cells, was not sensitive to HNE (Cheng et al., 2014).

When Ube2V2 was treated with HNE stoichiometrically, UbE2N activity was stimulated in vitro. HNEylation of Ube2V2 or Ube2V1 under T-REX conditions led to stimulation of UbE2N activity in cells. Remarkably, Ube2V2 T-REX showed selective DDR upregulation, but there was

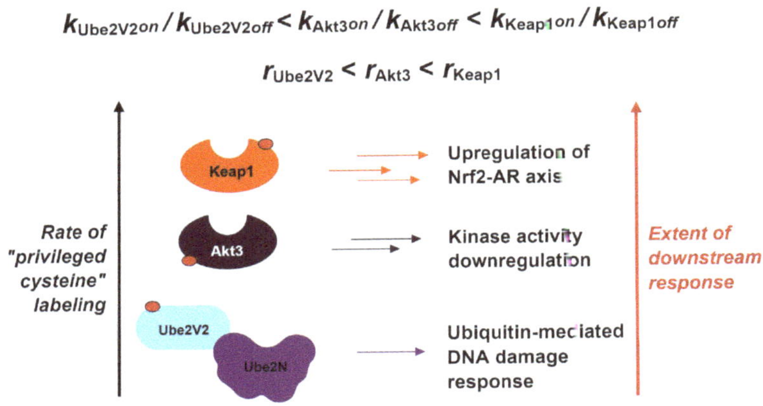

Fig. 7 T-REX ranks relative RES-sensitivity across RES-sensor POIs and G-REX unearths "Rosetta Stone" sensing and RES-signal translation. Differential RES sensitivity as a possible insight into the mechanisms of downstream signal propagation. (Top) Higher-ranking HNE sensors (i.e., those judged to have higher relative targeting efficiency/target occupancy derived by T-REX analysis such as Keap1) (Lin et al., 2015; Long & Aye, 2016; Parvez et al., 2015) and Akt3 (Long et al., 2017; Parvez et al., 2016), directly influence downstream responses as "first-responder" sensors. (Bottom) By contrast, lower-ranking sensors such as Ube2V2 mediate sensing through activation of a binding-partner (Ube2N in this case) in an allosteric fashion (Zhao et al., 2018).

no change in NFKB signaling. The opposite outcome was found for Ube2V1. We also identified the residue modified under T-REX, and we showed that mutation of this residue (to serine) prevented HNEylation under T-REX and downstream signaling was ablated when T-REX was replicated on this Cys-to-Ser mutant. We thus established an "in trans," or "homoallosteric mechanism" to regulate UbE2N activity that is marshaled by HNEylation of the binding-parter Ube2V2 or Ube2V1.

5.1 Rosetta Stone proteins

Our investigations into Ube2V2/Ube2V1 were to the best of our knowledge the first time HNE had been found to regulate activity of an enzyme through such a "trans" or homoallosteric mechanism (Fig. 7). Although this finding clearly complicates redox signaling significantly, we feel that ultimately such complex allosteric regulation mechanisms are to be expected of endogenous signaling pathways. Based on the data, proteins such as Ube2V2/Ube2V1 were best referred to as "Rosetta Stone" proteins. We designate this term to describe a protein that senses an electrophile signal rapidly, but then the RES signal cross-talks with a canonical PTM-code (here, ubiquitin signaling) in propagating pathway activity/function.

In the case of Ube2V2/Ube2V1, one additional key aspect is that these two proteins are catalytically-defunct E2 ligases, with no enzymatic function. However, Ube2V2/Ube2V1 are significantly more electrophile-sensitive than UbE2N, which has an intrinsically nucleophilic thiol residue that is reportedly sensitive to synthetic electrophilic inhibitors. This finding underscores how proteins with catalytically-essential cysteines in active sites may not be optimal for covalent targeting, but can be controlled through RES-modification-dependent allosteric regulation of their binding partners.

5.2 Implications of Rosetta Stone proteins

We hope more Rosetta Stone proteins will be unveiled in the future. However, we are also cognizant of the issues that Rosetta Stone proteins raise for understanding the impacts of redox signaling. It is a natural assumption that proteins that are RES sensors will undergo a change to their canonical activity/function upon RES-modification. For instance, we have shown that HNEylation of PTEN and Akt3 dominantly downregulate the activity of these two enzymes (Long et al., 2017; Parvez et al., 2016). Most other groups have made the same assumption. However, what has not been extensively considered is how interactomes and potential allosteric interactions could be affected by RES-sensing.

Of course, T-REX can decipher if there are any changes in activity and if that can lead to changes in canonical signaling. But it is possible that there are RES-orchestrated changes to non-canonical functions of the protein that are difficult to decipher. These non-canonical changes could also be evaluated by T-REX. However, it is difficult to decide what changes there could be. Possibly the coupling of T-REX with next-generation-sequencing (NGS) methods, or performing T-REX on catalytically-deactivated mutants would be some ways to investigate to what extent non-canonical protein functions are associated with redox signaling. Alternatively, coupling T-REX to various affinity proteomics methods may also be illuminating.

The Rosetta Stone concept also provides one more interesting insight that we are keen to pursue. Rosetta Stone proteins must sense and signal in a single step. In our discussion, we conceptually broke down sensing and signaling for simplicity. However, it is likely that sensing occurs because the protein has an inherent affinity for the electrophile and this affinity (which is often used in biology as a way to drive protein conformational/association changes) in turn drives pathway modulation (Liu et al., 2019). Thus we propose that there is a synergism between sensing and signaling that assists in the development and evolution of Rosetta Stone proteins. We have disclosed models for how such proteins could come about and we are trying to investigate these processes currently.

5.3 Study of electrophile signaling—especially Rosetta Stones—is important for drug discovery and understanding target engagement

We, and others, have stressed that RES-signaling does indeed have high utility and relevance to human health. This is principally because numerous drugs are now in development or have been approved that have appendages similar to endogenous electrophile signals. Examples include: broad-spectrum covalent drugs, such as dimethyl fumarate (Tecfidera) (Blewett et al., 2016); targeted covalent inhibitors, such as ibrutinib (Gayko et al., 2015); and numerous drugs in development or previously in development, such as sulforaphane/benzyl isothiocyanate (Lawson et al., 2017), curcumin (Amalraj, Pius, Gopi, & Gopi, 2017), and quercetin (Anand, Arulmoli, & Parasuraman, 2016). We thus believe it is critical to study native electrophiles and particularly to develop methods both to identify targets *and* to study impacts of precise, individual-protein-specific signal-processing mechanisms. Since clinically-validated drug mechanisms remain quite limited (Pammolli, Magazzini, & Riccaboni, 2011), we believe that finding proteins that use their intrinsic chemistry (i.e., non-enzyme-assisted RES-sensing/-modifications) to regulate key cellular processes may open the playing field further. Importantly, in cases where Rosetta Stone proteins are allosteric regulators, our data have revealed that they can modify their partner protein's activity positively. Such a mode-of-action is still not generally accessible for pharmaceutical communities.

6. Conclusion

Our understanding of redox signaling is improving. With excellent HT-methods not only to ID sensitive proteins, but also interrogate the signaling ramifications of these non-enzymatic events, the field is moving forward energetically. We also hope that the field adopts a more quantitative and metered approach to studying these reactive signals, so that we can develop quantitative understanding of all aspects of this very exciting area.

Acknowledgments

Novartis Foundation for Medical-Biological Research (Switzerland); National Centre of Competence in Research (NCCR): Chemical Biology (Swiss National Science Foundation); US National Institutes of Health (NIH) (1DP2GM114850); and Swiss Federal Institute of Technology Lausanne (EPFL) (to Y.A.); and US National Science Foundation (NSF) Graduate Research Fellowship Program (DGE-1650441) (to D.A.U.) are acknowledged for support. The G-REX platform has been filed for patent application by Cornell University (USA); the authors declare no conflict of interest outside of this.

References

Amalraj, A., Pius, A., Gopi, S., & Gopi, S. (2017). Biological activities of curcuminoids, other biomolecules from turmeric and their derivatives—A review. *Journal of Traditional and Complementary Medicine*, 7(2), 205–233. https://doi.org/10.1016/j.jtcme.2016.05.005.

Anand, D. A. V., Arulmoli, R., & Parasuraman, S. (2016). Overviews of biological importance of quercetin: A bioactive flavonoid. *Pharmacognosy Reviews*, *10*(20), 84–89. https://doi.org/10.4103/0973-7847.194044.

Beavers, W. N., Rose, K. L., Galligan, J. J., Mitchener, M. M., Rouzer, C. A., Tallman, K. A., et al. (2017). Protein modification by endogenously generated lipid electrophiles: Mitochondria as the source and target. *ACS Chemical Biology*, *12*(8), 2062–2069. https://doi.org/10.1021/acschembio.7b00480.

Blatnik, M., Frizzell, N., Thorpe, S. R., & Baynes, J. W. (2008). Inactivation of glyceraldehyde-3-phosphate dehydrogenase by fumarate in diabetes: Formation of S-(2-succinyl)cysteine, a novel chemical modification of protein and possible biomarker of mitochondrial stress. *Diabetes*, *57*(1), 41–49. https://doi.org/10.2337/db07-0838.

Blewett, M. M., Xie, J., Zaro, B. W., Backus, K. M., Altman, A., Teijaro, J. R., et al. (2016). Chemical proteomic map of dimethyl fumarate-sensitive cysteines in primary human T cells. *Science Signaling*, *9*(445), rs10. https://doi.org/10.1126/scisignal.aaf7694.

Bollong, M. J., Lee, G., Coukos, J. S., Yun, H., Zambaldo, C., Chang, J. W., et al. (2018). A metabolite-derived protein modification integrates glycolysis with KEAP1-NRF2 signalling. *Nature*, *562*(7728), 600–604. https://doi.org/10.1038/s41586-018-0622-0.

Carol, R. J., & Dolan, L. (2006). The role of reactive oxygen species in cell growth: Lessons from root hairs. *Journal of Experimental Botany*, *57*(8), 1829–1834. https://doi.org/10.1093/jxb/erj201.

Castro, J. P., Jung, T., Grune, T., & Siems, W. (2017). 4-Hydroxynonenal (HNE) modified proteins in metabolic diseases. *Free Radical Biology & Medicine*, *111*, 309–315. https://doi.org/10.1016/j.freeradbiomed.2016.10.497.

Chen, Y., Liu, Y., Lan, T., Qin, W., Zhu, Y., Qin, K., et al. (2018). Quantitative profiling of protein carbonylations in ferroptosis by an aniline-derived probe. *Journal of the American Chemical Society*, *140*(13), 4712–4720. https://doi.org/10.1021/jacs.8b01462.

Cheng, J., Fan, Y. H., Xu, X., Zhang, H., Dou, J., Tang, Y., et al. (2014). A small-molecule inhibitor of UBE2N induces neuroblastoma cell death via activation of p53 and JNK pathways. *Cell Death & Disease*, *5*, e1079. https://doi.org/10.1038/cddis.2014.54.

Dinkova-Kostova, A. T., Holtzclaw, W. D., & Kensler, T. W. (2005). The role of Keap1 in cellular protective responses. *Chemical Research in Toxicology*, *18*(12), 1779–1791. https://doi.org/10.1021/tx050217c.

Fan, X., Zhang, J., Theves, M., Strauch, C., Nemet, I., Liu, X., et al. (2009). Mechanism of lysine oxidation in human lens crystallins during aging and in diabetes. *The Journal of Biological Chemistry*, *284*(50), 34618–34627. https://doi.org/10.1074/jbc.M109.032094.

Friedersdorf, M. B., & Keene, J. D. (2014). Advancing the functional utility of PAR-CLIP by quantifying background binding to mRNAs and lncRNAs. *Genome Biology*, *15*, R2. https://doi.org/10.1186/gb-2014-15-1-r2.

Gatenby, R. A., & Gillies, R. J. (2004). Why do cancers have high aerobic glycolysis? *Nature Reviews. Cancer*, *4*(11), 891–899. https://doi.org/10.1038/nrc1478.

Gayko, U., Fung, M., Clow, F., Sun, S., Faust, E., Price, S., et al. (2015). Development of the Bruton's tyrosine kinase inhibitor ibrutinib for B cell malignancies. *Annals of the New York Academy of Sciences*, *1358*, 82–94. https://doi.org/10.1111/nyas.12878.

Hacker, S. M., Backus, K. M., Lazear, M. R., Forli, S., Correia, B. E., & Cravatt, B. F. (2017). Global profiling of lysine reactivity and ligandability in the human proteome. *Nature Chemistry*, *9*(12), 1181–1190. https://doi.org/10.1038/nchem.2826.

Huang, N., Lee, I., Marcotte, E. M., & Hurles, M. E. (2010). Characterising and predicting haploinsufficiency in the human genome. *PLoS Genetics*, *6*(10), e1001154. https://doi.org/10.1371/journal.pgen.1001154.

Jastroch, M., Divakaruni, A. S., Mookerjee, S., Treberg, J. R., & Brand, M. D. (2010). Mitochondrial proton and electron leaks. *Essays in Biochemistry*, *47*, 53–67. https://doi.org/10.1042/bse0470053.

Kang, S. W., Lee, S., & Lee, E. K. (2015). ROS and energy metabolism in cancer cells: Alliance for fast growth. *Archives of Pharmacal Research*, *38*(3), 338–345. https://doi.org/10.1007/s12272-015-0550-6.

Kumari, S., Badana, A. K., Gavara, M. M., Gugalavath, S., & Malla, R. (2018). Reactive oxygen species: A key constituent in cancer survival. *Biomarker Insights*, *13*, 1177271918755391. https://doi.org/10.1177/1177271918755391.

Lawson, A. P., Bak, D. W., Shannon, D. A., Long, M. J. C., Vijaykumar, T., Yu, R., et al. (2017). Identification of deubiquitinase targets of isothiocyanates using SILAC-assisted quantitative mass spectrometry. *Oncotarget*, *8*(31), 51296–51316. https://doi.org/10.18632/oncotarget.17261.

Lawson, A. P., Long, M. J. C., Coffey, R. T., Qian, Y., Weerapana, E., El Oualid, F., et al. (2015). Naturally occurring isothiocyanates exert anticancer effects by inhibiting deubiquitinating enzymes. *Cancer Research, 75*(23), 5130–5142. https://doi.org/10.1158/0008-5472.CAN-15-1544.

Li, X. H., & Babu, M. M. (2018). Human diseases from gain-of-function mutations in disordered protein regions. *Cell, 175*(1), 40–42. https://doi.org/10.1016/j.cell.2018.08.059.

Lin, H. Y., Haegele, J. A., Disare, M. T., Lin, Q., & Aye, Y. (2015). A generalizable platform for interrogating target- and signal-specific consequences of electrophilic modifications in redox-dependent cell signaling. *Journal of the American Chemical Society, 137*(19), 6232–6244. https://doi.org/10.1021/ja5132648.

Liu, X., Long, M. J. C., & Aye, Y. (2019). Proteomics and beyond: Cell decision-making shaped by reactive electrophiles. *Trends in Biochemical Sciences, 44*(1), 75–89. https://doi.org/10.1016/j.tibs.2018.09.014.

Long, M. J., & Aye, Y. (2016). The die is cast: Precision electrophilic modifications contribute to cellular decision making. *Chemical Research in Toxicology, 29*(10), 1575–1582. https://doi.org/10.1021/acs.chemrestox.6b00261.

Long, M. J. C., Parvez, S., Zhao, Y., Surya, S. L., Wang, Y., Zhang, S., et al. (2017). Akt3 is a privileged first responder in isozyme-specific electrophile response. *Nature Chemical Biology, 13*(3), 333–338. https://doi.org/10.1038/nchembio.2284.

Long, M. J. C., Urul, D. A., Chawla, S., Lin, H. Y., Zhao, Y., Haegele, J. A., et al. (2018). Precision electrophile tagging in *Caenorhabditis elegans*. *Biochemistry, 57*(2), 216–220. https://doi.org/10.1021/acs.biochem.7b00642.

McMahon, M., Lamont, D. J., Beattie, K. A., & Hayes, J. D. (2010). Keap1 perceives stress via three sensors for the endogenous signaling molecules nitric oxide, zinc, and alkenals. *Proceedings of the National Academy of Sciences of the United States of America, 107*(44), 18838–18843. https://doi.org/10.1073/pnas.1007387107.

Menard, L., Maughan, D., & Vigoreaux, J. (2014). The structural and functional coordination of glycolytic enzymes in muscle: Evidence of a metabolon? *Biology (Basel), 3*(3), 623–644. https://doi.org/10.3390/biology3030623.

Owen, O. E., Kalhan, S. C., & Hanson, R. W. (2002). The key role of anaplerosis and cataplerosis for citric acid cycle function. *The Journal of Biological Chemistry, 277*(34), 30409–30412. https://doi.org/10.1074/jbc.R200006200.

Pammolli, F., Magazzini, L., & Riccaboni, M. (2011). The productivity crisis in pharmaceutical R&D. *Nature Reviews. Drug Discovery, 10*(6), 428–438. https://doi.org/10.1038/nrd3405.

Park, P. J. (2009). ChIP-seq: Advantages and challenges of a maturing technology. *Nature Reviews. Genetics, 10*(10), 669–680. https://doi.org/10.1038/nrg2641.

Parvez, S., Fu, Y., Li, J., Long, M. J., Lin, H. Y., Lee, D. K., et al. (2015). Substoichiometric hydroxynonenylation of a single protein recapitulates whole-cell-stimulated antioxidant response. *Journal of the American Chemical Society, 137*(1), 10–13. https://doi.org/10.1021/ja5084249.

Parvez, S., Long, M. J., Lin, H. Y., Zhao, Y., Haegele, J. A., Pham, V. N., et al. (2016). T-REX on-demand redox targeting in live cells. *Nature Protocols, 11*(12), 2328–2356. https://doi.org/10.1038/nprot.2016.114.

Parvez, S., Long, M. J. C., Poganik, J. R., & Aye, Y. (2018). Redox signaling by reactive electrophiles and oxidants. *Chemical Reviews, 118*(18), 8798–8888. https://doi.org/10.1021/acs.chemrev.7b00698.

Pizzimenti, S., Toaldo, C., Pettazzoni, P., Dianzani, M. U., & Barrera, G. (2010). The "two-faced" effects of reactive oxygen species and the lipid peroxidation product 4-hydroxynonenal in the hallmarks of cancer. *Cancers (Basel), 2*(2), 338–363. https://doi.org/10.3390/cancers2020338.

Poganik, J. R., & Aye, Y. (2018). Weighing up the selenocysteome uncovers new sec-rets. *Cell Chemical Biology*, *25*(11), 1315–1317. https://doi.org/10.1016/j.chembiol.2018.10.027.

Qian, Y., & Weerapana, E. (2017). A quantitative mass-spectrometry platform to monitor changes in cysteine reactivity. In H. S. Overkleeft, B. I. Florea, & J. M. Walker (Eds.), *Activity-based proteomics: Methods and protocols* (pp. 11–22). New York, NY: Humana Press.

Sallmyr, A., Fan, J., & Rassool, F. V. (2008). Genomic instability in myeloid malignancies: Increased reactive oxygen species (ROS), DNA double strand breaks (DSBs) and error-prone repair. *Cancer Letters*, *270*(1), 1–9. https://doi.org/10.1016/j.canlet.2008.03.036.

Ségalat, L. (2007). Loss-of-function genetic diseases and the concept of pharmaceutical targets. *Orphanet Journal of Rare Diseases*, *2*(1), 30. https://doi.org/10.1186/1750-1172-2-30.

She, R., Chakravarty, A. K., Layton, C. J., Chircus, L. M., Andreasson, J. O., Damaraju, N., et al. (2017). Comprehensive and quantitative mapping of RNA-protein interactions across a transcribed eukaryotic genome. *Proceedings of the National Academy of Sciences of the United States of America*, *114*(14), 3619–3624. https://doi.org/10.1073/pnas.1618370114.

Spitzer, J., Hafner, M., Landthaler, M., Ascano, M., Farazi, T., Wardle, G., et al. (2014). PAR-CLIP (photoactivatable ribonucleoside-enhanced crosslinking and immunoprecipitation): A step-by-step protocol to the transcriptome-wide identification of binding sites of RNA-binding proteins. *Methods in Enzymology*, *539*, 113–161. https://doi.org/10.1016/B978-0-12-420120-0.00008-6.

Tomoshige, S., Naito, M., Hashimoto, Y., & Ishikawa, M. (2015). Degradation of HaloTag-fused nuclear proteins using bestatin-HaloTag ligand hybrid molecules. *Organic & Biomolecular Chemistry*, *13*(38), 9746–9750. https://doi.org/10.1039/c5ob01395j.

Wang, C., Weerapana, E., Blewett, M. M., & Cravatt, B. F. (2014). A chemoproteomic platform to quantitatively map targets of lipid-derived electrophiles. *Nature Methods*, *11*(1), 79–87. https://doi.org/10.1038/nmeth.2759.

Weerapana, E., Wang, C., Simon, G. M., Richter, F., Khare, S., Dillon, M. B., et al. (2010). Quantitative reactivity profiling predicts functional cysteines in proteomes. *Nature*, *468*(7325), 790–795. https://doi.org/10.1038/nature09472.

Whitaker-Menezes, D., Martinez-Outschoorn, U. E., Flomenberg, N., Birbe, R. C., Witkiewicz, A. K., Howell, A., et al. (2011). Hyperactivation of oxidative mitochondrial metabolism in epithelial cancer cells in situ: Visualizing the therapeutic effects of metformin in tumor tissue. *Cell Cycle*, *10*(23), 4047–4064. https://doi.org/10.4161/cc.10.23.18151.

Wong, C. M., Bansal, G., Marcocci, L., & Suzuki, Y. J. (2012). Proposed role of primary protein carbonylation in cell signaling. *Redox Report*, *17*(2), 90–94. https://doi.org/10.1179/1351000212Y.0000000007.

Xu, G., Liu, Y., Mayank, K. M., & Sayre, L. M. (1999). Rapid cross-linking of proteins by 4-ketoaldehydes and 4-hydroxy-2-alkenals does not arise from the lysine-derived monoalkylpyrroles. *Chemical Research in Toxicology*, *12*, 855–861. https://doi.org/10.1021/tx990056a.

Zhao, Y., Long, M. J. C., Wang, Y., Zhang, S., & Aye, Y. (2018). Ube2V2 is a Rosetta Stone bridging redox and ubiquitin codes, coordinating DNA damage responses. *ACS Central Science*, *4*(2), 246–259. https://doi.org/10.1021/acscentsci.7b00556.

Zhou, D., Shao, L., & Spitz, D. R. (2014). Reactive oxygen species in normal and tumor stem cells. *Advances in Cancer Research*, *122*, 1–67. https://doi.org/10.1016/B978-0-12-420117-0.00001-3.

CHAPTER TWELVE

Building artificial genetic circuits to understand protein function

Louis H. Scott[a,b], **James C. Mathews**[a,b], **Aleksandra Filipovska**[a,b,c], **Oliver Rackham**[a,d,e,*]

[a]Harry Perkins Institute of Medical Research, Nedlands, WA, Australia
[b]The University of Western Australia Centre for Medical Research, Crawley, WA, Australia
[c]School of Molecular Sciences, The University of Western Australia, Crawley, WA, Australia
[d]School of Pharmacy and Biomedical Sciences, Curtin University, Bentley, WA, Australia
[e]Curtin Health Innovation Research Institute, Curtin University, Bentley, WA, Australia
*Corresponding author: e-mail address: oliver.rackham@curtin.edu.au

Contents

Abstract

Intrinsic protein properties that may not be apparent by only examining three-dimensional structures can be revealed by careful analysis of mutant protein variants. Deep mutational scanning is a technique that allows the functional analysis of millions of protein variants in a single experiment. To enable this high-throughput technique, the mutant genotype of protein variants must be coupled to a selectable function. This chapter outlines how artificial genetic circuits in the yeast *Saccharomyces cerevisiae* can maintain the genotype-phenotype link, thus enabling the general application of this approach. To do this, we describe how to engineer genetic selections in yeast, methods to construct mutant libraries, and how to analyze sequencing data. We investigate the structure-function relationships of the antimicrobial resistance protein TetX to illustrate this process. In doing so, we demonstrate that deep mutational scanning is a powerful method to dissect the importance of individual residues for the inactivation of antibiotic analogues, with consequences for the rational design of new drugs to combat antimicrobial resistance.

Methods in Enzymology, Volume 633
ISSN 0076-6879
https://doi.org/10.1016/bs.mie.2019.11.003

1. Introduction

Dissecting the relationship between a protein's structure and its activity is vital for understanding how basic biological processes occur and has allowed us to improve upon or evolve new enzymatic reactions and develop new drugs (Bornscheuer et al., 2012; Lutz, 2010; Steiner & Schwab, 2012; van Montfort & Workman, 2017). Three-dimensional snapshots of a protein's structure in distinct states as shown from X-ray crystallography (Shi, 2014) or cryo-electron microscopy (Cheng, Grigorieff, Penczek, & Walz, 2015) can illustrate conspicuous interactions of residues and ligands. However, as proteins are not static structures, their important dynamic processes can be investigated using protein nuclear magnetic resonance spectroscopy (Marion, 2013). While these methods alone are required for elucidating a protein's three-dimensional structure, mutational studies are equally important for directly investigating the function of specific residues. For example, alanine scanning can establish the importance of individual amino acids to protein stability and function (Cunningham & Wells, 1989; Morrison & Weiss, 2001). Alternatively, mutagenesis to the binding pocket or around the catalytic residues of enzymes can give insight into the specifics of catalysis, or a glimpse into possible promiscuity that can be appropriated into new functions (Gupta & Varadarajan, 2018). While these methods rely on analytical validation to link mutated genotype to phenotype, advances in synthetic biology offer an alternative when convenient phenotypic readouts are not available.

Genetic circuits that follow computational logic, where experimental inputs are translated into selectable or screenable outputs, have been engineered in many model organisms (Brophy & Voigt, 2014; Filipovska & Rackham, 2008; Liu, Yuan, & Stewart, 2013; Weber & Fussenegger, 2010). Using genetic circuitry, if a specific input can be linked to the survival of a microbe, then a single cell can act as an independent computational laboratory; thus, within a single culture, tens of millions of experiments can be run and analyzed concurrently. Therefore, coupling the genotype of mutant protein variants to the survival of a microbe is a powerful technique to study mutations, obviating the need for laborious analytical screening. Deep sequencing of resulting mutant libraries can then generate large datasets to reveal intrinsic protein properties possibly too subtle to otherwise gauge, a technique known as "deep mutational scanning" (Araya & Fowler, 2011; Fowler et al., 2010; Fowler & Fields, 2014; Fowler, Stephany, & Fields, 2014). The yeast *Saccharomyces cerevisiae* is well suited as a microscopic laboratory for this purpose as it is easy to culture, easy to genetically modify, and has many well-characterized biological parts shown to follow computational logic when combined (Blount, Weenink, & Ellis, 2012).

This chapter will describe how to design genetic circuits in yeast to probe the structure and function relationships of a protein of interest. We will explain how variants in random mutant libraries can be selected for the gain, loss, or retention of protein activity, and how deep mutational scanning can reveal what and how prevalent these mutations are. We will illustrate this process by probing the structure and function relationships of residues within TetX, an emerging antimicrobial resistance threat that causes the enzymatic inactivation of all known tetracycline antibiotics (Scott, Mathews, Flematti, Filipovska, & Rackham, 2018). Understanding the mutability of specific residues within a protein has broad consequences for understanding how it functions; in the case of an antimicrobial resistance protein, this understanding can guide the development of vitally important new drugs.

2. General principles

We propose three general steps to use genetic circuits in yeast to probe the structure and function relationships of a protein, as illustrated in Fig. 1. These steps involve engineering a genetic selection for protein activity, making a library of protein variants, and then analyzing the genotype of escape mutants.

The first crucial step is to link protein function to cell survival. This requires careful consideration, as how to engineer this association may not be directly recognizable. Fortunately, yeast strains already engineered to assay macromolecular interactions (for example, yeast two-hybrid systems) can be appropriated for part of this genetic circuitry (Fields & Song, 1989; Licitra & Liu, 1996; Van Criekinge & Beyaert, 1999; Wallis, Filipovska, & Rackham, 2018; Wallis, Richman, Filipovska, & Rackham, 2018). In these strains, when an experimental macromolecular interaction occurs, a functional transcription factor is created that activates the expression of a reporter gene. Therefore, if the activity of a protein of interest can also be linked to the expression of such a transcription factor (or generally, a "master regulator") in one of these strains, then this activity should be selectable. A wealth of engineered yeast one-hybrid, two-hybrid and three-hybrid strains containing a multitude of different reporter genes capable of both

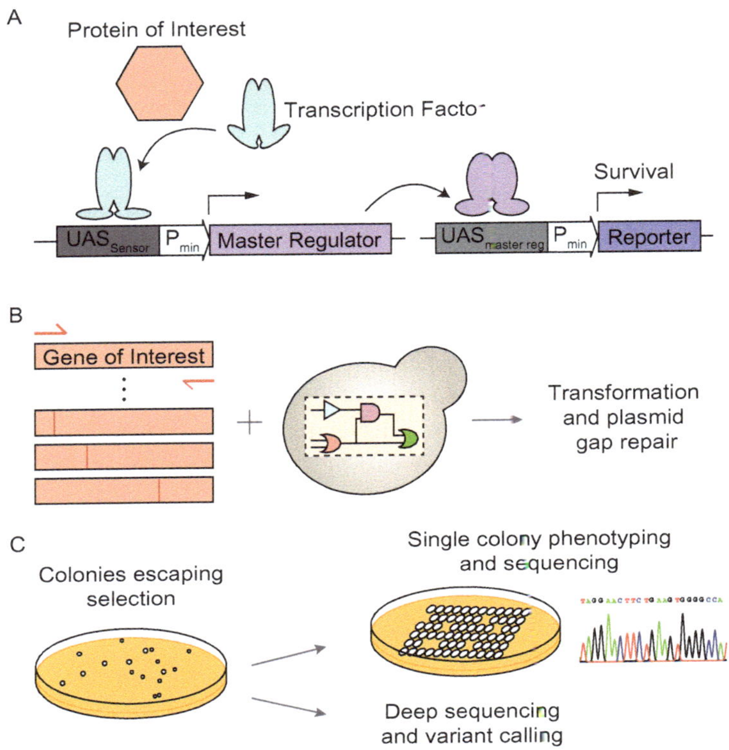

Fig. 1 The function of a protein of interest can be selected for if it activates a transcription factor that in turn activates a master regulator that turns on downstream reporter genes (A). A mutant library, generated for example by error-prone PCR, can be introduced into yeast engineered with genetic circuitry described in (A) and assembled *in vivo* with an expression vector by homologous recombination (B). Library transformants escaping selection can be analyzed as single clones by phenotypic screens in conjunction with Sanger sequencing, or all escape mutants can be analyzed in parallel with deep mutational scanning (C).

positive and negative selections is available, and this existing well-classified foundation can be used to build more complex genetic circuits to assay a specific protein function.

Second, a library of mutant proteins must be expressed in yeast. The mutational strategy chosen will be a trade-off between mutational depth versus amino acid sequence coverage and will need careful attention based on the size of the protein and on available structural or mutational information.

Third, mutants escaping selection should be analyzed. While very labor intensive, the analysis of individual escape mutants provides a direct link between genotype and phenotype and should be done with a manageable number of variants. Importantly, deep sequencing of all selected mutants will give a general view of the protein's mutational landscape. Using deep mutational scanning as a high-throughput approach can reveal intrinsic protein properties that would be missed by conventional sampling (Fowler & Fields, 2014).

3. Methodology

We recently used deep mutational scanning in yeast to investigate the structure and function of the antimicrobial resistance enzyme TetX (Scott et al., 2018). TetX is a flavin-dependent monooxygenase that can degrade all known tetracycline antibiotics (Yang et al., 2004). However, as not all tetracyclines are degraded at the same rate, there was potential to redesign current drugs to be better inhibitors and bypass this resistance mechanism. To investigate this, we designed genetic circuits that linked the function of TetX to the survival of yeast. Random mutant libraries of TetX were then selected for variants that could degrade tetracycline analogues or not. Finally, using deep mutational scanning, we investigated the structural requisites of TetX, dependent on the tetracycline analogues it could degrade. Combined with research into the structure of TetX and related enzymes (Moore, Hughes, & Wright, 2005; Park et al., 2017; Volkers et al., 2013; Volkers, Palm, Weiss, Wright, & Hinrichs, 2011), our study provides valuable information to be leveraged to rationally redesign tetracycline antibiotics to potentially overcome enzymatic destruction.

The next sections describe a methodological framework to study structure-function relationships of a protein using genetic circuits and deep mutational scanning in yeast. We describe this methodology using our recent study on the antimicrobial resistance protein TetX as example.

3.1 Genetic circuits to select for protein activity

3.1.1 A biosensor to select for TetX function

The following sections outline how an artificial genetic circuit can select for protein function in engineered yeast. As a starting point, one must test if the protein of interest is functionally expressed. Therefore, analytical techniques that give a direct readout on protein function should be used. Expression from a strong constitutive or inducible promoter in these initial experiments should ensure some detectable protein activity. Furthermore, codon usage optimization is recommended for all genes not originating from yeast—most gene synthesis companies offer this service. In our recent work, we found expression of the yeast codon usage optimized gene for TetX from the *ADH1* promoter on a 2-micron plasmid was sufficient to destroy the tetracycline anhydrotetracycline (ATC) in yeast cultures, as detected by high performance liquid chromatography-mass spectrometry (HPLC-MS) (Scott et al., 2018). While a similar expression configuration can be an initial foundation for circuit design, expression may need to be adjusted during the final library selection to ensure enough sensitivity to detect mutations that cause subtle changes in function. Many endogenous yeast promoters spanning orders of magnitude in activity have been characterized and can be appropriated for this purpose (Keren et al., 2013; Lee, DeLoache, Cervantes, & Dueber, 2015).

Once protein function in yeast is confirmed, the next crucial step is to link this activity to the expression of a "master regulator." A master regulator is simply a transcription factor that can activate downstream reporter genes that enable genetic selection. Many yeast-hybrid strains have reporter genes engineered to be activated by transcription factors, such as the yeast two-hybrid strain MaV204K that uses Gal4 to activate *HIS3* and *URA3* auxotrophic markers (Ito et al., 2000). These reporter genes can be selected for on media lacking histidine or uracil, but also against using 5-fluoroorotic acid (5-FOA) as it is transformed into a toxic product by *URA3*. Therefore, the function of TetX, or generally any other protein of interest, can be selected for or against if it can induce *GAL4* expression in MaV204K.

As TetX functions by degrading tetracyclines, a tetracycline inducible gene expression system was used to express the master regulator Gal4 in MaV204K (Scott et al., 2018). We built a genetic circuit where the reverse tetracycline-controlled transactivator (rtTA) can sense tetracyclines and bind to promoter elements upstream of the *GAL4* gene, encoding the Gal4 transcription factor. If this strain is cultured with tetracyclines, Gal4 expression is turned on, which in turn activates *HIS3* and *URA3* expression. If TetX is expressed in this strain, it will degrade the tetracycline inducer before it can stimulate Gal4 expression and downstream auxotrophic marker activation. Therefore, using this system, the function of TetX can be selected for or against dependent on the culture conditions. Integral to consistent results of such a selection is to use a centromeric plasmid that replicates as a single copy to avoid possible variability in Gal4 expression confounded by copy number.

Our initial exploratory experiments demonstrated that induction of Gal4 expression by ATC caused toxicity to yeast (Scott et al., 2018).

We hypothesized this may be caused by pleiotropic effects of Gal4 overexpression (Keren et al., 2016). One way to overcome this is to modulate the translation of the master regulator using strong and weak Kozak consensus sequences or start codons. This can easily be achieved by site directed mutagenesis without having to replace promoters (Chen, Lin, Chang, Yeh, & Wang, 2008). Subtle effects on translation can be made by altering the six nucleotides preceding the start codon, for example, to cytosines, thus creating a weak Kozak consensus sequence. Alternatively, changing the start codon from ATG to the less commonly used TTG can cause a more pronounced reduction in translation. Using this technique, a yeast strain conditionally expressing Gal4 with a TTG start codon enabled yeast to survive on media lacking histidine and uracil in the presence of ATC, thus creating cells that were dependent on tetracyclines for survival. When TetX was expressed in this strain it could inactivate ATC, thus prohibiting Gal4 expression from triggering of the nutritional marker expression cascade.

3.1.2 General genetic circuit design considerations

A transcriptional regulator, that operates through the tetracycline repressor (TetR), was used to sense the presence of tetracyclines in our recent work. Nevertheless, one is not constricted to this class of antibiotics as biosensors have been designed against a myriad of other targets (Dietrich, McKee, & Keasling, 2010). Therefore, if a biosensor exists against a protein or ligand of interest, then the structure and function of a protein that modulates it can be explored using the methodology presented in this chapter. Furthermore, the concept of utilizing a transcriptional regulator as a sensor is generally applicable to many other antibiotic and small molecule altering enzymes far beyond those already detectable by a biosensor. Most biosynthetic clusters encode a self-resistance mechanism that is transcriptionally regulated by the metabolite produced. This duality concept has been exploited to discover new natural products via genome mining (Yan et al., 2018). Furthermore, there are currently over 200,000 TetR family transcriptional regulators in public sequence databases that recognize many diverse small molecules, including numerous desirable natural products that are modified by a multitude of accessory proteins (Ramos et al., 2005). The structures of about 200 of these transcriptional regulators are known, and protein engineering approaches have shown that they can be redesigned to bind new ligands (Tang & Cirino, 2011; Tang et al., 2013). Therefore, the concept of linking the activity of a protein that alters a natural product, via a transcriptional regulator, to life-death selections in yeast should be generally applicable to most natural products of interest.

While Gal4 functions well in the context of available yeast-hybrid strains, toxicity from overexpression is a confounding factor that could be an issue (Keren et al., 2016). Many yeast two-hybrid strains have operator sites for the LexA repressor upstream of reporter genes; therefore, expressing a fusion of the LexA repressor with an activating domain could overcome potential pleotropic Gal4 effects (Fashena, Serebriiskii, & Golemis, 2000). Furthermore, expanding the repertoire of reporter strains to encompass fully orthogonal components will be highly beneficial for genetic circuit design. Strain engineering is becoming even easier with CRISPR-Cas9 systems (Jessop-Fabre et al., 2016), and future iterations should make use of the expanding collection of well-characterized genetic parts (Naseri et al., 2017).

In terms of practicalities, a biosensor should have a wide operational and dynamic range (see Fig. 2) (Dietrich et al., 2010; Rogers, Taylor, & Church, 2016). Typically, high molecular specificity should also be sought after (Lederer et al., 1996). However, the promiscuous nature of TetR used to induce tetracycline dependent Gal4 expression allowed the analysis of different inducers in our recent study, emphasizing that choice of parts during genetic circuit construction is highly dependent on the experimental question asked. Another important aspect for genetic circuit design is the ability for rapid prototyping, or failing quickly, as this will drastically decrease the

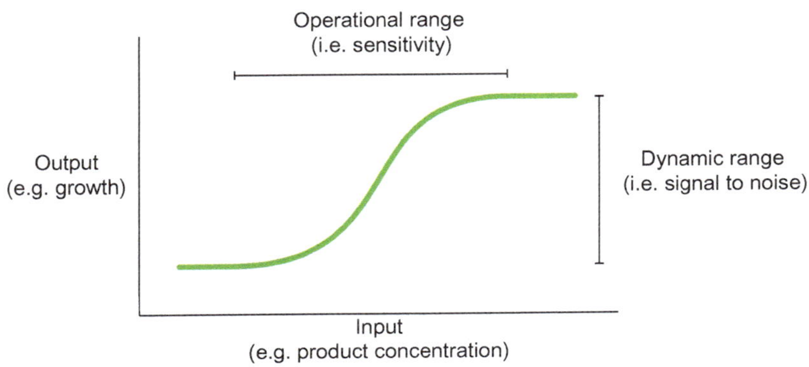

Fig. 2 Input concentrations over which the biosensor responds is the operational range. It spans the limit of detection to the saturation point. The dynamic range is the output to any given concentration of input. It can be measured in cell growth or fluorescence depending on the reporter gene(s) chosen. A large dynamic range will ensure the signal resulting from the experimental input can be detected over background noise.

design-build-test cycle required to engineer a suitable selection. To aid rapid prototyping, yeast survival assays are a quick semi-quantitative way to assess genetic circuit performance. However, thorough exploratory experiments are essential to large-scale library screening to understand how the selection mechanism will behave *in situ*. Consequently, more detailed assessment of short-listed selections should be done mimicking the final selection conditions.

3.2 Mutant library construction

Before constructing a mutant library, planning is required to ensure that it contains useful answers to your research questions. Choice of mutational strategy is a key decision when planning deep mutational scanning, with saturation or random mutagenesis being two broad classes of mutational strategy. Saturation mutagenesis is the exhaustive examination of amino acid changes within a discrete sequence region. This approach, enabled by efficient chemical synthesis of combinatorial DNA libraries (Medina-Cucurella et al., 2019) or by polymerase chain reaction (PCR) with trimer phosphoramidites (Filipovska, Razif, Nygård, & Rackham, 2011), offers comprehensive and largely unbiased exploration of the sequence space but is limited by practical constraints to targeting short protein sequences.

Random mutagenesis strategies include a variety of techniques that introduce DNA changes by chance. For example, mutations can be introduced by modified PCR (Leung, Chen, & Goeddel, 1989), by treatment with chemical or radiological agents (Bose, 2016), or by introducing target sequences to "mutator" strains of organisms with inherently low fidelity of DNA replication (Muteeb & Sen, 2010). Although libraries produced by random strategies will not be as comprehensive or controlled as saturation mutagenesis libraries, random strategies are often simpler to implement and can more easily target longer sequences of whole genes or even whole genomes. Furthermore, random mutagenesis also more closely approximates the conditions by which mutations arise in nature, and so is a powerful tool for evolving proteins with new or improved functions.

3.2.1 *Error-prone polymerase chain reaction libraries*

Error-prone polymerase chain reaction (EP-PCR) is a tried and tested method for introducing random mutations to full-length proteins where reaction conditions can be changed to fine tune the mutation rate. Overall mutation rate and library size are two key considerations controlling diversity within the library: if the mutation rate is too low and/or the library is too small, amino acid changes will not be sufficiently diverse to provide useful data (especially when synonymous mutations are expected). If the rate is too high, there will be many loss of function mutants carrying multiple mutations, making it difficult to attribute function loss to any one mutation. For these reasons, aiming for a mutation rate close to two substitutions per kilobase can maximize the number of transformants carrying proteins with just one productive amino acid change. At this rate, if mutation frequency is assumed to follow a Poisson distribution, roughly 14% of clones will be not mutated, roughly 27% will carry one mutation, roughly 27% will carry two mutations, and the remainder more than two. Although there are many methods for adjusting mutation rate, the simplest to modify is number of PCR cycles with an error-prone polymerase—this parameter reportedly titrating mutation rates between 0.01% and 10% per base position (Vanhercke, Ampe, Tirry, & Denolf, 2005).

Another key consideration for EP-PCR library design is reducing mutation bias. DNA polymerase enzymes are naturally biased toward certain nucleotide substitutions, notably AT → GC and AT → TA. Although a limited repertoire of base substitutions can theoretically still result in a very diverse library of amino acid substitutions, it is advisable to consider methods for reducing this bias to allow detection of rare mutant phenotypes. Common methods for controlling mutation bias are addition of Mn^{2+} up to 5 mM in the reaction buffer (Lin-Goerke, Robbins, & Burczak, 1997), changed ratio of deoxynucleoside triphosphates (dNTPs) (Cadwell & Joyce, 1992), and combining multiple polymerase enzymes with different mutational preferences (Vanhercke et al., 2005). For our example TetX library, we chose to experiment with a novel method for combatting mutational bias. By using an EP-PCR template obtained by fairly low fidelity gene fragment synthesis (7.52×10^{-4} errors per base position) additional non-polymerase biased mutations can be introduced into the library (Invitrogen, 2015).

When planning an EP-PCR library size target for a whole open reading frame, one should maximize the completeness of the library, so that it contains as many protein sequences as possible that differ from the wild type by a single amino acid. The proportion of these sequences present can be thought of as the "completeness" of the library, and the maximum number of single amino acid substitutions calculated by multiplying the length of the protein sequence by the 19 alternative amino acids possible at each position. Since the start codon ordinarily needs to be preserved, the amino acid sequence length can be reduced by one for this calculation. For example, at a rate of two mutations per kilobase for the *TETX* gene, an ideal EP-PCR library containing codons for all 7353 possible single amino acid substitutions would require the impractical 2×10^{11} yeast transformants to completely sample. Knowing that large-scale yeast transformations typically yield closer to 10^{7} transformants, the practical library completeness would be closer to 57% as calculated by the online tool PEDEL-AA (Firth & Patrick, 2008). It should also be considered at this point that library size targets are limited by the requirements of downstream sequencing experiments. When using next-generation sequencing technologies, it is advisable for each variant to be covered by at least 100 sequencing reads for accurate quantitative analysis (Fowler & Fields, 2014).

Selecting a method for incorporation of genes of interest into the host also requires some thought. One advantage of using yeast is the chance to harness its robust homologous recombination machinery to assemble vectors *in situ*. When using this technique, one should, however, be aware that some transformants may carry re-ligated vector, inflating the usable transformation efficiency (Chandran & Shapland, 2017). For our example TetX library, we transformed variant *TETX* gene sequences along with linear copies of a yeast shuttle vector to place *TETX* expression under control of the constitutive *ADH1* promotor.

3.2.2 Small-scale library testing

Small-scale pilot library experiments are recommended to investigate the quality of the library and determine the reagents required before up-scaling. Key considerations that can be optimized in a small-scale experiment include transformation efficiency, mutation rate, and mutation type. During small-scale testing, it is useful to test various ratios of vector and insert DNA for their transformation efficiencies to select conditions that maximize yield during larger experiments. For our TetX pilot study, insert to vector ratios of 2:1 and 1:1 were transformed into yeast by the lithium acetate/single-stranded carrier DNA/polyethylene glycol method (Gietz & Schiestl, 2007; Gietz & Woods, 2002), then inoculated as serial dilutions on agar plates selecting only for plasmid maintenance. Colony counts revealed that an insert to vector ratio of 2:1 yielded the most transformants in our experimental conditions. We recommend, however, testing even more ratios, for example, up to an insert to vector ratio of 5:1.

Using Sanger sequencing, small-scale transformation tests are an excellent opportunity to compare the achieved mutation rate with the original target. PCR conditions can then be adjusted to titrate this rate if necessary.

Sanger sequencing of yeast transformant colonies from our TetX pilot library found mutation rates of 1.5 nucleotide substitutions per kilobase, resulting in an average 1.2 mutations per gene. Encouragingly, no clones carrying empty re-ligated vector were detected.

As well as mutation rate, the type of mutations occurring in a small set of EP-PCR transformants will provide key information on the strength of your mutational strategy. For our example TetX library, it was found that for 25% of those mutants sequenced, insertions and deletions occurred in a single 9 bp adenine homopolymer region of the gene (Fig. 3). If we proceeded with large-scale experiments, these indels would result in many missense mutants that would be non-functional and escape selection—overwhelming rarer loss of function mutants that arose from point mutations. In this case, it was simple to redesign the synthetic *TETX* gene template with two AAA codons replaced by the synonymous AAG to disrupt the homopolymer. This example demonstrates the importance of small-scale tests before attempting more substantial EP-PCR protocols.

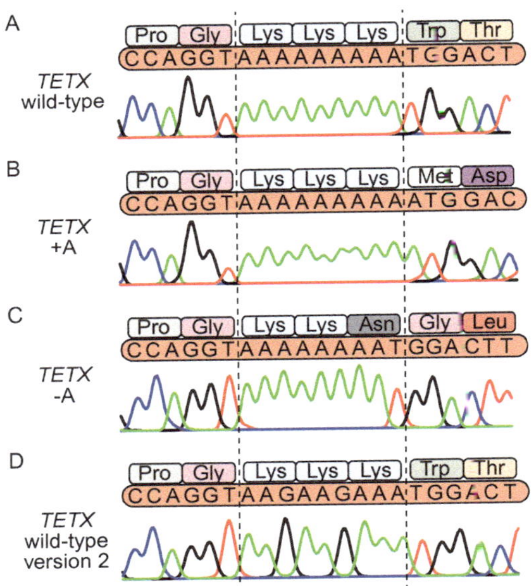

Fig. 3 Small-scale tests are crucial to the success of large-scale EP-PCR libraries. In this example, an unexpectedly high (25%) rate of insertions and deletions occurring in a single adenine homopolymer sequence (A) flooded the library with nonsense loss of function mutations (B and C). To correct this, it was trivial to replace two AAA codons with the synonymous AAG in the synthetic gene template (D).

3.2.3 Large-scale library screening

Large-scale library construction parameters should be extrapolated from results achieved during the small-scale pilot experiment. The small-scale experiment should give information on the lowest amount of DNA required to still deliver the most transformants possible that can be analyzed with downstream techniques. This is important as, although rare, increasing the DNA amount can lead to multiple variants being introduced to the same cell, decoupling the phenotype-genotype link. It is therefore recommended to scale up the whole transformation reaction, and not just the DNA amount. We recommend up-scaling using the lithium acetate/single-stranded carrier DNA/polyethylene glycol yeast transformation method which should yield tens of millions of clones (Gietz & Schiestl, 2007; Gietz & Woods, 2002). This process is highly contamination sensitive due to large volumes of rich yeast media undergoing repeated incubations, and the use of laminar air flow is recommended.

Once transformed, the library should be allowed to undergo additional doublings before plating on selective media. This is an important step that provides a clonal population of each mutant. Having multiple cells containing the same mutant ensures that even those that occur rarely in the population have a chance of being sampled. It also means that the entire diversity of the library can be captured on multiple selection conditions, and crucially on a non-selective condition to provide an un-selected representation of the library, which is necessary for calculating the enrichment of mutations. Additionally, multiple biological replicates should be constructed and selected. Comparing mutant enrichment between replicates using purpose-built software such as Enrich2 will ensure the planned library diversity is captured and can also decrease noise and false positives from inconsistent escape mutants (Rubin et al., 2017).

In our recent work, additional doublings of the library allowed TetX mutants that could not inactivate different tetracyclines (ATC or doxycycline (DOX)) to be selected. This was done using media lacking tyrosine and lysine, to select for TetX expression and biosensor plasmids, and lacking uracil and histidine, to select for reporter activation because of enzyme function loss. Furthermore, we also conducted some selection experiments using media lacking tyrosine and lysine, supplemented with ATC and 5-FOA—to select for functional TetX variants only—demonstrating the flexibility of this selection system (Scott et al., 2018). Aliquots of the library were also taken following rescue of transformants and after the final additional doublings, to calculate final transformation efficiency and library size by serial dilution survival assays on media selecting for plasmid maintenance only.

3.3 Mutant library analysis

3.3.1 *Single colony analysis*

In conjunction with large-scale library analysis using deep sequencing, it is useful to isolate mutant strains with interesting phenotypes by traditional replica plating techniques. Doing so will offer a chance to study these mutants' activity in greater detail and against a greater variety of substrates. The enrichment state of these phenotypes in deep sequencing data can also serve as validation for selection, sequencing, and analysis strategies. Isolated mutants with studied activity levels offered a valuable point of comparison with our *TETX* deep sequencing data, their enrichment in the larger data set confirmed the validity of the deep sequencing data and indicated the relative level of activity loss that corresponded to various enrichment values. Following selection from a portion of our TetX library, yeast colonies that escaped selection were inoculated on selective agar plates with alternative enzyme substrates, in this case ATC and DOX. By comparing growth between these plates, it was possible to identify mutant enzymes with a preference for one substrate. Sequencing these mutations and modeling them with molecular visualization software provided crucial structure-function insights (Scott et al., 2018). Care must be taken to retain part of each sampled colony on the original selection plate to ensure their representation during deep sequencing.

3.3.2 *Whole population analysis*

3.3.2.1 Generation of sequencing data

As it is not possible to individually sequence and phenotypically screen each mutant, deep sequencing the whole mutant population and comparing this with an unselected population can reveal which mutations are enriched after a selection condition. There are many sequencing technologies and protocols suitable for this purpose; however, choices here will be governed by availability, cost, and experimental needs. For shorter gene sequences or smaller libraries, Sanger sequencing many individual mutants in parallel remains a reliable and robust option. However, for longer gene sequences and larger populations containing rare mutant phenotypes, next-generation sequencing technologies are more appropriate. Of these technologies, sequencing by synthesis (Illumina), pyrosequencing (454), and sequencing by ligation (SOLiD) reportedly give the highest accuracy (99.9% for all three) (Liu et al., 2012).

For our example TetX library, we chose to use Illumina MiSeq with paired-end reads, due to its ready availability and low cost per base. Since deep mutational scanning is a quantitative method relying on detection of rare phenotypes, we aimed to maximize the coverage depth achieved in the *TETX* open reading frame using sub-amplicon sequencing. This is a technique where primers tagged with known Illumina compatible sequences are used to amplify a series of short products spanning the target (Hiatt, Patwardhan, Turner, Lee, & Shendure, 2010; Starita et al., 2013). These amplicons can then be applied directly to the flow cell as paired-end reads. For our example TetX library, we found that 250 bp sub-amplicon reads provided the greatest coverage depth and most accurate results. These reads were generated from mutants in an unselected condition, escape mutants from ATC and from DOX selections (selection for non-functional TetX), and escape mutants from selection with ATC and 5-FOA (selection for functional TetX).

3.3.2.2 Analysis of sequencing data

After optimization of a sequencing strategy and sequencing of your library, you will need software to convert raw read data into useful information about the mutants that make up your library. In general, the process of analysis will include pre-processing of reads for quality, alignment of reads to the reference DNA sequence, and finally calling mutations and calculating enrichment scores for mutations compared to an unselected population. Naturally, the choice of bioinformatics tool(s) will depend on the sequencing technology used, characteristics of the library, characteristics of the target gene(s), and the nature of your research questions. For many applications, obvious choices include the widely used open-source toolkits SAMtools (Li et al., 2009) and GATK (McKenna et al., 2010). If these platforms do not meet the specific needs of your experiment, there are specialized open-source tools available from code repositories developed by researchers who have needed to extend the reach of their deep sequencing and deep mutational scanning experiments. Enrich2 and DMStools are examples of purpose-built software that will treat sequencing data with a statistically principled approach to generate plots for deep mutational scanning data visualization (Bloom, 2015; Rubin et al., 2017). Recently, machine learning algorithms have been developed to incorporate variant function prediction of large mutant datasets (Gray, Hause, Luebeck, Shendure, & Fowler, 2018).

For our example TetX library, we performed initial read quality and adaptor trimming, as well as extraction and realignment of subamplicons, using the proprietary software Geneious (Kearse et al., 2012). Enrich2 was then used to calculate enrichment scores for mutations across the protein sequence, and these enrichment scores visualized graphically as well as mapped onto protein structure using PyMOL (Schrödinger LLC, 2018).

Mutational bias can also be investigated at this stage. When the frequency of all possible base substitutions in the *TETX* library was calculated, TA→CG transversions were most frequently observed, followed by GC→AT transversions (Fig. 4). This trend is consistent with the known mutation bias of Taq and with the relatively low GC content of the target gene (Vanhercke et al., 2005). Encouragingly, a low but consistent rate of most other substitutions was detected in the library, perhaps resulting from our novel approach to reduce mutation bias.

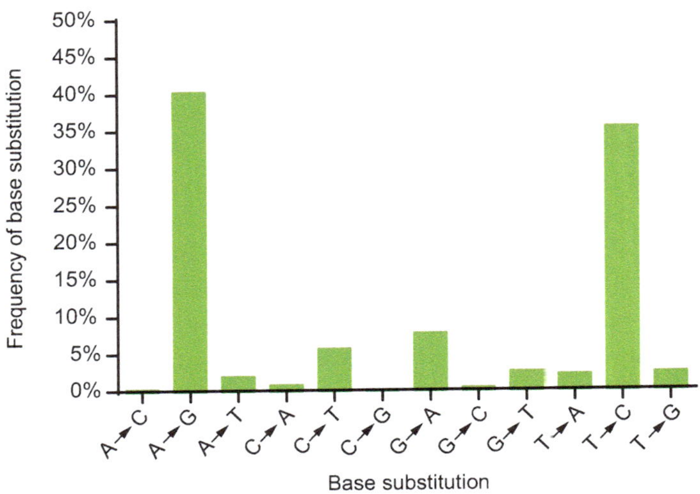

Fig. 4 Frequencies of base substitutions occurring in an EP-PCR library following selection with an artificial yeast genetic circuit are generally consistent with expectations for Taq polymerase. Low but consistent levels of rare mutations were attributable to the use of a low fidelity gene synthesis template for the reaction.

4. Summary

Artificial genetic circuits in yeast can provide a valuable tool to link mutant genotype to protein function. In doing so, the analysis of tens of millions of variants can be achieved in parallel, a technique known as deep mutational scanning. This chapter describes this process using the analysis of tetracycline resistance protein TetX variants, as an example. We describe in detail how to engineer genetic selections for protein function, how to create libraries of mutant proteins, and how deep sequencing can reveal the relevance of specific residues for protein activity. Investigation of the structure and function of TetX provides insights into how new drugs could be made to combat this resistance mechanism, illustrating the power of this methodology.

Acknowledgments

This work was supported by fellowships, scholarships and grants from the Australian Research Council (DP180101656 to A.F. and O.R.), the National Health and Medical Research Council (APP1154646 to A.F. and APP1154932 to O.R.), and the Cancer Council Western Australia (to A.F. and O.R.).

References

Araya, C. L., & Fowler, D. M. (2011). Deep mutational scanning: Assessing protein function on a massive scale. *Trends in Biotechnology*, *29*, 435–442.

Bloom, J. D. (2015). Software for the analysis and visualization of deep mutational scanning data. *BMC Bioinformatics*, *16*, 168.

Blount, B. A., Weenink, T., & Ellis, T. (2012). Construction of synthetic regulatory networks in yeast. *FEBS Letters*, *586*, 2112–2121.

Bornscheuer, U. T., Huisman, G. W., Kazlauskas, R. J., Lutz, S., Moore, J. C., & Robins, K. (2012). Engineering the third wave of biocatalysis. *Nature*, *485*, 185–194.

Bose, J. L. (2016). Chemical and UV mutagenesis. In J. Bose (Ed.), *Methods in molecular biology: Vol.1373. The genetic manipulation of Staphylococci* (pp. 111–115). New York, NY: Humana Press.

Brophy, J. A. N., & Voigt, C. A. (2014). Principles of genetic circuit design. *Nature Methods*, *11*, 508–520.

Cadwell, R. C., & Joyce, G. F. (1992). Randomization of genes by PCR mutagenesis. *PCR Methods and Applications*, *2*, 28–33.

Chandran, S., & Shapland, E. (2017). Efficient assembly of DNA using yeast homologous recombination (YHR). In R. Hughes (Ed.), *Methods in molecular biology: Vol. 1472. Synthetic DNA* (pp. 187–192). New York, NY: Humana Press.

Chen, S.-J., Lin, G., Chang, K.-J., Yeh, L.-S., & Wang, C.-C. (2008). Translational efficiency of a non-AUG initiation codon is significantly affected by its sequence context in yeast. *The Journal of Biological Chemistry*, *283*, 3173–3180.

Cheng, Y., Grigorieff, N., Penczek, P. A., & Walz, T. (2015). A primer to single-particle cryo-electron microscopy. *Cell*, *161*, 438–449.

Cunningham, B., & Wells, J. (1989). High-resolution epitope mapping of hGH-receptor interactions by alanine-scanning mutagenesis. *Science*, *244*, 1081–1085.

Dietrich, J. A., McKee, A. E., & Keasling, J. D. (2010). High-throughput metabolic engineering: Advances in small-molecule screening and selection. *Annual Review of Biochemistry*, *79*, 563–590.

Fashena, S. J., Serebriiskii, I. G., & Golemis, E. A. (2000). LexA-based two-hybrid systems. *Methods in Enzymology*, *328*, 14–26.

Fields, S., & Song, O. (1989). A novel genetic system to detect protein–protein interactions. *Nature*, *340*, 245–246.

Filipovska, A., & Rackham, O. (2008). Building a parallel metabolism within the cell. *ACS Chemical Biology*, *3*, 51–63.

Filipovska, A., Razif, M. F. M., Nygård, K. K. A., & Rackham, O. (2011). A universal code for RNA recognition by PUF proteins. *Nature Chemical Biology*, *7*, 425–427.

Firth, A. E., & Patrick, W. M. (2008). GLUE-IT and PEDEL-AA: New programmes for analyzing protein diversity in randomized libraries. *Nucleic Acids Research*, *36*, W281–W285.

Fowler, D. M., Araya, C. L., Fleishman, S. J., Kellogg, E. H., Stephany, J. J., Baker, D., et al. (2010). High-resolution mapping of protein sequence-function relationships. *Nature Methods*, *7*, 741–746.

Fowler, D. M., & Fields, S. (2014). Deep mutational scanning: A new style of protein science. *Nature Methods*, *11*, 801–807.

Fowler, D. M., Stephany, J. J., & Fields, S. (2014). Measuring the activity of protein variants on a large scale using deep mutational scanning. *Nature Protocols*, *9*, 2267–2284.

Gietz, R. D., & Schiestl, R. H. (2007). Large-scale high-efficiency yeast transformation using the LiAc/SS carrier DNA/PEG method. *Nature Protocols*, *2*, 38–41.

Gietz, R. D., & Woods, R. A. (2002). Transformation of yeast by lithium acetate/single-stranded carrier DNA/polyethylene glycol method. *Methods in Enzymology*, *350*, 87–96.

Gray, V. E., Hause, R. J., Luebeck, J., Shendure, J., & Fowler, D. M. (2018). Quantitative missense variant effect prediction using large-scale mutagenesis data. *Cell Systems*, *6*. 116–124.e3.

Gupta, K., & Varadarajan, R. (2018). Insights into protein structure, stability and function from saturation mutagenesis. *Current Opinion in Structural Biology*, *50*, 117–125.

Hiatt, J. B., Patwardhan, R. P., Turner, E. H., Lee, C., & Shendure, J. (2010). Parallel, tag-directed assembly of locally derived short sequence reads. *Nature Methods*, *7*, 119–122.

Invitrogen. (2015). *Evaluation of linear synthetic DNA fragments from separate suppliers*. Thermo Fisher Scientific.

Ito, T., Tashiro, K., Muta, S., Ozawa, R., Chiba, T., Nishizawa, M., et al. (2000). Toward a protein-protein interaction map of the budding yeast: A comprehensive system to examine two-hybrid interactions in all possible combinations between the yeast proteins. *Proceedings of the National Academy of Sciences of the United States of America*, *97*, 1143–1147.

Jessop-Fabre, M. M., Jakočiūnas, T., Stovicek, V., Dai, Z., Jensen, M. K., Keasling, J. D., et al. (2016). EasyClone-MarkerFree: A vector toolkit for marker-less integration of genes into *Saccharomyces cerevisiae* via CRISPR-Cas9. *Biotechnology Journal*, *11*, 1110–1117.

Kearse, M., Moir, R., Wilson, A., Stones-Havas, S., Cheung, M., Sturrock, S., et al. (2012). Geneious basic: An integrated and extendable desktop software platform for the organization and analysis of sequence data. *Bioinformatics*, *28*, 1647–1649.

Keren, L., Hausser, J., Lotan-Pompan, M., Vainberg Slutskin, I., Alisar, H., Kaminski, S., et al. (2016). Massively parallel interrogation of the effects of gene expression levels on fitness. *Cell*, *166*, 1282–1294.e18.

Keren, L., Zackay, O., Lotan-Pompan, M., Barenholz, U., Dekel, E., Sasson, V., et al. (2013). Promoters maintain their relative activity levels under different growth conditions. *Molecular Systems Biology*, *9*, 701.

Lederer, T., Kintrup, M., Takahashi, M., Sum, P. E., Ellestad, G. a., & Hillen, W. (1996). Tetracycline analogs affecting binding to Tn10-Encoded Tet repressor trigger the same mechanism of induction. *Biochemistry*, *35*, 7439–7446.

Lee, M. E., DeLoache, W. C., Cervantes, B., & Dueber, J. E. (2015). A highly characterized yeast toolkit for modular, multipart assembly. *ACS Synthetic Biology*, *4*, 975–986.

Leung, D. W., Chen, E., & Goeddel, D. V. (1989). A method for random mutagenesis of a defined DNA segment using a modified polymerase chain reaction. *Techniques*, *1*, 11–15.

Li, H., Handsaker, B., Wysoker, A., Fennell, T., Ruan, J., Homer, N., et al. (2009). The sequence alignment/map format and SAMtools. *Bioinformatics*, *25*, 2078–2079.

Licitra, E. J., & Liu, J. O. (1996). A three-hybrid system for detecting small ligand-protein receptor interactions. *Proceedings of the National Academy of Sciences of the United States of America*, *93*, 12817–12821.

Lin-Goerke, J. L., Robbins, D. J., & Burczak, J. D. (1997). PCR-based random mutagenesis using manganese and reduced dNTP concentration. *BioTechniques*, *23*, 409–412.

Liu, L., Li, Y., Li, S., Hu, N., He, Y., Pong, R., et al. (2012). Comparison of next-generation sequencing systems. *Journal of Biomedicine and Biotechnology*, *2012*, 1–11.

Liu, W., Yuan, J. S., & Stewart, C. N., Jr. (2013). Advanced genetic tools for plant biotechnology. *Nature Reviews Genetics*, *14*, 781–793.

Lutz, S. (2010). Beyond directed evolution—semi-rational protein engineering and design. *Current Opinion in Biotechnology*, *21*, 734–743.

Marion, D. (2013). An introduction to biological NMR spectroscopy. *Molecular & Cellular Proteomics: MCP*, *12*, 3006–3025.

McKenna, A., Hanna, M., Banks, E., Sivachenko, A., Cibulskis, K., Kernytsky, A., et al. (2010). The genome analysis toolkit: A MapReduce framework for analyzing next-generation DNA sequencing data. *Genome Research*, *20*, 1297–1303.

Medina-Cucurella, A. V., Steiner, P. J., Faber, M. S., Beltrán, J., Borelli, A. N., Kirby, M. B., et al. (2019). User-defined single pot mutagenesis using unamplified oligo pools. *Protein Engineering, Design and Selection*, *32*, 41–45.

Moore, I. F., Hughes, D. W., & Wright, G. D. (2005). Tigecycline is modified by the flavin-dependent monooxygenase TetX. *Biochemistry*, *44*, 11829–11835.

Morrison, K. L., & Weiss, G. A. (2001). Combinatorial alanine-scanning. *Current Opinion in Chemical Biology*, *5*, 302–307.

Muteeb, G., & Sen, R. (2010). Random mutagenesis using a mutator strain. In J. Braman (Ed.), *Methods in molecular biology (Methods and protocols)*: *Vol. 634. In vitro mutagenesis protocols* (pp. 411–419). Totowa, NJ: HumanaPress.

Naseri, G., Balazadeh, S., Machens, F., Kamranfar, I., Messerschmidt, K., & Mueller-Roeber, B. (2017). Plant-derived transcription factors for orthologous regulation of gene expression in the yeast *Saccharomyces cerevisiae*. *ACS Synthetic Biology*, *6*, 1742–1756.

Park, J., Gasparrini, A. J., Reck, M. R., Symister, C. T., Elliott, J. L., Vogel, J. P., et al. (2017). Plasticity, dynamics, and inhibition of emerging tetracycline resistance enzymes. *Nature Chemical Biology*, *13*, 730–736.

Ramos, J. L., Martínez-Bueno, M., Molina-Henares, A. J., Terán, W., Watanabe, K., Zhang, X., et al. (2005). The TetR family of transcriptional repressors. *Microbiology and Molecular Biology Reviews: MMBR*, *69*, 326–356.

Rogers, J. K., Taylor, N. D., & Church, G. M. (2016). Biosensor-based engineering of biosynthetic pathways. *Current Opinion in Biotechnology*, *42*, 84–91.

Rubin, A. F., Gelman, H., Lucas, N., Bajjalieh, S. M., Papenfuss, A. T., Speed, T. P., et al. (2017). A statistical framework for analyzing deep mutational scanning data. *Genome Biology*, *18*, 150.

Schrödinger LLC. (2018). *The PyMOL molecular graphics system*. New York, NY: Schrödinger, LLC.

Scott, L. H., Mathews, J. C., Flematti, G. R., Filipovska A., & Rackham, O. (2018). An artificial yeast genetic circuit enables deep mutational scanning of an antimicrobial resistance protein. *ACS Synthetic Biology*, 7, 1907–1917.

Shi, Y. (2014). A glimpse of structural biology through X-ray crystallography. *Cell*, *159*, 995–1014.

Starita, L. M., Pruneda, J. N., Lo, R. S., Fowler, D. M., Kim, H. J., Hiatt, J. B., et al. (2013). Activity-enhancing mutations in an E3 ubiquitin ligase identified by high-throughput mutagenesis. *Proceedings of the National Academy of Sciences of the United States of America*, *110*, E1263–E1272.

Steiner, K., & Schwab, H. (2012). Recent advances in rational approaches for enzyme engineering. *Computational and Structural Biotechnology Journal*, *2*, e201209010.

Tang, S.-Y., & Cirino, P. C. (2011). Design and application of a mevalonate-responsive regulatory protein. *Angewandte Chemie International Edition*, *50*, 1084–1086.

Tang, S.-Y., Qian, S., Akinterinwa, O., Frei, C. S., Gredell, J. A., & Cirino, P. C. (2013). Screening for enhanced triacetic acid lactone production by recombinant *Escherichia coli* expressing a designed triacetic acid lactone reporter. *Journal of the American Chemical Society*, *135*, 10099–10103.

Van Criekinge, W., & Beyaert, R. (1999). Yeast two-hybrid: State of the art. *Biological Procedures Online*, *2*, 1–38.

van Montfort, R. L. M., & Workman, P. (2017). Structure-based drug design: Aiming for a perfect fit. *Essays in Biochemistry*, *61*, 431–437.

Vanhercke, T., Ampe, C., Tirry, L., & Denolf, P. (2005). Reducing mutational bias in random protein libraries. *Analytical Biochemistry*, *339*, 9–14.

Volkers, G., Damas, J. M., Palm, G. J., Panjikar, S., Soares, C. M., & Hinrichs, W. (2013). Putative dioxygen-binding sites and recognition of tigecycline and minocycline in the tetracycline-degrading monooxygenase TetX. *Acta Crystallographica. Section D, Biological Crystallography*, *69*, 1758–1767.

Volkers, G., Palm, G. J., Weiss, M. S., Wright, G. D., & Hinrichs, W. (2011). Structural basis for a new tetracycline resistance mechanism relying on the TetX monooxygenase. *FEBS Letters*, *585*, 1061–1066.

Wallis, C. P., Filipovska, A., & Rackham, O. (2018). A modified yeast three-hybrid system enabling both positive and negative selections. *Biotechnology Letters*, *40*, 1127–1134.

Wallis, C. P., Richman, T. R., Filipovska, A., & Rackham, O. (2018). Tighter ligand binding can compensate for impaired stability of an RNA-binding protein. *ACS Chemical Biology*, *13*, 1499–1505.

Weber, W., & Fussenegger, M. (2010). Synthetic gene networks in mammalian cells. *Current Opinion in Biotechnology*, *21*, 690–696.

Yan, Y., Liu, Q., Zang, X., Yuan, S., Bat-Erdene, U., Nguyen, C., et al. (2018). Resistance-gene-directed discovery of a natural-product herbicide with a new mode of action. *Nature*, *559*, 415–418.

Yang, W., Moore, I. F., Koteva, K. P., Bareich, D. C., Hughes, D. W., & Wright, G. D. (2004). TetX is a flavin-dependent monooxygenase conferring resistance to tetracycline antibiotics. *The Journal of Biological Chemistry*, *279*, 52346–52352.

CHAPTER THIRTEEN

Methods for studying human sirtuins with activity-based chemical probes

Song Zheng, Jessica Wohlfahrt, Ian Cohen, Yana Cen*,†
Department of Pharmaceutical Sciences, Albany College of Pharmacy and Health Sciences, Colchester, VT, United States
*Corresponding author: e-mail address: ceny2@vcu.edu

Contents

Abstract

Sirtuins are unique posttranslational modification enzymes that utilize NAD^+ as the co-substrate to remove acyl groups from lysine residues. The deacylation events result in profound biological consequences, from transcription silencing to metabolism regulation. This article focuses on a newly developed technology using activity-based chemical probes to report sirtuin functional state in various settings. These chemical probes, thioacyllysine peptides carrying photo-cross-linker as well as bioorthogonal functionality, target the active site of sirtuins to form stalled reaction intermediate.

† Current Address: Department of Medicinal Chemistry, Virginia Commonwealth University, Richmond, VA 23219, United States

Methods in Enzymology, Volume 633
ISSN 0076-6879
https://doi.org/10.1016/bs.mie.2019.11.004

Subsequently, the probe forms covalent adduct with the protein through photo-crosslinking. Ultimately, the active sirtuin can be visualized via "click" chemistry-mediated conjugation to a fluorescent tag. Here, we describe the labeling protocols on recombinant protein, whole cell lysate, and *in situ* labeling.

1. Introduction

Sirtuins, also called class III HDACs, are NAD^+-dependent protein deacetylases found in diverse organisms from archaea to human (Blander & Guarente, 2004; Sauve, Wolberger, Schramm, & Boeke, 2006). Sirtuin activities are upregulated by calorie restriction (CR) in multiple species such as yeast, worms, flies and mice (Cohen et al., 2004; Lin, Ford, Haigis, Liszt, & Guarente, 2004; Rogina & Helfand, 2004; Tissenbaum & Guarente, 2001). More importantly CR causes extended lifespan and improved health profiles in the same species (Cohen et al., 2004; Lin et al., 2004; Rogina & Helfand, 2004; Tissenbaum & Guarente, 2001). The notion that sirtuin activity mediates the critical health and longevity benefits of CR has been supported by studies showing that increased gene dosage of sirtuins increases lifespan in model organisms. In addition, deletion of sirtuin activity abrogates the longevity effects provided by CR (Lee, Lee, Lee, & Min, 2019). Hence, sirtuins are sometimes referred as "longevity proteins" (Sack & Finkel, 2012).

In human, there are seven sirtuin isoforms (SIRT1–7) with diverse subcellular localizations and biological functions (Haigis & Sinclair, 2010). They have been implicated in the regulation of a myriad of cellular events including gene silencing (Brachmann et al., 1995), DNA repair (Ghosh, Liu, Wang, Hao, & Zhou, 2015), mitochondrial biogenesis (Yuan et al., 2016; Zhao et al., 2016) as well as cell cycle control (Zhang et al., 2019). Studies from the past few years revealed that sirtuins are complex enzymes with diverse catalytic activities. For example, in addition to being a weak deacetylase, SIRT5 also harbors robust desuccinylase, demalonylase, and deglutarylase activities (Du et al., 2011; Tan et al., 2014). Through these novel posttranslational modifications, SIRT5 regulates proteins and enzymes involved in carbohydrate and fatty acid metabolism, ammonia detoxification, and ketone body production (Kumar & Lombard, 2018). On the other hand, SIRT6 prefers to remove long-chain fatty acyl groups from lysines (Feldman, Baeza, & Denu, 2013; Jiang et al., 2013). These pleiotropic enzymatic activities give SIRT6 its far-reaching functions in genomic stability through the regulation of DNA repair (Mao et al., 2011; Van Meter et al., 2016) and telomere integrity (Tennen, Bua, Wright, & Chua, 2011) and in metabolism through the control of glucose and lipid homeostasis (Cui et al., 2017; Zhong et al., 2010). Due to their importance in various biological processes, human sirtuins have been intensely pursued as therapeutic targets for the treatment of metabolic disease (Kane & Sinclair, 2018), cancer (Carafa, Altucci, & Nebbioso, 2019) and other age-related disorders (Grabowska, Sikora, & Bielak-Zmijewska, 2017).

It is essential to elucidate the deacetylation landscape and to assess the functional state of each individual sirtuin for a better understanding of their physiological and pathological roles. Recently, we developed an assay to directly characterize sirtuin functions in biological samples using activity-based chemical probes (Graham, Rymarchyk, Wood, & Cen, 2018). The chemical structures of the first-generation probes are shown in Fig. 1. As a common feature, these probes carry the thioacyl groups which can target specifically the active site of sirtuins, forming stalled thioimidate intermediates (Fig. 2). The presence of a photocrosslinkable group, benzophenone, allows the formation of a covalent adduct upon photo irradiation. The terminal alkyne group, a bioorthogonal functionality, enables the conjugation of the protein-probe complex to an azido-fluorophore via copper (I)-catalyzed [3+2]-cycloaddition (also known as the "click" reaction) for in-gel visualization (Fig. 2). This labeling protocol can accurately report the active sirtuin content in complex proteomes. The detection and profiling of enzyme activities can be accomplished in one setup that cannot be

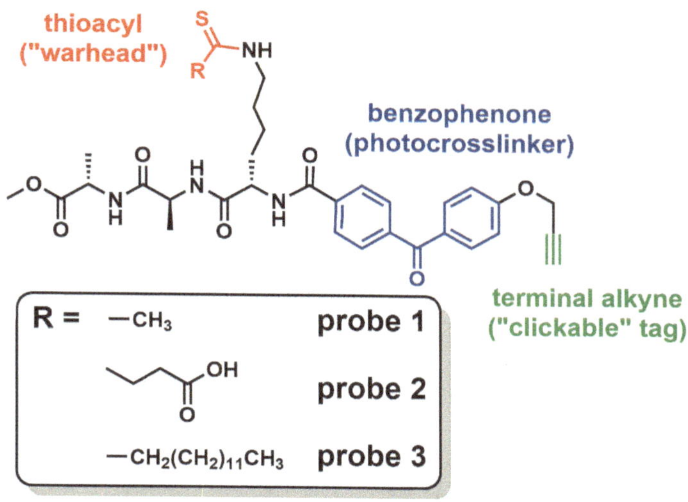

Fig. 1 Chemical structure of the probes.

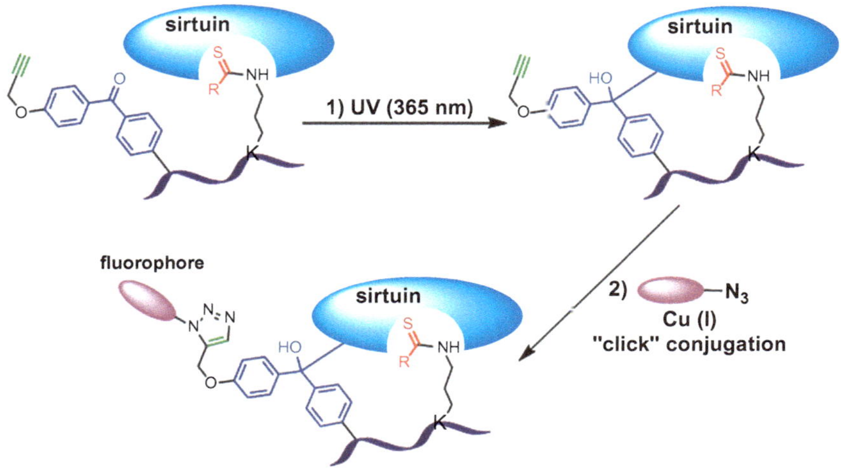

Fig. 2 General scheme of the labeling protocol.

directly assayed by other existing methods for activity screenings. This workflow represents a valuable new approach for functional genomics study under physiologically relevant conditions.

In this chapter, we first detail the expression and purification of several recombinant human sirtuins. The labeling procedures for recombinant enzymes are then described. We also describe the preparation of whole cell lysate and the subsequent labeling with the chemical probe. Finally, the *in situ* labeling of SIRT2 is introduced in which the enzymatic activity is directly interrogated in live cells in response to NAD^+ concentration change, a factor that is known to alter sirtuin activity.

2. Recombinant sirtuin expression and purification

Plasmids for SUMO-SIRT2 (38–356), SIRT5 (34–302), SIRT6 (1–314) and ULP-1 are generous gifts from Dr. Hening Lin at Cornell University. The identity of the protein is further confirmed by trypsin digestion followed by LC-MS/MS analysis. Protein concentrations are determined by Bradford assay.

2.1 SIRT2

2.1.1 Procedures for SUMO-SIRT2 expression

1. Transform the pET28a-SUMO-SIRT2 plasmid into Rosetta™ (DE3) competent cells. Plate the cells on LB plates with 50 μg/mL Kanamycin and 20 μg/mL Chloramphenicol.

2. Select single colony to inoculate overnight starter culture with 10 mL of LB media containing 50 μg/mL Kanamycin and 20 μg/mL Chloramphenicol in a 37 °C shaker.
3. Inoculate 2 L of LB media containing 50 μg/mL Kanamycin and 20 μg/mL Chloramphenicol with the overnight starter culture in a 37 °C shaker until OD_{600} reaches 0.6.
4. Induce protein expression with 20 μM of IPTG. Incubate cells in a shaker at 15 °C and 200 rpm for 20 h.
5. Harvest the cells by centrifugation at 4 °C and 8000 rpm for 5 min. Re-suspend the cell pellet in 40 mL of lysis buffer (20 mM Tris-HCl, 500 mM NaCl, 5 mM imidazole, 10 mM $MgCl_2$, 5% glycerol at pH 7.5) with protease inhibitor (without EDTA).
6. Add lysozyme (50 mg) to the cell suspension and incubate it on ice for 30 min. Sonicate the solution on ice for 30 s, 4 times, with 1 min cool-down on ice in between the sonication steps.
7. Remove cell debris by centrifugation at 4 °C and 13,000 rpm for 60 min. Incubate the supernatant with Ni-NTA agarose beads (1.5 mL per 2 L cells, pre-equilibrated with lysis buffer) at 4 °C for 2 h.
8. Load the protein bound Ni-NTA beads onto an Econo-column® and wash twice with 20 mL of lysis buffer containing 25 mM imidazole.
9. Elute the protein with elution buffers (50, 100, 150, 200, 250, 300, 350 and 500 mM imidazole in the lysis buffer, 3 mL each).
10. Dialyze the purified protein against 20 mM Tris-HCl, 500 mM NaCl and 5% glycerol at pH 7.5. Concentrate the protein using Amicon® Ultra-4 centrifugal filter unit.

2.1.2 Procedures for ULP-1 expression

1. Transform the pET28a-ULP-1 plasmid into Rosetta™ (DE3) competent cells. Plate the cells on LB plates with 50 μg/mL Kanamycin and 20 μg/mL Chloramphenicol.
2. Select single colony to inoculate overnight starter culture with 10 mL of LB media containing 50 μg/mL Kanamycin and 20 μg/mL Chloramphenicol in a 37 °C shaker.
3. Inoculate 2 L of LB media containing 50 μg/mL Kanamycin and 20 μg/mL Chloramphenicol with the overnight starter culture in a 37 °C shaker until OD_{600} reaches 0.6.
4. Induce protein expression with 20 μM of IPTG and incubate cells in a shaker at 15 °C and 200 rpm for 20 h.

5. Harvest the cells by centrifugation at 4 °C and 8000 rpm for 5 min. Re-suspend the cell pellet in 40 mL of lysis buffer (20 mM Tris-HCl, 500 mM NaCl, 5 mM imidazole, 10 mM $MgCl_2$, 5% glycerol at pH 7.5) with protease inhibitor (without EDTA).
6. Add lysozyme (50 mg) to the cell suspension and incubate it on ice for 30 min. Sonicate the solution on ice for 30 s, 4 times, with 1 min cool-down on ice in between the sonication steps.
7. Remove cell debris by centrifugation at 4 °C and 13,000 rpm for 60 min. Incubate the supernatant with pre-equilibrated Ni-NTA agarose beads (1.5 mL per 2 L cells) at 4 °C for 2 h.
8. Load the protein bound Ni-NTA beads onto an Econo-column® and wash twice with 20 mL of lysis buffer containing 25 mM imidazole. Elute the protein with elution buffers (50, 100, 150, 200, 250, 300, 350 and 500 mM imidazole in the lysis buffer, 3 mL each).
9. Dialyze the purified protein against 20 mM Tris-HCl, 500 mM NaCl and 5% glycerol at pH 7.5. Concentrate the protein using Amicon® Ultra-4 centrifugal filter unit.

2.1.3 Procedures for SIRT2 purification

The His_6-SUMO tag of SIRT2 is removed by ULP-1.

1. Incubate purified SUMO-SIRT2 with ULP-1 in 20 mM Tris-HCl, 500 mM NaCl and 5% glycerol at pH 7.5 for 2 h at 4 °C.
2. Incubate the protein mixture with pre-equilibrated Ni-NTA resins at 4 °C for 2 h. Load the resins onto an Econo-column® and collect the flow-through (contains tag-removed SIRT2).

2.2 SIRT5

Procedures for SIRT5 expression and purification

1. Transform the pET28a-SIRT5 plasmid into Rosetta™ (DE3) competent cells. Plate the cells on LB plates with 100 μg/mL Ampicillin and 20 μg/mL Chloramphenicol.
2. Select single colony to inoculate overnight starter culture with 10 mL of LB media containing 100 μg/mL Ampicillin and 20 μg/mL Chloramphenicol in a 37 °C incubator.
3. Inoculate 2 L of LB media containing 100 μg/mL Ampicillin and 20 μg/mL Chloramphenicol with the overnight starter culture in a 37 °C shaker until OD_{600} reaches 0.6. Induce protein expression with 100 μM of IPTG and incubate cells in a shaker at 15 °C and 200 rpm for 20 h.

4. Harvest the cells by centrifugation at 4 °C and 8000 rpm for 5 min. Re-suspend the cell pellet in 40 mL of lysis buffer (20 mM Tris-HCl, 500 mM NaCl, 5 mM imidazole, 10 mM $MgCl_2$, 5% glycerol at pH 8) with EDTA-free protease inhibitor.
5. Add lysozyme (50 mg) to the cell suspension and incubate it on ice for 30 min. Sonicate the solution on ice for 30 s, 4 times, with 1 min cool-down on ice in between the sonication steps.
6. Remove cell debris by centrifugation at 4 °C and 13,000 rpm for 60 min. Incubate the supernatant with Ni-NTA agarose beads pre-equilibrated with lysis buffer at 4 °C for 2 h.
7. Load the Ni-NTA beads onto an Econo-column® and wash twice with 20 mL of lysis buffer containing 25 mM imidazole. Elute the protein with elution buffers (50, 100, 150, 200, 250, 300, 350 and 500 mM imidazole in the lysis buffer, 3 mL each).
8. Dialyze the purified protein against 20 mM Tris-HCl, 500 mM NaCl and 5% glycerol at pH 8. Concentrate the protein using Amicon® Ultra-4 centrifugal filter unit.

2.3 SIRT6

Procedures for SIRT6 expression and purification

1. Transform the pET28a-SIRT6 plasmid into Rosetta™ (DE3) competent cells. Plate the cells on LB plates with 50 μg/mL Kanamycin and 20 μg/mL Chloramphenicol.
2. Select single colony to inoculate overnight starter culture with 10 mL of 2xYT media containing 50 μg/mL Kanamycin and 20 μg/mL Chloramphenicol in a 37 °C shaker.
3. Inoculate 2 L of 2xYT media containing 50 μg/mL Kanamycin and 20 μg/mL Chloramphenicol with the overnight starter culture in a 37 °C shaker until OD_{600} reaches 0.6. Induce protein expression with 200 μM of IPTG and incubate cells in a shaker at 16 °C and 200 rpm for 20 h.
4. Harvest the cells by centrifugation at 4 °C and 8000 rpm for 5 min. Re-suspend the cell pellet in 40 mL of lysis buffer (20 mM Tris-HCl, 500 mM NaCl, 5 mM imidazole, 10 mM $MgCl_2$, 5% glycerol at pH 7.2) with EDTA-free protease inhibitor.
5. Add lysozyme (50 mg) to the cell suspension and incubate it on ice for 30 min. Sonicate the solution on ice for 30 s, 4 times, with 1 min cool-down on ice in between the sonication steps.

6. Remove cell debris by centrifugation at 4 °C and 13,000 rpm for 60 min. Incubate the supernatant with Ni-NTA agarose beads pre-equilibrated with lysis buffer at 4 °C for 2 h.
7. Load the Ni-NTA beads onto an Econo-column® and wash twice with 20 mL of lysis buffer containing 25 mM imidazole. Elute the protein with elution buffers (50, 100, 150, 200, 250, 300, 350 and 500 mM imidazole in the lysis buffer, 3 mL each).
8. Dialyze the purified protein against 20 mM Tris-HCl, 500 mM NaCl and 5% glycerol at pH 7.2. Concentrate the protein using Amicon® Ultra-4 centrifugal filter unit.

3. Activity-based labeling of recombinant sirtuin

The following is a standard protocol for labeling recombinant human sirtuins with activity-based chemical probes. The exogenously expressed and purified sirtuins are incubated with NAD^+ and the probe. Following photo irradiation and copper-mediated "click" conjugation to a fluorophore, the active sirtuin content can be visualized on a SDS-PAGE gel via fluorescence scanning. This protocol can be applied to either time-dependent (Graham et al., 2018) or concentration-dependent (Fig. 3) labeling. In the latter case, the labeling efficiency should directly correlate with the concentrations of the probe. No appreciable labeling is detected for the sample with no probe. The optimal labeling is achieved by using the minimum concentration of the probe that reaches the signal plateau.

1. Add the protein and other reagents in a 0.7 mL Eppendorf tube: purified recombinant sirtuin (final concentration: 10 μM), NAD^+ (final concentration: 500 μM) and the activity-based probe (probe 1 for SIRT2, probe

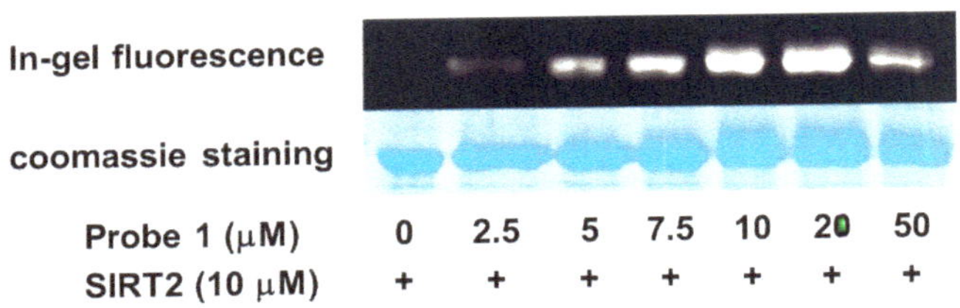

Fig. 3 Concentration-dependent labeling of recombinant SIRT2 with probe 1. SIRT2 (10 μM) was incubated with 500 μM NAD^+ and various concentrations of probe 1 (0, 2.5, 5, 7.5, 10, 20 and 50 μM) at 37 °C for 10 min. The samples were then subjected to the UV irradiation and "click" conjugation process as described in Section 3. SIRT2 can be robustly labeled by probe 1 in a dose-dependent manner. Coomassie staining serves as a loading control.

2 for SIRT5 and probe 3 for SIRT6). Add 100 mM phosphate buffer at pH 7.5 to make the final volume of the reaction 20 μL. Incubate the sample at 37 °C for 10 min, allowing the formation of the thioimidate intermediate.

2. Transfer the sample to a clear-bottom 96-well plate. Place the plate on ice, and irradiate the sample with a hand-held UV lamp at 365 nm for 1 h at 4 °C.
3. At the meantime, prepare the "click" reaction stocks:
   a. $CuSO_4$ (50 mM in distilled water, freshly prepared)
   b. Tris(2-carboxyethyl)phosphine (TCEP, 100 mM in distilled water, freshly prepared)
   c. Tris[(1-benzyl-1H-1,2,3-triazol-4-yl)methyl]amine (TBTA, 30 mM in DMSO/*t*BuOH (3/1, v/v))
   d. TAMRA-azide (5 mM in DMSO)
4. Transfer the sample back to the Eppendorf tube. Add the ingredients of the "click" reaction to the sample to the following final concentrations: $CuSO_4$ 2 mM, TCEP 4 mM, TBTA 1.25 mM, TAMRA-azide 200 μM. Incubate the sample on a microshaker at 250 rpm and room temperature in the dark for 30 min.
5. Add 8 μL of 4 × loading dye to the reaction and boil it at 95 °C for 10 min.
6. Spin down the sample briefly and load it onto a 10% SDS-PAGE gel submerged in 1 × Tris-Glycine SDS running buffer. Place the electrophoresis cell in the dark to minimize fluorophore decomposition. Run the gel for 1 h at 120 V.
7. Destain the gel for 4 h at room temperature in methanol/distilled water/acetic acid (4/5/1, v/v/v).
8. Scan the gel using Biorad Versa Doc 4000 MP with excitation 532 nm, 580 nm cutoff filter, and 30 nm band-pass.
9. After the desired fluorescent images are acquired, stain the gel with Coomassie blue for loading control.

4. Labeling sirtuin in whole cell lysate

The activity-based chemical probes selectively interact with the active enzyme, leading to efficient labeling of the target protein. The probes we developed also demonstrate isoform selectivity with probe 1 prefers SIRT2, probe 2 is specific for SIRT5, and probe 3 is selective for SIRT6. This enables the labeling of one particular sirtuin isotype in a complex proteome such as whole cell lysate. The cell lysate can be incubated with the probe, followed by photocrosslinking and "click" coupling with azido-containing fluorophores. The labeling pattern and efficiency can then be easily evaluated by in-gel fluorescence scanning.

4.1 Cell culturing and lysis

1. Grow human embryonic kidney (HEK293) cells in Dulbecco's modified Eagle medium (DMEM) supplemented with 10% fetal bovine serum (FBS), 100 U/mL penicillin, and 100 mg/mL streptomycin in a 150 cm^2 cell culture flask to 70–80% confluency.
2. Harvest cells by repetitively pipetting up and down, and then centrifuge at 4 °C and 1000 × *g* for 10 min.
3. Remove the supernatant, and re-suspend the cells in PBS followed by centrifuging at 4 °C and 1500 × *g* for 10 min. Repeat this step three times.
4. Remove the supernatant and re-suspend the cell pellet in 4 mL of RIPA lysis buffer supplemented with protease inhibitor cocktail. Gently shake the tube and incubate on ice for 15 min.
5. Centrifuge the lysed cells at 4 °C and 14,000 × *g* for 15 min.
6. Recover the supernatant, and quantify the protein concentration by Bradford assay. Dilute the cell lysate with 50 mM Tris-HCl buffer to a final concentration of 2 mg protein/mL.

4.2 Labeling whole cell lysate

1. Add the cell lysate and other ingredients to a 0.7 mL Eppendorf tube
   a. 7.5 μL of cell lysate (15 μg of protein)
   b. 1 μL of 10 mM NAD^+ stock (final concentration: 500 μM)
   c. 0.2 μL of 1 mM probe stock (final concentration: 10 μM)
2. Add 100 mM phosphate buffer at pH 7.5 to make the final volume of the reaction 20 μL. Incubate the sample at 37 °C for 10 min.
3. Transfer the sample to a clear-bottom 96-well plate. Place the plate on ice, and irradiate the sample with a hand-held UV lamp at 365 nm for 1 h at 4 °C.
4. Transfer the sample back to the Eppendorf tube. Add the "click" reaction cocktail to the sample ($CuSO_4$ 2 mM, TCEP 4 mM, TBTA 1.25 mM, TAMRA-azide 200 μM). Gently agitate the sample on a microshaker at 250 rpm and room temperature in the dark for 30 min.

5. Add 8 μL of 4× loading dye to the reaction and boil it at 95 °C for 10 min.
6. Spin down the sample briefly and load it onto a 10% SDS-PAGE gel submerged in 1× Tris-Glycine SDS running buffer. Place the electrophoresis cell in the dark to minimize fluorophore decomposition. Run the gel for 1 h at 120 V.
7. Destain the gel for 4 h at room temperature in methanol/distilled water/acetic acid (4/5/1, v/v/v).
8. Scan the gel using Biorad Versa Doc 4000 MP with excitation 532 nm, 580 nm cutoff filter, and 30 nm band-pass.
9. After the desired fluorescent images are acquired, stain the gel with Coomassie blue for loading control.

5. *In situ* labeling of human SIRT2

Our recent study (Graham et al., 2018) indicates that probe 1 (Fig. 1) is a cell permeable sirtuin inhibitor, a feature allows it to be used in *in situ* labeling. The dynamic change of enzyme activity under different cellular conditions can thus be captured and compared. Ample literature precedent has suggested that NAD^+ concentration change can impact sirtuin activity (Skoge, Dolle, & Ziegler, 2014; Yang et al., 2007), but not sirtuin protein abundance. The intracellular NAD^+ level can be significantly reduced by the pharmacological inhibition of NAD^+ biosynthetic pathways. Subsequently, sirtuin activity is reduced as reflected by the elevated acetylation level of endogenous sirtuin substrates (Skoge et al., 2014; Van Gool et al., 2009). This activity change in response to NAD^+ availability can then be directly assessed through *in situ* labeling using probe 1 (Graham et al., 2018). Indeed, in the NAD^+ depletion scenario (Fig. 4, FK866 treatment), the labeling of SIRT2 is drastically decreased compared to the control sample despite the fact that SIRT2 protein level remains constant as determined by Western blot (Fig. 4, WB).

5.1 Transfection and FK866 treatment

SIRT2 Flag was a generous gift from Eric Verdin (Addgene plasmid no. 13813) (North, Marshall, Borra, Denu, & Verdin, 2003). FK866 is a nanomolar inhibitor of nicotinamide phosphoribosyltransferase (NAMPT) (Hasmann & Schemainda, 2003), the rate-limiting enzyme of the NAD^+ salvage pathway (Fig. 5). Intracellular NAD^+ level is decreased upon FK866 treatment (Graham et al., 2018; Skoge et al., 2014). The following

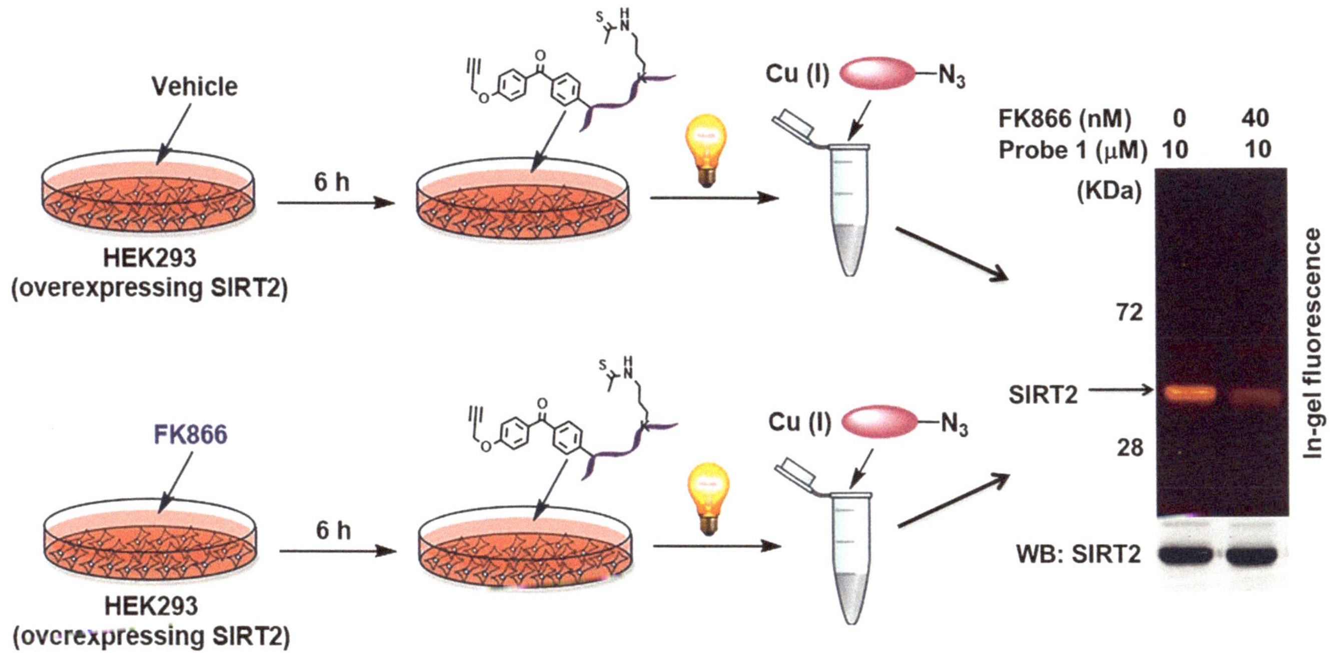

Fig. 4 *In situ* labeling of SIRT2. HEK293 cells were transfected to transiently overexpress human SIRT2. Cells were treated with either vehicle (DMSO) or 40 nM FK866 for 6 h. The medium was removed. After thorough rinsing, the culture was replenished with fresh medium supplemented with 10 μM probe 1. The cells were irradiated on ice at 365 nm for 1 h before the cell lysate was isolated and "click" conjugated to a fluorophore as described in Section 5.2.

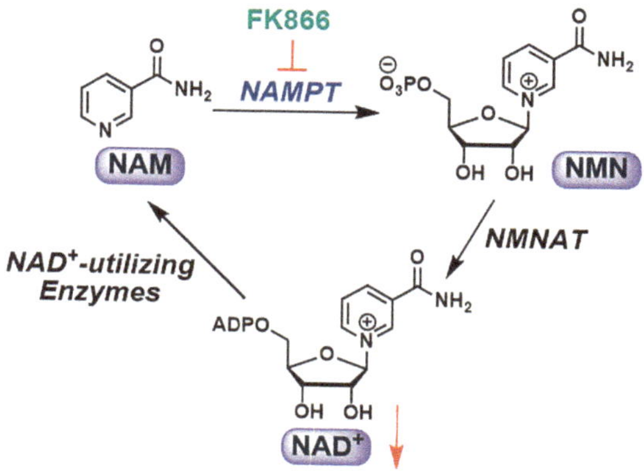

Fig. 5 FK866 inhibits NAD$^+$ biosynthesis. In mammalian cells, NAD$^+$ can be broken down to nicotinamide (NAM) by NAD$^+$-utilizing enzymes such as sirtuins, CD38 and PARP. NAM is recycled to nicotinamide mononucleotide (NMN) by nicotinamide phosphoribosyltransferase (NAMPT), the rate-limiting enzyme of the NAD$^+$ salvage pathway. Ultimately, NMN is fully incorporated into NAD$^+$ by NMN adenylyltransferase (NMNAT). FK866, a potent and selective NAMPT inhibitor, can significantly reduce intracellular NAD$^+$ level.

protocol is used in a 6-well plate format. All the amounts and numbers are given on a per well basis.

1. Culture HEK293 cells in DMEM/10% FBS without antibiotics to 90–95% confluency.
2. One hour before the transfection, replace the growth medium with fresh DMEM alone (be careful not to disturb the cells).
3. Dilute 4 μg of plasmid into 250 μL of Opti-MEM medium, mix gently.
4. Dilute 10 μL of Lipofectamine 2000 into 250 μL of Opti-MEM medium. Incubate at room temperature for 5 min.
5. Combine the diluted plasmid with diluted Lipofectamine. Mix gently and incubate at room temperature for 20 min.
6. Add the plasmid mixture dropwise into the well, mix gently.
7. Incubate cells at 37 °C for 24 h.
8. Add either vehicle (DMSO) or FK866 to a final concentration of 40 nM and culture for additional 6 h.

5.2 *In situ* labeling of SIRT2

After being cultured in the presence or absence of FK866, the SIRT2-overexpressing cells are treated with probe 1, followed by in cell photocrosslinking, cell lysis and the "click" conjugation to the fluorescent reporter.

1. Remove the medium. Add fresh DMEM and mix gently by rocking the plate back and forth. Repeat this step three times.
2. Remove the medium. Add fresh DMEM supplemented with 10 μM of probe 1.
3. Incubate the cells at 37 °C for 1 h.
4. Place the cell culture plate on ice, and irradiate the sample with a hand-held UV lamp at 365 nm for 1 h at 4 °C.
5. Harvest cells by repetitively pipetting up and down, and then centrifuge at 4 °C and 1000 × *g* for 10 min.
6. Remove the supernatant, and re-suspend the cells in PBS followed by centrifuging at 4 °C and 1500 × *g* for 10 min. Repeat this step three times.
7. Remove the supernatant and re-suspend the cell pellet in 100 μL of RIPA lysis buffer supplemented with protease inhibitor cocktail. Gently shake the tube and incubate on ice for 15 min.
8. Centrifuge the lysed cells at 4 °C and 14,000 × *g* for 15 min.
9. Recover the supernatant, and quantify the protein concentration by Bradford assay.
10. Add the cell lysate (15 μg of protein) to a 0.7 mL Eppendorf tube. Add 100 mM phosphate buffer at pH 7.5 to make the final volume of the reaction 20 μL.
11. Add the "click" reaction cocktail to the sample ($CuSO_4$ 2 mM, TCEP 4 mM, TBTA 1.25 mM, TAMRA-azide 200 μM). Gently agitate the sample on a microshaker at 250 rpm and room temperature in the dark for 30 min.
12. Add 8 μL of 4 × loading dye to the reaction and boil it at 95 °C for 10 min.
13. Load the sample onto a 10% SDS-PAGE gel. Run the gel in the dark at 120 V for 1 h.
14. Destain the gel for 4 h at room temperature in methanol/distilled water/acetic acid (4/5/1, v/v/v).
15. Acquire the fluorescence images using a Biorad Versa Doc 4000 MP imager.
16. Determine SIRT2 protein level by western blot using anti-SIRT2 antibody from Santa Cruz Biotechnology.

5.3 Determination of NAD^+ concentration

The cellular NAD^+ concentration can be assessed right after the FK866 treatment using a continuous and sensitive fluorescence intensity assay as

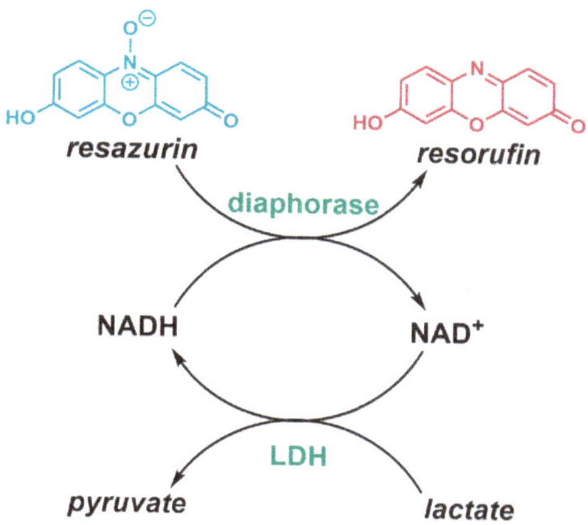

Fig. 6 Cycling assay for NAD^+ measurement.

described previously (Fulco et al., 2008). In this assay, the concentration of NAD^+ is measured using a coupled enzyme system with lactate dehydrogenase (LDH) and diaphorase (Fig. 6).

5.3.1 Procedures for sample preparation

1. Harvest cells by repetitively pipetting up and down, and then centrifuge at 4 °C and 1000 × *g* for 10 min.
2. Remove the supernatant, and re-suspend the cells in PBS followed by centrifuging at 4 °C and 1500 × *g* for 10 min. Repeat this step three times.
3. Remove the supernatant and re-suspend the cell pellet in 30 μL of ice-cold 7% perchloric acid.
4. Vortex the sample for 30 s and then sonicate on ice for 5 min. Repeat this step three times.
5. Centrifuge the sample at 14,000 × *g* and 4 °C for 3 min.
6. Transfer the clear supernatant to another Eppendorf tube, and neutralize it to pH 7 with 3 M of NaOH and 1 M phosphate buffer (pH 9). Neutralized sample can be used in the following steps for NAD^+ measurement.

5.3.2 Procedures for NAD^+ cycling assay

1. Prepare the following stock solutions for the measurement:
   a. 100 mM Tris-HCl, pH 7.5
   b. 250 mM $MgCl_2$
   c. 466.7 mM KCl
   d. 27 mM resazurin

e. 0.1 U/μL LDH

f. 100 mM lactate

2. Combine the ingredients of the cycling assay buffer. For each 150 μL reaction, combine 37.5 μL of Tris-HCl, 3 μL of $MgCl_2$, 16.1 μL of KCl, 0.3 μL of resazurin, 0.6 μL of LDH, 3.4 μL of lactate, and 59.1 μL of distilled water.
3. Load the neutralized samples and NAD^+ standards in a black 96-well plate. Load 2 and 4 μL of the sample. Load NAD^+ standard to the following final concentrations (for 150 μL reaction): 0, 50 nM, 100 nM, 200 nM, 400 nM, 500 nM, 1 μM and 2 μM. Add distilled water to all the samples to bring the total volume up to 10 uL.
4. Add 120 μL of the cycling assay buffer to each well.
5. Add 20 μL of 0.035 U/μL of diaphorase to each well using a multichannel pipette to initiate the cycling reactions.
6. Record the fluorescence intensity change using a microplate reader with excitation at 530 nm and emission at 580 nm at 27 s intervals for 15 min.
7. Plot the readings at different time points to acquire slopes. Create the NAD^+ standard curve by plotting slopes as a function of their concentrations.
8. Compare the slope of the sample with that of the NAD^+ standard curve to determine the NAD^+ concentration in each well. Adjust the final concentration of the sample by multiplying by the dilution factor. The final NAD^+ concentration is expressed as pmol/10^6 cells or pmol/mg protein.

6. Conclusion

A set of powerful small molecules that can directly interrogate the functional state of sirtuins in their native matrix has been developed. They demonstrate good labeling efficiency and isoform selectivity. They label individual recombinant sirtuin, one particular sirtuin isotype in protein mixture, and also report the sirtuin activity change in complex cellular setting when subjected to different stimuli. In this chapter, we have discussed the expression and isolation of recombinant sirtuins, cell culturing and cell lysate preparation. More importantly, we have highlighted the labeling assays for different biological samples ranging from recombinant protein to live cell labeling. Application of the above-mentioned chemical probes and associated methods will facilitate the elucidation of sirtuin activity in specific physiological and pathological processes.

Acknowledgments

This work was supported by 1R15GM123393 from NIH/NIGMS (to Y.C.), Student Summer Research Award from Albany College of Pharmacy and Health Sciences (to S.Z.), and Graduate Research Assistantship from Albany College of Pharmacy and Health Sciences (to J.W.).

References

Blander, G., & Guarente, L. (2004). The Sir2 family of protein deacetylases. *Annual Review of Biochemistry*, *73*, 417–435. https://doi.org/10.1146/annurev.biochem.73.011303.073651.

Brachmann, C. B., Sherman, J. M., Devine, S. E., Cameron, E. E., Pillus, L., & Boeke, J. D. (1995). The SIR2 gene family, conserved from bacteria to humans, functions in silencing, cell cycle progression, and chromosome stability. *Genes & Development*, *9*(23), 2888–2902. https://doi.org/10.1101/gad.9.23.2888.

Carafa, V., Altucci, L., & Nebbioso, A. (2019). Dual tumor suppressor and tumor promoter action of Sirtuins in determining malignant phenotype. *Frontiers in Pharmacology*, *10*, 38. https://doi.org/10.3389/fphar.2019.00038.

Cohen, H. Y., Miller, C., Bitterman, K. J., Wall, N. R., Hekking, B., Kessler, B., et al. (2004). Calorie restriction promotes mammalian cell survival by inducing the SIRT1 deacetylase. *Science*, *305*(5682), 390–392. https://doi.org/10.1126/science.1099196.

Cui, X., Yao, L., Yang, X., Gao, Y., Fang, F., Zhang, J., et al. (2017). SIRT6 regulates metabolic homeostasis in skeletal muscle through activation of AMPK. *American Journal of Physiology Endocrinology and Metabolism*, *313*(4), E493–E505. https://doi.org/10.1152/ajpendo.00122.2017.

Du, J., Zhou, Y., Su, X., Yu, J. J., Khan, S., Jiang, H., et al. (2011). Sirt5 is a NAD-dependent protein lysine demalonylase and desuccinylase. *Science*, *334*(6057), 806–809. https://doi.org/10.1126/science.1207861.

Feldman, J. L., Baeza, J., & Denu, J. M. (2013). Activation of the protein deacetylase SIRT6 by long-chain fatty acids and widespread deacylation by mammalian sirtuins. *The Journal of Biological Chemistry*, *288*(43), 31350–31356. https://doi.org/10.1074/jbc.C113.511261.

Fulco, M., Cen, Y., Zhao, P., Hoffman, E. P., McBurney, M. W., Sauve, A. A., et al. (2008). Glucose restriction inhibits skeletal myoblast differentiation by activating SIRT1 through AMPK-mediated regulation of Nampt. *Developmental Cell*, *14*(5), 661–673. https://doi.org/10.1016/j.devcel.2008.02.004.

Ghosh, S., Liu, B., Wang, Y., Hao, Q., & Zhou, Z. (2015). Lamin a is an endogenous SIRT6 activator and promotes SIRT6-mediated DNA repair. *Cell Reports*, *13*(7), 1396–1406. https://doi.org/10.1016/j.celrep.2015.10.006.

Grabowska, W., Sikora, E., & Bielak-Zmijewska, A. (2017). Sirtuins, a promising target in slowing down the ageing process. *Biogerontology*, *18*(4), 447–476. https://doi.org/10.1007/s10522-017-9685-9.

Graham, E., Rymarchyk, S., Wood, M., & Cen, Y. (2018). Development of activity-based chemical probes for human Sirtuins. *ACS Chemical Biology*, *13*(3), 782–792. https://doi.org/10.1021/acschembio.7b00754.

Haigis, M. C., & Sinclair, D. A. (2010). Mammalian sirtuins: Biological insights and disease relevance. *Annual Review of Pathology*, *5*, 253–295. https://doi.org/10.1146/annurev.pathol.4.110807.092250.

Hasmann, M., & Schemainda, I. (2003). FK866, a highly specific noncompetitive inhibitor of nicotinamide phosphoribosyltransferase, represents a novel mechanism for induction of tumor cell apoptosis. *Cancer Research*, *63*(21), 7436–7442. Retrieved from https://www.ncbi.nlm.nih.gov/pubmed/14612543.

Jiang, H., Khan, S., Wang, Y., Charron, G., He, B., Sebastian, C., et al. (2013). SIRT6 regulates TNF-alpha secretion through hydrolysis of long-chain fatty acyl lysine. *Nature*, *496*(7443), 110–113. https://doi.org/10.1038/nature12038.

Kane, A. E., & Sinclair, D. A. (2018). Sirtuins and NAD(+) in the development and treatment of metabolic and cardiovascular diseases. *Circulation Research*, *123*(7), 868–885. https://doi.org/10.1161/CIRCRESAHA.118.312498.

Kumar, S., & Lombard, D. B. (2018). Functions of the sirtuin deacylase SIRT5 in normal physiology and pathobiology. *Critical Reviews in Biochemistry and Molecular Biology*, *53*(3), 311–334. https://doi.org/10.1080/10409238.2018.1458071.

Lee, S. H., Lee, J. H., Lee, H. Y., & Min, K. J. (2019). Sirtuin signaling in cellular senescence and aging. *BMB Reports*, 52(1), 24–34. Retrieved from https://www.ncbi.nlm.nih.gov/pubmed/30526767.

Lin, S. J., Ford, E., Haigis, M., Liszt, G., & Guarente, L. (2004). Calorie restriction extends yeast life span by lowering the level of NADH. *Genes & Development*, *18*(1), 12–16. https://doi.org/10.1101/gad.1164804.

Mao, Z., Hine, C., Tian, X., Van Meter, M., Au, M., Vaidya, A., et al. (2011). SIRT6 promotes DNA repair under stress by activating PARP1. *Science*, *332*(6036), 1443–1446. https://doi.org/10.1126/science.1202723.

North, B. J., Marshall, B. L., Borra, M. T., Denu, J. M., & Verdin, E. (2003). The human Sir2 ortholog, SIRT2, is an NAD+-dependent tubulin deacetylase. *Molecular Cell*, 11(2), 437–444. Retrieved from https://www.ncbi.nlm.nih.gov/pubmed/12620231.

Rogina, B., & Helfand, S. L. (2004). Sir2 mediates longevity in the fly through a pathway related to calorie restriction. *Proceedings of the National Academy of Sciences of the United States of America*, *101*(45), 15998–16003. https://doi.org/10.1073/pnas.0404184101.

Sack, M. N., & Finkel, T. (2012). Mitochondrial metabolism, sirtuins, and aging. *Cold Spring Harbor Perspectives in Biology*, *4*(12). https://doi.org/10.1101/cshperspect.a013102.

Sauve, A. A., Wolberger, C., Schramm, V. L., & Boeke, J. D. (2006). The biochemistry of sirtuins. *Annual Review of Biochemistry*, *75*, 435–465. https://doi org/10.1146/annurev.biochem.74.082803.133500.

Skoge, R. H., Dolle, C., & Ziegler, M. (2014). Regulation of SIRT2-dependent alpha-tubulin deacetylation by cellular NAD levels. *DNA Repair (Amst)*, *23*, 33–38. https://doi.org/10.1016/j.dnarep.2014.04.011.

Tan, M., Peng, C., Anderson, K. A., Chhoy, P., Xie, Z., Dai, L., et al. (2014). Lysine glutarylation is a protein posttranslational modification regulated by SIRT5. *Cell Metabolism*, *19*(4), 605–617. https://doi.org/10.1016/j.cmet.2014.03.014.

Tennen, R. I., Bua, D. J., Wright, W. E., & Chua, K. F. (2011). SIRT6 is required for maintenance of telomere position effect in human cells. *Nature Communications*, *2*, 433. https://doi.org/10.1038/ncomms1443.

Tissenbaum, H. A., & Guarente, L. (2001). Increased dosage of a sir-2 gene extends lifespan in *Caenorhabditis elegans*. *Nature*, *410*(6825), 227–230. https://doi.org/10.1038/35065638.

Van Gool, F., Galli, M., Gueydan, C., Kruys, V., Prevot, P. P., Bedalov, A., et al. (2009). Intracellular NAD levels regulate tumor necrosis factor protein synthesis in a sirtuin-dependent manner. *Nature Medicine*, *15*(2), 206–210. https://doi.org/10.1038/nm.1906.

Van Meter, M., Simon, M., Tombline, G., May, A., Morello, T. D., Hubbard, B. P., et al. (2016). JNK phosphorylates SIRT6 to stimulate DNA double-strand break repair in response to oxidative stress by recruiting PARP1 to DNA breaks. *Cell Reports*, *16*(10), 2641–2650. https://doi.org/10.1016/j.celrep.2016.08.006.

Yang, H., Yang, T., Baur, J. A., Perez, E., Matsui, T., Carmona, J. J., et al. (2007). Nutrient-sensitive mitochondrial NAD+ levels dictate cell survival. *Cell*, *130*(6), 1095–1107. https://doi.org/10.1016/j.cell.2007.07.035.

Yuan, Y., Cruzat, V. F., Newsholme, P., Cheng, J., Chen, Y., & Lu, Y. (2016). Regulation of SIRT1 in aging: Roles in mitochondrial function and biogenesis. *Mechanisms of Ageing and Development*, *155*, 10–21. https://doi.org/10.1016/j.mad.2016.02.003.

Zhang, W., Feng, Y., Guo, Q., Guo, W., Xu, H., Li, X., et al. (2019). SIRT1 modulates cell cycle progression by regulating CHK2 acetylation-phosphorylation. *Cell Death and Differentiation*. https://doi.org/10.1038/s41418-019-0369-7.

Zhao, H. C., Ding, T., Ren, Y., Li, T. J., Li, R., Fan, Y., et al. (2016). Role of Sirt3 in mitochondrial biogenesis and developmental competence of human in vitro matured oocytes. *Human Reproduction*, *31*(3), 607–622. https://doi.org/10.1093/humrep/dev345.

Zhong, L., D'Urso, A., Toiber, D., Sebastian, C., Henry, R. E., Vadysirisack, D. D., et al. (2010). The histone deacetylase Sirt6 regulates glucose homeostasis via Hif1alpha. *Cell*, *140*(2), 280–293. https://doi.org/10.1016/j.cell.2009.12.041.

CHAPTER FOURTEEN

Site-directed labeling of β-arrestin with monobromobimane for measuring their interaction with G protein-coupled receptors

Ashish Srivastava, Mithu Baidya, Hemlata Dwivedi-Agnihotri, Arun K. Shukla*

Department of Biological Sciences and Bioengineering, Indian Institute of Technology, Kanpur, India
*Corresponding author: e-mail address: arshukla@iitk.ac.in

Contents

Abstract

β-arrestins (βarrs) are multifunctional proteins that interact with activated and phosphorylated G protein-coupled receptors (GPCRs) to regulate their signaling and trafficking. Understanding the intricate details of GPCR-βarr interaction continues to be a key research area in the field of GPCR biology. Bimane fluorescence spectroscopy has been one of the key approaches among a broad range of methods employed to study GPCR-βarr interaction using purified and reconstituted system. Here, we present a step-by-step protocol for labeling βarrs with monobromobimane (mBBr) in a site-directed fashion for measuring their interaction with GPCRs and the resulting conformational changes. This simple protocol can be directly applied to other protein-protein interaction modules as well for measuring interactions and conformational changes in reconstituted systems in vitro.

Methods in Enzymology, Volume 633
ISSN 0076-6879
https://doi.org/10.1016/bs.mie.2019.11.009

1. Introduction

Interaction of βarrs with activated and phosphorylated GPCRs is a central mechanism for exerting a precise control on downstream signaling (Kang, Tian, & Benovic, 2014; Lefkowitz & Shenoy, 2005; Luttrell et al., 2018). In the classical paradigm, βarrs sterically hinder the coupling and activation of heterotrimeric G-proteins to the receptors, while also directing internalization of the activated receptors thereby reducing their surface levels (DeWire, Ahn, Lefkowitz, & Shenoy, 2007; Reiter, Ahn, Shukla, & Lefkowitz, 2012). In addition, a critical contribution of βarrs in GPCR mediated signaling has also been established for a large number of GPCRs (DeWire et al., 2007; Reiter et al., 2012; Shukla, Singh, & Ghosh, 2014). More recent paradigm of sustained GPCR signaling, for example, from endosomal compartments also involves an important role of βarrs, potentially via formation of GPCR-G-protein-βarr supercomplexes (Ranjan, Dwivedi, Baidya, Kumar, & Shukla, 2017; Thomsen et al., 2016). Thus, understanding the intricate details and molecular mechanisms of GPCR-βarr interaction is an important area in GPCR biology research.

GPCR-βarr interaction typically depends on two distinct activation-dependent features of GPCRs, first, receptor phosphorylation and second, conformational change in the transmembrane domain (Chen, Iverson, & Gurevich, 2018; Gurevich & Gurevich, 2004, 2018; Ranjan et al., 2017; Shukla et al., 2013, 2014). Receptor attached phosphates primarily interact with the positively charged residues on the N-domain of βarrs while the opening on the intracellular side of the receptor transmembrane domain accommodates finger loop of βarrs (Chen et al., 2018; Gurevich & Gurevich, 2018; Shukla et al., 2013, 2014). A number of methods have been employed to measure GPCR-βarr interaction in cellular context and in vitro. For example, confocal microscopy, enzyme complementation, FRET, BRET and Tango assay are widely used to measure the recruitment of βarrs to GPCRs with spatio-temporal resolution in cellular context. Methods like co-immunoprecipitation, ELISA, isothermal calorimetry and fluorescence spectroscopy are most commonly used to measure this interaction in vitro using purified components.

Fluorescence spectroscopy, especially the one using monobromobimane (mBBr) labeled βarrs, has been extensively employed to report the interaction of GPCRs with βarrs as well as to measure the conformational changes in βarrs arising upon their interaction with the receptors (Ghosh et al., 2019; Kumari et al., 2016, 2017; Rahmeh et al., 2012; Sente et al., 2018; Staus et al., 2018). mBBr is an environmentally sensitive fluorophore that exhibits a change in its fluorescence intensity and emission maxima upon alteration of its physico-chemical environment (Chinn, Pigiet, & Fahey, 1986; Mansoor & Farrens, 2004; Mansoor, McHaourab, & Farrens, 2002). Site-directed labeling of βarrs has allowed deciphering the engagement of various loops in βarrs with the receptor, the contribution of receptor phosphorylation and activation in their interaction, and conformational differences among βarr isoforms (Ghosh et al., 2017, 2019; Kumari et al., 2016, 2017; Rahmeh et al., 2012; Staus et al., 2018). These studies involving GPCRs and βarrs have been designed based on extensive studies on rhodopsin-visual-arrestin system described in the literature (Sommer & Farrens, 2006; Sommer, Farrens, McDowell, Weber, & Smith, 2007; Sommer, Smith, & Farrens, 2005; Sommer, Smith, & Farrens, 2006).

Here, we present a step-by-step protocol for labeling βarrs in their finger loop using monobromobimane based on our previous studies (Kumari et al., 2016, 2017) (Fig. 1). The method presented here should allow other researchers to establish this technique in their laboratory for investigating GPCR-βarr interaction, and moreover, it should also be adaptable to other protein–protein interaction systems involved in cellular signaling (Fig. 2).

2. Materials and methods

2.1 Chemicals, reagents and equipments

Monobromobimane (Sigma, Cat#B4380)
Fluorometer (e.g., Perkin Elmer, LS-55)
Quartz microcuvette (Perkin Elmer)
E. coli BL21 (DE3) (NEB, Cat#C25271)
LB media
Buffers (Tris-HCL and HEPES buffer)
Isopropyl β-D-1-thiogalactopyranoside (IPTG) (SRL, Cat#67208)
Lysozyme (SRL, Cat#45822)
DMSO (Dimethyl Sulfoxide)
Sonicator
Ultracentrifuge
Econo-column (Biorad Cat# 7374251)
Glutathione resin (Genescript, Cat#L0026)
Thrombin, Bovine (Merck, Cat#605157)
PD10 desalting column (GE healthcare Cat# 17085101)

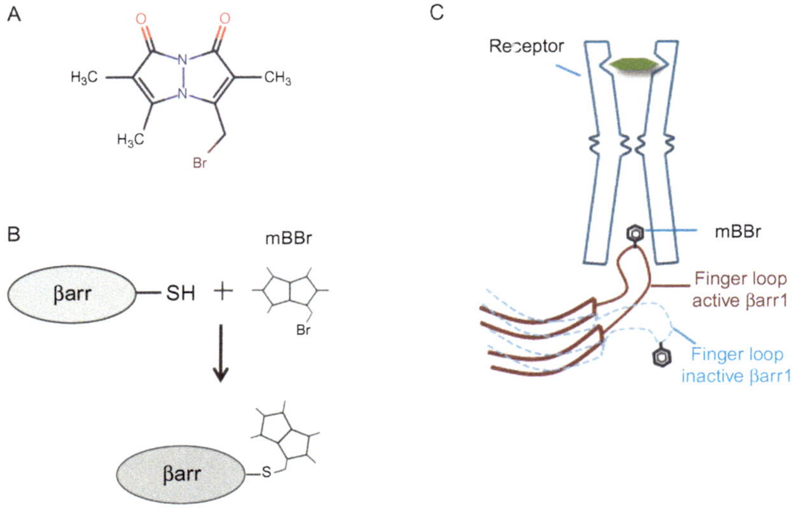

Fig. 1 Site-specific labeling of proteins with monobromobimane (mBBr) for biophysical studies. (A) Chemical structure of mBBr drawn using Marvin JS online tool. (B) A schematic representation of typical bimane-labeling using free thiol group on the protein of interest. (C) Schematic representation of bimane fluorescence assay as a readout of the core interaction between a GPCR and βarr1 labeled in its finger loop as a case example. Here, mBBr is conjugated to Cys^{68} position in the finger loop of βarr1. Upon interaction of βarr1 with activated GPCR, the finger loop interacts with the receptor-core leading to a change in mBBr fluorescence due to a change in its chemical environment. *This figure is adapted from Kumari, P., Srivastava, A., Banerjee, R., Ghosh, E., Gupta, P., Ranjan, R., Chen, X., Gupta, B., Gupta, C., Jaiman, D., Shukla, A.K. (2016). Functional competence of a partially engaged GPCR-beta-arrestin complex.* Nature Communications, 7, *13416.*

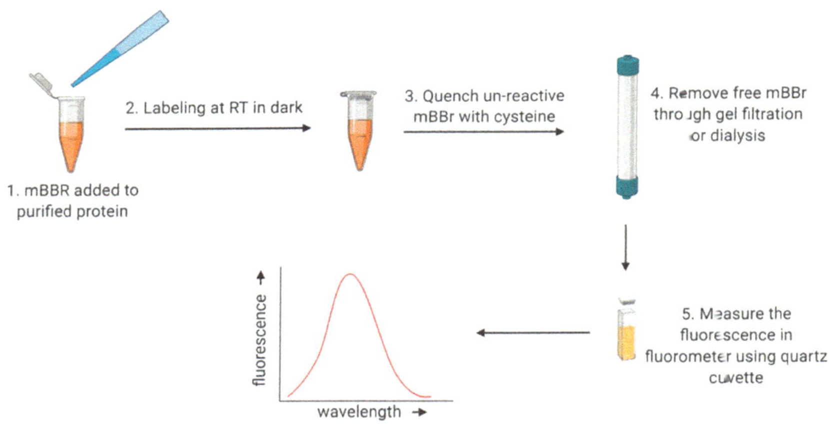

Fig. 2 A schematic flow diagram for step-by-step labeling of proteins with mBBr. Purified protein of interest is incubated with mBBr solution under light-protected conditions for specified period of time followed by quenching the reaction with L-cysteine. Subsequently, unreacted mBBr can be separated either by size-exclusion chromatography or dialysis. The labeling efficiency of the mBBr-labeled protein can be measured with a standard fluorometer using experimental settings described in the text.

2.2 Expression and purification of βarr constructs for bimane labeling

This protocol is based on our previous studies designed for site-directed labeling of βarrs in their finger loop, i.e., Leu^{68} in βarr1 and Leu^{69} in βarr2 (Kumari et al., 2016, 2017; Sente et al., 2018).

2.2.1. Synthesize or generate by site-directed mutagenesis βarr expression constructs where all the natural cysteines have been mutated to Ala/Val followed by substitution of Leu^{68}/Leu^{69} with cysteine. We have used pGEX-4T based vector system for expression of βarrs in *E. coli* (BL21-DE3), however, other expression vectors should work as well.

2.2.2. Express and purify single-Cys variant of βarrs following standard protocol using glutathione sepharose affinity chromatography and on-column cleavage of βarrs with thrombin as described previously (Shukla et al., 2013).

2.2.2a. Transform βarr expression plasmid in BL21 (DE3) competent cells, inoculate and grow primary culture overnight at 37 °C. Next day, inoculate secondary culture and grow until OD_{600} reaches 0.6–0.8.

2.2.2b. Add 0.1 mM IPTG to the secondary culture and let it grow for 16 h at 18 °C. Afterward, harvest the cell pellet, lyse in binding buffer (25 mM Tris, pH 8.5, 150 mM NaCl, 1 mM EDTA, 2 mM DTT, 1 mM PMSF, 2 mM benzamidine) by sonication and incubate the supernatant with pre-equilibrated glutathione sepharose (GS) resin overnight at 4 °C.

2.2.2c. Subsequently, wash the GS-resin extensively with wash buffer (25 mM Tris, pH 8.5, 150 mM NaCl) and incubate it with thrombin for 2 h at room temperature with gentle shaking. Collect the flow through containing βarr and estimate its purity on SDS-PAGE.

2.2.2d. βarr can be further purified using anion-exchange chromatography or size-exclusion chromatography as described previously (Shukla et al., 2013).

Caution! Purity of the protein of interest should be better than 80–85% in order to avoid non-specific labeling of impurities which may otherwise interfere with subsequent fluorescence spectroscopy data collection and interpretation.

2.2.3. We typically validate the functionality of bimane-labeled βarrs using co-immunoprecipitation assay involving a phosphopeptide

(V2Rpp) and active-conformation of βarr1 specific antibody fragment (Fab30). However, other approaches such as direct interaction with activated and phosphorylated receptors or other interaction partners can also be used to establish the functionality of βarrs after bimane-labeling.

2.3 Site directed labeling of βarrs with monobromobimane (mBBr)

Here, we describe a protocol to label purified βarrs at targeted cysteine residues using thiol-reactive monobromobimane reagent (Fig. 1).

2.3.1. Exchange the buffer of purified protein with labeling buffer (20 mM HEPES, pH 7.5, 150 mM NaCl) and concentrate to about 2.0 mg/mL. Alternatively, if the protein is purified in this buffer, it can be directly used for bimane-labeling.

2.3.2. Prepare a fresh stock of mBBr at a concentration of 100 mM in DMSO and add it to the purified protein at a molar ratio of 1:10. Incubate the reaction for 1 h in dark at room temperature. Alternatively, bimane labeling can also be performed at 4 °C or on ice depending on the stability of the protein and the labeling pattern.

Caution! Extended labeling time may cause protein aggregation and therefore, it is important to optimize the labeling condition for the protein of interest. A systematic optimization of protein to mBBr molar ratio, incubation temperature and time are recommended. It is also important to protect the reaction from light as it may affect mBBr stability and labeling efficiency.

2.3.3. Quench the labeling reaction by adding 100 mM L-cysteine. Subsequently, centrifuge the sample at 14,000 × *g* for 30 min to remove any aggregated protein.

2.3.4. Unreacted mBBr can be removed by desalting on PD-10 columns, standard size exclusion chromatography or dialysis in appropriate buffer.

2.3.5. Fluorescence of bimane-labeled βarrs can be measured using a standard Fluorescence spectrophotometer in photon counting mode using excitation and emission band pass of 5 nm. For labeled proteins, it is advisable to read full scan (200–800 nm) of excitation and emission spectrum to precisely measure the excitation and emission maxima, respectively. In our experiments, we have typically

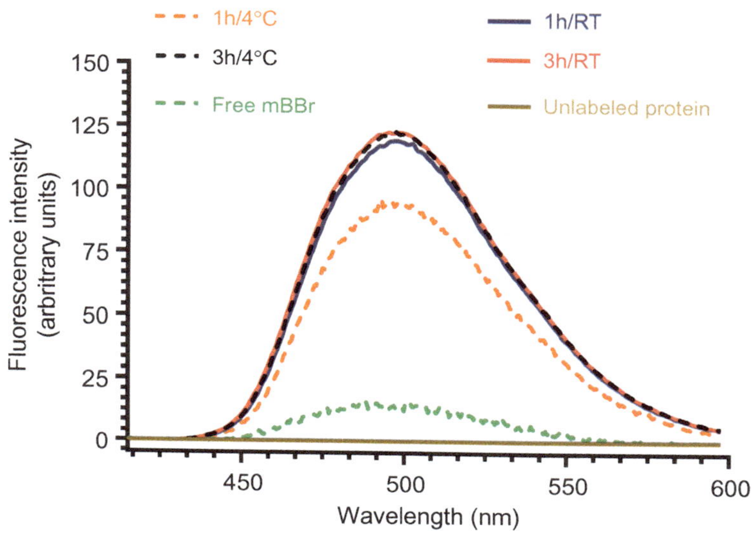

Fig. 3 A typical fluorescence scan profile of bimane-labeled βarr1. Purified βarr1 was labeled with mBBr under indicated experimental conditions followed by fluorescence scan using a standard LS-55 fluorometer in photon counting mode. Fluorescence scan profile of free mBBr and that of non-labeled βarr1 is presented as control. RT = room temperature. *This figure is adapted and modified from a previous study (Kumari, P., Srivastava, A., Banerjee, R., Ghosh, E., Gupta, P., Ranjan, R., Chen, X., Gupta, B., Gupta, C., Jaiman, D., Shukla, A.K. (2016). Functional competence of a partially engaged GPCR-beta-arrestin complex.* Nature Communications, *7, 13416.*

used an excitation wavelength of 397 nm and emission wavelength range from 415 to 600 nm with scan speed of 50 nm/min using a Perkin Elmer, LS-55 fluorometer (Fig. 3).

Caution! Slit width throughout the path of the light can greatly affect the sensitivity of the experiment. It is advisable to optimize this setting for individual labeled protein and must be kept constant throughout for a set of experiment.

2.3.6. Bimane fluorescence intensity should be corrected for background fluorescence by measuring the fluorescence intensity of just buffer (blank) and subtracting it from the measured values.

2.3.7. Bimane-labeled protein can be flash frozen with 10% glycerol and stored in small aliquots at −80 °C for several weeks.

2.4 Calculation of bimane-labeling efficiency

Bimane-labeling efficiency can be measured in terms of moles of bimane fluorophore per mole of protein of interest using the following formula:

$$\frac{Ax}{\in} \times \frac{\text{MW of protein}}{\text{Mg protein/mL}} = \frac{\text{moles of fluorophore}}{\text{moles of protein}}$$

Ax = fluorescence value at emission maximum wavelength. ϵ = molar extinction coefficient of dye fluorophore.

In our experiments, the labeling efficiency of βarrs was typically in the range of 80–85%.

3. Potential problems, troubleshooting and additional considerations

3.1. If the labeling efficiency is sub-optimal, a number of troubleshooting steps can be implemented.

- **3.1.1.** Carefully assess the buffer conditions of the protein and carry out additional dialysis steps or size exclusion chromatography.
- **3.1.2.** It is advisable to use freshly prepared mBBr solution as stored stocks may not be very stable over time. Additionally, multiple freeze-thaws of mBBr stocks should be avoided.
- **3.1.3.** If the protein of interest exhibits aggregation or precipitation during labeling reaction, the molar ratio of mBBr:protein should be titrated and optimized accordingly.

3.2. It is advisable to avoid excess labeling of protein with mBBr as it may alter the structural and functional properties of the protein. It is also advisable to validate the functionality of mBBr-labeled protein before using it in subsequent experiments.

Acknowledgments

The research program in Dr. Shukla's laboratory is supported by an Intermediate Fellowship of the Wellcome Trust/DBT India Alliance Fellowship (grant number IA/I/14/1/501285) awarded to A.K.S.; the Science and Engineering Research Board (SERB) (EMR/2017/003804); Innovative Young Biotechnologist Award from the Department of Biotechnology (DBT) (BT/08/IYBA/2014-3); and the Indian Institute of Technology, Kanpur. A.K.S. is an Intermediate Fellow of Wellcome Trust/DBT India Alliance, EMBO Young Investigator, and Joy Gill Chair Professor. H.D.-A. and A.S. are supported by DBT-BioCaRE (BT/PR31791/BIC/101/1228/2019) and Early Career

Fellowship of the Wellcome Trust/DBT India Alliance Fellowship (grant number IA/E/17/1/503687) respectively.

References

Chen, Q., Iverson, T. M., & Gurevich, V. V. (2018). Structural basis of arrestin-dependent signal transduction. *Trends in Biochemical Sciences*, *43*(6), 412–423.

Chinn, P. C., Pigiet, V., & Fahey, R. C. (1986). Determination of thiol proteins using monobromobimane labeling and high-performance liquid chromatographic analysis: Application to *Escherichia coli* thioredoxin. *Analytical Biochemistry*, *159*(1), 143–149.

DeWire, S. M., Ahn, S., Lefkowitz, R. J., & Shenoy, S. K. (2007). Beta-arrestins and cell signaling. *Annual Review of Physiology*, *69*, 483–510.

Ghosh, E., Dwivedi, H., Baidya, M., Srivastava, A., Kumari, P., Stepniewski, T., et al. (2019). Conformational sensors and domain swapping reveal structural and functional differences between beta-arrestin isoforms. *Cell Reports*, *28*(13), 3287–3299 e6.

Ghosh, E., Srivastava, A., Baidya, M., Kumari, P., Dwivedi, H., Nidhi, K., et al. (2017). A synthetic intrabody-based selective and generic inhibitor of GPCR endocytosis. *Nature Nanotechnology*, *12*(12), 1190–1198.

Gurevich, V. V., & Gurevich, E. V. (2004). The molecular acrobatics of arrestin activation. *Trends in Pharmacological Sciences*, *25*(2), 105–111.

Gurevich, V. V., & Gurevich, E. V. (2018). GPCRs and signal transducers: Interaction stoichiometry. *Trends in Pharmacological Sciences*, *39*(7), 672–684.

Kang, D. S., Tian, X., & Benovic, J. L. (2014). Role of beta-arrestins and arrestin domain-containing proteins in G protein-coupled receptor trafficking. *Current Opinion in Cell Biology*, *27*, 63–71.

Kumari, P., Srivastava, A., Banerjee, R., Ghosh, E., Gupta, P., Ranjan, R., et al. (2016). Functional competence of a partially engaged GPCR-beta-arrestin complex. *Nature Communications*, 7, 13416.

Kumari, P., Srivastava, A., Ghosh, E., Ranjan, R., Dogra, S., Yadav, P. N., et al. (2017). Core engagement with beta-arrestin is dispensable for agonist-induced vasopressin receptor endocytosis and ERK activation. *Molecular Biology of the Cell*, *28*(8), 1003–1010.

Lefkowitz, R. J., & Shenoy, S. K. (2005). Transduction of receptor signals by beta-arrestins. *Science*, *308*(5721), 512–517.

Luttrell, L. M., Wang, J., Plouffe, B., Smith, J. S., Yamani, L., Kaur, S., et al. (2018). Manifold roles of beta-arrestins in GPCR signaling elucidated with siRNA and CRISPR/Cas9. *Science Signaling*, *11*(549), 1–22.

Mansoor, S. E., & Farrens, D. L. (2004). High-throughput protein structural analysis using site-directed fluorescence labeling and the bimane derivative (2-pyridyl)dithiobimane. *Biochemistry*, *43*(29), 9426–9438.

Mansoor, S. E., McHaourab, H. S., & Farrens, D. L. (2002). Mapping proximity within proteins using fluorescence spectroscopy. A study of T4 lysozyme showing that tryptophan residues quench bimane fluorescence. *Biochemistry*, *41*(8), 2475–2484.

Rahmeh, R., Damian, M., Cottet, M., Orcel, H., Mendre, C., Durroux, T., et al. (2012). Structural insights into biased G protein-coupled receptor signaling revealed by fluorescence spectroscopy. *Proceedings of the National Academy of Sciences of the United States of America*, *109*(17), 6733–6738.

Ranjan, R., Dwivedi, H., Baidya, M., Kumar, M., & Shukla, A. K. (2017). Novel structural insights into GPCR-beta-arrestin interaction and signaling. *Trends in Cell Biology*, *27*(11), 851–862.

Reiter, E., Ahn, S., Shukla, A. K., & Lefkowitz, R. J. (2012). Molecular mechanism of beta-arrestin-biased agonism at seven-transmembrane receptors. *Annual Review of Pharmacology and Toxicology*, *52*, 179–197.

Sente, A., Peer, R., Srivastava, A., Baidya, M., Lesk, A. M., Balaji, S., et al. (2018). Molecular mechanism of modulating arrestin conformation by GPCR phosphorylation. *Nature Structural & Molecular Biology, 25*(6), 538–545.

Shukla, A. K., Manglik, A., Kruse, A. C., Xiao, K., Reis, R. I., Tseng, W. C., et al. (2013). Structure of active beta-arrestin-1 bound to a G-protein-coupled receptor phosphopeptide. *Nature, 497*(7447), 137–141.

Shukla, A. K., Singh, G., & Ghosh, E. (2014). Emerging structural insights into biased GPCR signaling. *Trends in Biochemical Sciences, 39*(12), 594–602.

Shukla, A. K., Westfield, G. H., Xiao, K., Reis, R. I., Huang, L. Y., Tripathi-Shukla, P., et al. (2014). Visualization of arrestin recruitment by a G-protein-coupled receptor. *Nature, 512*(7513), 218–222.

Sommer, M. E., & Farrens, D. L. (2006). Arrestin can act as a regulator of rhodopsin photochemistry. *Vision Research, 46*(27), 4532–4546.

Sommer, M. E., Farrens, D. L., McDowell, J. H., Weber, L. A., & Smith, W. C. (2007). Dynamics of arrestin-rhodopsin interactions: Loop movement is involved in arrestin activation and receptor binding. *The Journal of Biological Chemistry, 282*(35), 25560–25568.

Sommer, M. E., Smith, W. C., & Farrens, D. L. (2005). Dynamics of arrestin-rhodopsin interactions: Arrestin and retinal release are directly linked events. *The Journal of Biological Chemistry, 280*(8), 6861–6871.

Sommer, M. E., Smith, W. C., & Farrens, D. L. (2006). Dynamics of arrestin-rhodopsin interactions: Acidic phospholipids enable binding of arrestin to purified rhodopsin in detergent. *The Journal of Biological Chemistry, 281*(14), 9407–9417.

Staus, D. P., Wingler, L. M., Choi, M., Pani, B., Manglik, A., Kruse, A. C., et al. (2018). Sortase ligation enables homogeneous GPCR phosphorylation to reveal diversity in beta-arrestin coupling. *Proceedings of the National Academy of Sciences of the United States of America, 115*(15), 3834–3839.

Thomsen, A. R. B., Plouffe, B., Cahill, T. J., 3rd, Shukla, A. K., Tarrasch, J. T., Dosey, A. M., et al. (2016). GPCR-G protein-beta-arrestin super-complex mediates sustained G protein signaling. *Cell, 166*(4), 907–919.

CHAPTER FIFTEEN

Reversible biotinylation of purified proteins for measuring protein–protein interactions

Hemlata Dwivedi-Agnihotri, Ashish Srivastava, Arun K. Shukla*
Department of Biological Sciences and Bioengineering, Indian Institute of Technology, Kanpur, India
*Corresponding author: e-mail address: arshukla@iitk.ac.in

Contents

Abstract

Measuring protein–protein interactions using purified proteins in vitro is one of the most frequently used approach to understand the biochemical and mechanistic details of cellular signaling pathways. Typically, affinity tags are genetically fused to proteins of interest, and they are used to capture and detect them. However, in some cases, fusion of bulky affinity tags might present a significant limitation in these experiments, especially if the regions in close proximity of tags are involved in protein–protein interactions. Here, we present a step-by-step protocol for an alternative approach that involves reversible biotinylation of purified proteins using a simple chemical-conjugation of cleavable biotin moiety. Biotinylated proteins can be directly used as bait for selective immobilization on solid support for measuring protein–protein interactions. Furthermore, biotinylation of protein of interest also allows specific detection in standard biochemical assays. This simple, straightforward and modular protocol can be directly adapted and applied to facilitate the detection of novel protein–protein interactions as well as measuring apparent affinities of such interactions.

Methods in Enzymology, Volume 633
ISSN 0076-6879
https://doi.org/10.1016/bs.mie.2019.11.008

1. Introduction

The process of cellular signaling involves a large number of transient, non-covalent protein–protein interaction networks which are essential for signal-recognition and propagation in cellular context (Braun & Gingras, 2012; De Las Rivas & Fontanillo, 2010; Nooren & Thornton, 2003). Protein–protein interactions are typically tightly regulated with respect to their spatio-temporal patterns and exquisite selectivity, which in turn determine the range and duration of signaling events in cells (De Las Rivas & Fontanillo, 2010; Scott & Pawson, 2009; Yang, Wagner, & Beli, 2015). For example, upon agonist-activation, G protein-coupled receptors (GPCRs) undergo a conformational change followed by coupling to heterotrimeric G proteins (Gilman, 1987; Maguire, Van Arsdale, & Gilman, 1976). This leads to GDP/GTP exchange on Gα sub-unit and dissociation of Gα from Gβγ sub-units followed by activation of their downstream effectors such as adenylyl cyclase and ion channels (Bockaert & Pin, 1999; Gilman, 1987). Subsequently, GPCRs are phosphorylated by GRKs (GPCR kinases) and other kinases which in turn promote the recruitment of multifunctional proteins called arrestins (Inglese, Freedman, Koch, & Lefkowitz, 1993; Ranjan, Dwivedi, Baidya, Kumar, & Shukla, 2017). Arrestins typically block G-protein coupling through steric hindrance on one hand, and mediate receptor endocytosis on the other via nucleating the assembly of the components of clathrin coated endocytosis machinery such as clathrin and adaptin (Freedman & Lefkowitz, 1996; Goodman et al., 1996; Kang, Tian, & Benovic, 2014).

There are a large number of methods available to measure protein–protein interaction in cellular context such as Bioluminescence Resonance Energy Transfer (BRET), Fluorescence Resonance Energy Transfer (FRET), Proximity Ligation assay (PSA) etc. (Berggard, Linse, & James, 2007; Miura, 2018; Phizicky & Fields, 1995). In vitro detection and characterization of protein–protein interaction can be carried out using label free approaches such as Isothermal Calorimetry (ITC) and Surface Plasmon Resonance (SPR) among others (Lin & Wu, 2019; Nguyen, Park, Kang, & Kim, 2015). While these methods yield immensely useful information on precise affinity of interactions, thermodynamic parameters and interaction stoichiometry, they require sophisticated instrumentation and expertise. As an alternative approach, ELISA and co-immunoprecipitation (co-IP) based assays are more frequently used across a large number of laboratories for qualitative assessment of protein–protein interaction (Lequin, 2005; Lin & Lai, 2017).

Typically, proteins of interest are genetically tagged with affinity tags at either the N- or the C-terminus, followed by their expression and purification. Subsequently, either affinity resins (e.g., Ni-NTA for Histidine tag) or antibody-based approaches (e.g., FLAG M2 antibody agarose for FLAG tag) can be used to capture and detect their interaction using standard ELISA and co-IP assays. In some cases, however, genetic fusion of affinity tags may compromise the activity and functionality of proteins of interest and thereby, limits the utility of this approach. Moreover, this type of approach cannot be employed for proteins isolated from their native sources. Although, using antibodies against proteins of interest provide an alternative strategy in ELISA and co-IP assays, suitable antibodies may not always be available for this purpose. Therefore, a simple, modular and adaptable strategy to capture and detect purified proteins can be of significant interest to many laboratories engaged in protein biochemistry research.

Here, we present a step-by-step protocol for biotinylating purified proteins via chemical conjugation of biotin reagents which can significantly facilitate the detection and characterization of protein–protein interactions in vitro. Considering its small size, biotin-conjugation should not typically interfere with the biological activity of the proteins and it offers a modular approach for labeling the proteins without any genetic modifications. This protocol is based on our previously published proof-of-principle studies using biotinylation of several proteins involved in GPCR signaling and regulatory paradigms (Ghosh et al., 2017, 2019; Kumari et al., 2016, 2017).

2. Materials and methods

2.1 Reagents for biotin labeling of purified proteins

EZ-Link Sulfo-NHS-SS-Biotin (Thermofisher Scientific; cat no. 21331 or A39258)
DMSO (Dimethyl sulfoxide)
Purified protein of interest to be biotinylated
Tris–HCl
HEPES (4-(2-hydroxyethyl)-1-piperazineethanesulfonic acid)
NaCl
PD-10 Desalting Column, with Sephadex G-25 resin, 1.0–2.5 mL samples
Vivaspin protein concentrator spin columns

Slide-A-Lyzer™ Dialysis Cassettes, 3.5K MWCO, 3 mL
Syringe (5 mL)
Regular micro-tips (1000, 200, 20 μL)
Microcentrifuge tubes (1.5, 2 mL)
Centrifuge tubes (15 mL)
Glycerol (for freezing protein sample)
Bradford reagent (Sigma; cat no. B6916).

2.2 Reagents for capture and detection of biotinylated proteins

Streptavidin beads
Magnetic separation rack
DTT (Dithiothreitol)
Pre-cast or manually prepared acrylamide gels for SDS-PAGE
Suitable power supply and SDS-PAGE apparatus
Protein molecular weight marker
Chemicals for different buffers (Tris–HCl, Glycine, Methanol, Tween-20, Hydrochloric acid)
Coomassie Brilliant blue
Suitable power supply and Western blot apparatus for Semi-Dry transfer
PVDF membrane for protein transfer
Bovine Serum Albumin (For membrane blocking)
Streptavidin–horseradish peroxidase antibody (Genscript)
Enhanced chemi-luminescence (ECL) reagents
Chemi-Doc imaging system
96-well Immuno plates
H_2SO_4
3,3,5,5-Tetramethylbenzidine or TMB substrate
Plate reader (Victor 4X multimode plate reader, Perkin Elmer)

2.3 Method

2.3.1 Biotin labeling of purified proteins

2.3.1.1 Preparing the working stock solution of the biotinylation reagent

There are several biotinylation reagents commercially available with or without cleavable biotin group, different spacer arms and functional groups. The protocol described below is based on our previous studies with Sulfo-NHS-SS-Biotin reagent that has a disulfide bond in its spacer arm allowing cleavage by reducing agents (Fig. 1).

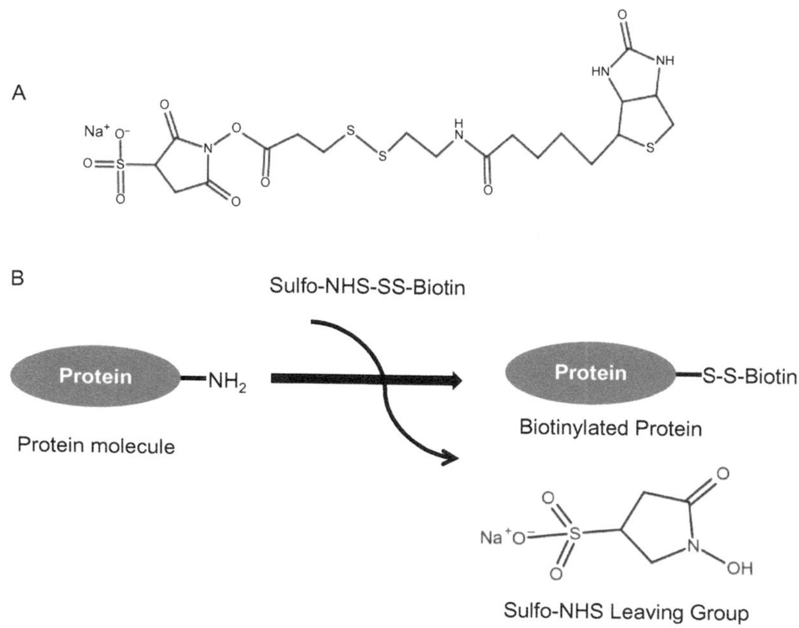

Fig. 1 Chemical structure of Sulfo-NHS-SS-biotin and a schematic of a typical biotinylation reaction. (A) The Sulfo-NHS-SS-Biotin molecule is a primary amine-reactive, thiol cleavable molecule with long spacer arm of 24.3 Å. (B) The N-terminus primary amine and epsilon amino group of lysine residues react with ester group of Sulfo-NHS-SS-biotin molecule to form an amide bond, which covalently links biotin to the protein of interest. The cleaved Sulfo-NHS group can be subsequently separated though size exclusion chromatography or dialysis.

2.3.1.1.a. Take out the EZ-Link Sulfo-NHS-SS-Biotin from −20 °C freezer and equilibrate it to room temperature before opening the vial for weighing. Alternatively, no-weigh vials can also be used to avoid repeated freeze thaw of powder stock of the reagent.

Caution!: Sulfo-NHS-SS-Biotin is hygroscopic in nature. Hence, it is important to bring down the reagent to ambient temperature before opening the vial to avoid moisture retention and tightly cap the vial after usage. It is advisable to store it with desiccant.

2.3.1.1.b. Weigh a small amount of reagent, adequate enough for the labeling reaction. Prepare a 10 mM stock in DMSO. An aqueous stock can also be prepared as Sulfo-NHS-SS-Biotin is water soluble.

Caution!: The NHS ester group hydrolyzes readily and therefore, the aqueous solution should always be prepared fresh prior to use and residual reconstituted reagent should be discarded. The DMSO stock, however, can be stored in small aliquots for up to a couple of months at −80°C without a significant loss in labeling efficiency. Repeated freeze–thaw cycle of the stock solution should be avoided.

2.3.2 Preparing the protein sample for biotinylation

2.3.2.1. Determine the concentration of the protein to be labeled using any standard assay. Typically, the concentration should be >1 mg/mL for efficient labeling. If the concentration of the protein sample is low, it can be concentrated using VivaSpin protein concentrator spin columns. Protein of interest should be purified to a reasonable level (e.g., >90%) in order to avoid labeling of non-specific impurities which may interfere with subsequent experiments.

Caution!: The protein should be resuspended in a buffer that is free of primary amines (Tris or Glycine) as they will react with the biotinylation reagent and therefore, impair the efficiency of the biotinylation reaction. If the protein buffer has amines, then either dialyze in compatible buffer (PBS or HEPES-NaCl) or consider performing size exclusion chromatography using PD-10 desalting columns. The buffer should also be free of any reducing agent to prevent cleavage of disulfide bond. The ester bond is labile at higher pH (10 − 11), and therefore, the optimal pH range is typically between 7 and 9.

2.3.3 Biotinylation of the protein of interest

2.3.3.1 Incubate the protein sample, preferably in a volume of less than 3 mL with 20-fold molar excess of the biotinylation reagent for 30–60 min at room temperature. Alternatively, the labeling reaction can be carried out for longer time periods at 4°C or on ice.

Note: It is advisable that the efficiency of biotinylation is compared and optimized using a range of concentrations of the biotinylation reagent, for example, 1:5, 1:10, 1:20, 1:50 and 1:100 M ratios. In our studies, we have observed that proteins are efficiently biotinylated at a 20-fold molar excess. The labeling efficacy depends on the accessibility of the amine-group at the

N-terminus of the protein and the number of exposed Lysine/Arginine residues that harbor free amine-group for biotinylation using NHS reagent.

Note: In addition to NHS-Biotin reagent, a range of other biotinylation reagents are also commercially available which label the proteins of interest through other functional groups such as Iodoacetyl-LC-Biotin (sulfhydryl-reactive groups), Amine-PEG3-Biotin (Carboxyl reactive group) etc.

2.3.3.2 After the completion of the biotinylation reaction, quench the unreacted biotinylation reagent with 50 mM Tris, pH 8.5. Typically, 1/10th of the reaction volume is sufficient to quench the free biotinylation reagent.

2.3.4 Removal of the excess biotinylation reagent

The unreacted biotinylation reagent can be removed either through desalting/gel-filtration column or dialysis.

2.3.4.1 Removal of unreacted biotinylation reagent through buffer exchange on PD-10 desalting columns

2.3.4.1a Remove the bottom seal of the PD-10 column, place it on the column stand and let the storage buffer pass out by gravity flow.

2.3.4.1b Equilibrate the PD-10 column with approximately 25 mL of PD-10 buffer (20 mM HEPES, pH 7.4, 100 mM NaCl) and discard the flow through.

2.3.4.1c Load the biotinylated protein sample on top of the resin bed and collect the flow through. Typically, the sample is loaded in a volume of 2.5 mL, followed by the collection of flow-through and subsequent elution in 3.5 mL with elution buffer in a falcon tube placed below the column. In case, the sample volume is less than 2.5 mL, add protein sample first, let it pass through, and then add buffer to complete the total volume of 2.5 mL. Afterwards, collect seven fractions of 0.5 mL each in microcentrifuge tubes, and assess the presence of protein in the eluted fractions using Bradford dye through colorimetric reaction. Pool the eluted fractions containing the biotinylated protein and store as required for subsequent experiments.

Note: Instead of PD-10 based procedure, a standard size exclusion chromatography may also be used to remove unreacted biotinylation reagent.

2.3.4.2 Removal of unreacted biotinylation reagent through dialysis

2.3.4.2a Take a dialysis cassette and hydrate it by immersing in dialysis buffer (20 mM HEPES, pH 7.5 and 100 mM NaCl) for about 2–5 min.

2.3.4.2b From one end of the cassette, inject the sample using a 24 gauge syringe. Withdraw the air from the dialysis cassette while the needle is still inside so that the sample touches maximum surface of the membrane. Insert the cassette into the groove of the buoy and immerse in dialysis buffer. Dialyze for 12–16 h at 4°C, preferably with replacement of dialysis buffer once at 6–8 h to maximize the dialysis efficiency.

Caution!: It is important to inject or withdraw the sample carefully from the dialysis cassette to avoid membrane rupture. It is advisable to check the cassette for leakage by injecting sterilized water.

2.3.4.2c Collect the sample from dialysis cassette by first inflating it by injecting air into the dialysis bag from one of the side ports using a syringe so that the sample collects at one place and can be sucked into the syringe.

2.3.4.2d Measure the concentration of biotinylated protein using standard assays, prepare small aliquots depending on subsequent experiments, flash-freeze the aliquots in liquid nitrogen and store at −80 °C.

2.3.5 Detection of biotinylated proteins using Western blotting

Biotinylated proteins can be detected using standard Western blotting assay using HRP-coupled streptavidin or anti-biotin antibodies.

2.3.5.1 Separate the biotinylated proteins using standard SDS-PAGE followed by transfer on PVDF/nitrocellulose membrane.

2.3.5.2 After incubating the membrane with 5% BSA in 1 × TBST for 1 h at room temperature to block non-specific sites, incubate it further in HRP-coupled streptavidin solution (final concentration 0.2 μg/mL) for 2 h at room temperature or overnight at 4°C.

2.3.5.3 Wash the membrane 3 times (5 min each) with 1 × TBST. Add 1–2 mL of ECL (Enhanced Chemiluminescence) substrate sufficient enough to cover the entire membrane and visualize the proteins using a standard CCD camera attached to the ChemiDoc imaging system.

Note: A typical example of detection of biotinylated adaptin (β-appendage of Adaptin2, 592-951 amino acid residues) is presented in Fig. 2A.

2.3.6 Detection of biotinylated proteins using ELISA

Biotinylated proteins can also be detected using standard ELISA assay by means of HRP-coupled streptavidin or anti-biotin antibodies.

2.3.6.1 Immobilize varying amounts of biotinylated proteins on MaxiSorp plates via adsorption. As a negative control, also immobilize non-biotinylated protein at an amount equal to the maximal amount of biotinylated protein. Incubate for 1 h at room temperature.

2.3.6.2 Block the non-specific binding sites in the well using 200 μL of 1% BSA (prepared in 20 mM HEPES, pH 7.5, 100 mM NaCl) in each well for 1 h at room temperature.

2.3.6.3 Incubate the wells with 100 μL of HRP-coupled streptavidin (final concentration of 0.2 μg/mL) for 1 h at room temperature. Subsequently, wash the wells 3–5 times with 100 μL of washing buffer (20 mM HEPES, pH 7.5, 100 mM NaCl, 0.01% MNG). Detect the biotinylated protein by adding 50 μL of TMB-ELISA

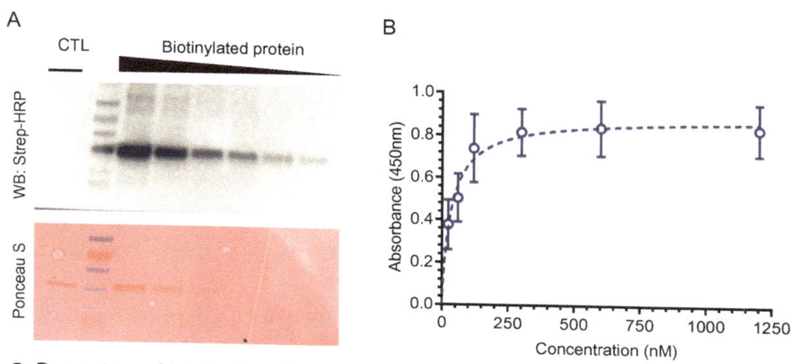

Fig. 2 Detection of biotinylated protein using Western Blotting and ELISA approach. (A) Varying concentrations of purified and biotinylated β-appendage of adaptin2 (residues from 592 to 951) was separated on SDS-PAGE followed by Western blot based detection using HRP-coupled anti-streptavidin antibody. Non-biotinylated protein was used as negative control (CTL). The upper blot shows the detection of biotinylated protein in a dose-dependent manner (detected up to 1/50th of the highest concentration). Non-biotinylated protein was not detected under the same experimental conditions. The lower panel shows Ponceau S staining of the blot to visualize the total protein loaded on the gel. Due to low sensitivity of Ponceau S, lower amounts of biotinylated adaptin2 are not visible in the lower panel. (B) Biotinylated proteins can also be detected using ELISA based approach as presented here using biotinylated clathrin (terminal-domain) as a case example. Varying amounts of biotinylated protein were immobilized in MaxiSorp 96-well plate followed by blocking of potential non-specific binding sites in the wells with 1% BSA. Subsequently, immobilized protein was detected using HRP-coupled anti-streptavidin antibody as indicated in the main text. Non-biotinylated protein was used as negative control and the signals were normalized.

substrate per well and quench the reaction by adding 50 μL of 2 M H_2SO_4 in each well. Upon addition of TMB-ELISA, a blue color will develop which turns yellow after quenching the reaction with H_2SO_4.

2.3.6.4 Measure the absorbance at 450 nM using a Plate reader (Victor 4X multimode plate reader, Perkin Elmer).

Note

(i) Blocking with BSA or non-fat dry milk reduces signal to noise ratio by blocking non-specific sites. It is advisable to use 200 μL of blocking agent per well to minimize the non-specific binding surface.

(ii) It is important to quench the colorimetric reaction before the signal reaches saturation which can otherwise negatively impact accurate estimation of biotinylated protein.

(iii) It is important to systematically optimize the dilution of HRP-coupled streptavidin to allow the detection of biotinylated protein with suitable signal to noise ratio.

(iv) A typical example of detection of biotinylated Clathrin terminal-domain is presented in Fig. 2B.

(v) In experiments where the bait protein is immobilized on ELISA plate, its interaction with the protein of interested, which is biotinylated, can be detected using HRP-coupled streptavidin following a protocol mentioned above.

2.3.7 *Capture and immobilization of biotinylated proteins using magnetic streptavidin beads*

Biotinylated proteins can be captured on either standard streptavidin agarose/sepharose beads or magnetic beads coated with streptavidin for subsequent use in protein–protein interaction studies. Here, we present a protocol for capturing biotinylated proteins using magnetic streptavidin beads.

2.3.7.1 Equilibrate magnetic streptavidin beads in equilibration buffer (20 mM HEPES, pH 7.5, 100 mM NaCl) by three cycles of suspension and precipitation using magnetic separation stand.

2.3.7.2 Incubate approximately 5 μg of biotinylated protein in a volume of 100 μL with the pre-equilibrated beads for 10–15 min at room temperature with gentle shaking. Typically, 50 μL of the original magnetic streptavidin bead slurry should be sufficient to capture 5–10 μg of biotinylated proteins.

2.3.7.3 Use the magnetic separation rack to collect the flow-through and take out 20 μL for running on SDS-PAGE.

2.3.7.4 Wash the beads 2–3 times with 100 μL of the washing buffer (20 mM HEPES, pH 7.5, 100 mM NaCl) and collect wash fractions.

2.3.7.5 Elute the bound protein in 100 μL elution buffer (20 mM HEPES, pH 7.5, 100 mM NaCl, 100 mM DTT) by incubating the beads in this buffer for 5–10 min at room temperature with gentle shaking followed by collection of flow through.

2.3.7.6 Analyze the capture and elution of biotinylated proteins by running the samples collected from every step on SDS-PAGE and visualize using Coomassie brilliant blue staining.

Note

(i) Before incubating protein sample with beads in 2.3.7.1, take out 20 μL of the sample and add reducing SDS-PAGE sample running dye to it. Samples should be collected at every step to run SDS-PAGE in order to determine the extent of biotinylation.

(ii) A typical schematic of the capture and elution strategy is presented in Fig. 3A. An example of capture and elution of biotinylated Clathrin terminal-domain is presented in Fig. 3B.

3. Potential problems, troubleshooting guidelines and additional considerations

3.1 Inefficient labeling of protein of interest may arise from a number of reasons including a sub-optimal pH of the buffer, presence of primary-amines and reducing agent in the buffer, or the quality of the biotinylation reagent stock.

3.1.1 The optimal pH range of the buffer is between 7 and 9, and therefore, the purified protein should be prepared in a buffer of this pH range.

3.1.2 Purified protein should not have any contamination of primary amines or reducing agents in the buffer, and therefore, the protein should be purified in appropriate buffer or buffer exchange should be carried out effectively.

3.1.3 Although solution stocks of biotinylated reagents prepared in DMSO can be stored at −80°C, it is advisable to prepare fresh stock every time just before use to avoid any issues with degradation of the biotinylation reagent.

3.2 It should be noted that the protocol presented here is not designed for site-directed biotinylation, and therefore, the functionality and activity

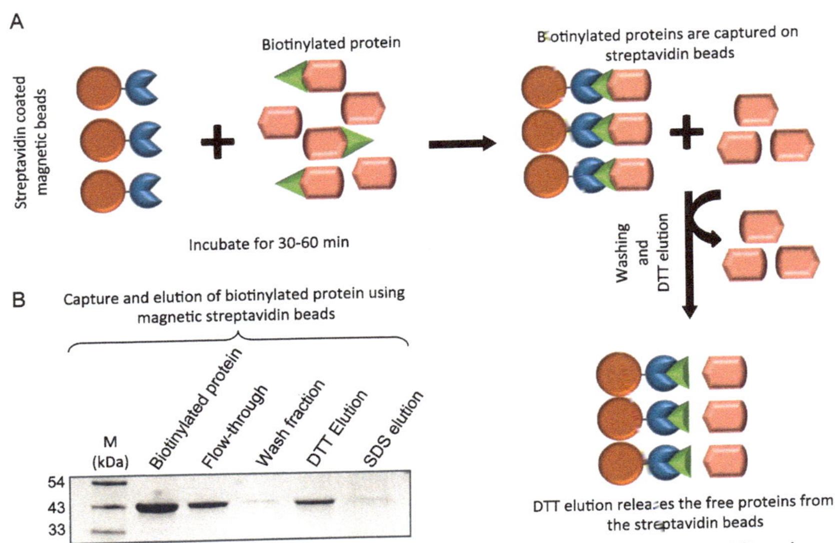

Fig. 3 Capturing biotinylated proteins using magnetic Streptavidin beads. (A) A schematic flow-diagram describing the process of capturing biotinylated proteins using magnetic streptavidin beads and subsequent elution using a reducing agent. (B) A typical profile of capture and elution experiment using biotinylated clathrin terminal-domain as a case example. Samples from different steps were separated on SDS-PAGE followed by Coomassie staining for visualization. SDS elution is carried out to test whether any biotinylated protein is still present on the beads. Densitometry-based quantification can be performed to measure the apparent capture and elution efficiency.

of biotinylated proteins should be characterized before using them in subsequent experiments. It is advisable to optimize the biotinylation conditions in such a way that the stoichiometry of biotinylation is typically 1:1, i.e., 1 biotin molecule per molecule of protein.

Acknowledgments

The research program in Dr. Shukla's laboratory is supported by an Intermediate Fellowship of the Wellcome Trust/DBT India Alliance Fellowship (grant number IA/I/14/1/501285) awarded to A.K.S.; the Science and Engineering Research Board (SERB) (EMR/2017/003804); Innovative Young Biotechnologist Award from the Department of Biotechnology (DBT) (BT/08/IYBA/2014-3); and the Indian Institute of Technology, Kanpur. A.K.S. is an Intermediate Fellow of Wellcome Trust/DBT India Alliance, EMBO Young Investigator, and Joy Gill Chair Professor. H.D.-A. and A.S. are supported by National Post-Doctoral Fellowship of SERB (PDF/2016/002930) and Early Career Fellowship of the Wellcome Trust/DBT India Alliance Fellowship (grant number IA/I/14/1/501285), respectively.

References

Berggard, T., Linse, S., & James, P. (2007). Methods for the detection and analysis of protein-protein interactions. *Proteomics*, 7(16), 2833–2842.

Bockaert, J., & Pin, J. P. (1999). Molecular tinkering of G protein-coupled receptors: An evolutionary success. *The EMBO Journal*, *18*(7), 1723–1729.

Braun, P., & Gingras, A. C. (2012). History of protein-protein interactions: From egg-white to complex networks. *Proteomics*, *12*(10), 1478–1498.

De Las Rivas, J., & Fontanillo, C. (2010). Protein-protein interactions essentials: Key concepts to building and analyzing interactome networks. *PLoS Computational Biology*, *6*(6), e1000807.

Freedman, N. J., & Lefkowitz, R. J. (1996). Desensitization of G protein-coupled receptors. *Recent Progress in Hormone Research*, *51*, 319–351. discussion 352-3.

Ghosh, E., Dwivedi, H., Baidya, M., Srivastava, A., Kumari, P., Stepniewski, T., et al. (2019). Conformational sensors and domain swapping reveal structural and functional differences between beta-arrestin isoforms. *Cell Reports*, *28*(13), 3287–3299 e6.

Ghosh, E., Srivastava, A., Baidya, M., Kumari, P., Dwivedi, H., Nidhi, K., et al. (2017). A synthetic intrabody-based selective and generic inhibitor of GPCR endocytosis. *Nature Nanotechnology*, *12*(12), 1190–1198.

Gilman, A. G. (1987). G proteins: Transducers of receptor-generated signals. *Annual Review of Biochemistry*, *56*, 615–649.

Goodman, O. B., Jr., Krupnick, J. G., Santini, F., Gurevich, V. V., Penn, R. B., Gagnon, A. W., et al. (1996). Beta-arrestin acts as a clathrin adaptor in endocytosis of the beta2-adrenergic receptor. *Nature*, *383*(6599), 447–450.

Inglese, J., Freedman, N. J., Koch, W. J., & Lefkowitz, R. J. (1993). Structure and mechanism of the G protein-coupled receptor kinases. *The Journal of Biological Chemistry*, *268*(32), 23735–23738.

Kang, D. S., Tian, X., & Benovic, J. L. (2014). Role of beta-arrestins and arrestin domain-containing proteins in G protein-coupled receptor trafficking. *Current Opinion in Cell Biology*, *27*, 63–71.

Kumari, P., Srivastava, A., Banerjee, R., Ghosh, E., Gupta, P., Ranjan, R., et al. (2016). Functional competence of a partially engaged GPCR-beta-arrestin complex. *Nature Communications*, 7, 13416.

Kumari, P., Srivastava, A., Ghosh, E., Ranjan, R., Dogra, S., Yadav, P. N., et al. (2017). Core engagement with beta-arrestin is dispensable for agonist-induced vasopressin receptor endocytosis and ERK activation. *Molecular Biology of the Cell*, *28*(8), 1003–1010.

Lequin, R. M. (2005). Enzyme immunoassay (EIA)/enzyme-linked immunosorbent assay (ELISA). *Clinical Chemistry*, *51*(12), 2415–2418.

Lin, J. S., & Lai, E. M. (2017). Protein-protein interactions: Co-immunoprecipitation. *Methods in Molecular Biology*, *1615*, 211–219.

Lin, K., & Wu, G. (2019). Isothermal titration calorimetry assays to measure binding affinities in vitro. *Methods in Molecular Biology*, *1893*, 257–272.

Maguire, M. E., Van Arsdale, P. M., & Gilman, A. G. (1976). An agonist-specific effect of guanine nucleotides on binding to the beta adrenergic receptor. *Molecular Pharmacology*, *12*(2), 335–339.

Miura, K. (2018). An overview of current methods to confirm protein-protein interactions. *Protein and Peptide Letters*, *25*(8), 728–733.

Nguyen, H. H., Park, J., Kang, S., & Kim, M. (2015). Surface plasmon resonance: A versatile technique for biosensor applications. *Sensors (Basel)*, *15*(5), 10481–10510.

Nooren, I. M., & Thornton, J. M. (2003). Diversity of protein-protein interactions. *The EMBO Journal*, *22*(14), 3486–3492.

Phizicky, E. M., & Fields, S. (1995). Protein-protein interactions: Methods for detection and analysis. *Microbiological Reviews*, *59*(1), 94–123.

Ranjan, R., Dwivedi, H., Baidya, M., Kumar, M., & Shukla, A. K. (2017). Novel structural insights into GPCR-beta-arrestin interaction and signaling. *Trends in Cell Biology*, *27*(11), 851–862.

Scott, J. D., & Pawson, T. (2009). Cell signaling in space and time: Where proteins come together and when they're apart. *Science*, *326*(5957), 1220–1224.

Yang, J., Wagner, S. A., & Beli, P. (2015). Illuminating spatial and temporal organization of protein interaction networks by mass spectrometry-based proteomics. *Frontiers in Genetics*, *6*, 344.

FSC
www.fsc.org
MIX
Papier | Fördert gute Waldnutzung
FSC® C083411

Kolibri 360 GmbH
Ferdinand-Jühlke-Straße 7
99095 Erfurt, Deutschland
produktsicherheit@kolibri360.de